Giovanni Battista Guccia

Benedetto Bongiorno • Guillermo P. Curbera

Giovanni Battista Guccia

Pioneer of International Cooperation in Mathematics

 Springer

Benedetto Bongiorno
Università degli Studi di Palermo
Palermo, Italy

Guillermo P. Curbera
Universidad de Sevilla
Sevilla, Spain

ISBN 978-3-319-78666-7   ISBN 978-3-319-78667-4   (eBook)
https://doi.org/10.1007/978-3-319-78667-4

Library of Congress Control Number: 2018944622

Mathematics Subject Classification (2010): 01A55, 01A60, 01A70, 01A74

Printed on acid-free paper

This Springer imprint is published by the registered company Springer International Publishing AG part of Springer Nature.
The registered company address is: Gewerbestrasse 11, 6330 Cham, Switzerland

Introduction

May 9, 1943, was a day of memorable tragedy for Palermo. At noon, the air alarms announced that a bombardment was about to start. This was not a novelty. After the Allied victory over the Africa Corps of Marshal Erwin Rommel in El Alamein at the end of 1942, the city had been suffering regular bombardments in preparation for the Allied landing in Sicily. The military targets of the bombings were Palermo's port facilities, the railway and the roads. Targeting had been poor however, as German defenses on the nearby Mount Pellegrino forced the Allied aircraft to fly at high altitudes. It was at this time that the Allied forces adopted the new and devastating military strategy of carpet-bombing. The human cost of the bombardment was relatively low, as many inhabitants had fled to the countryside months before, yet the destruction of Palermo's housing was significant. Figures indicate that forty percent of Palermo's housing was destroyed. The attack was repeated that night. In addition to housing, many public buildings were damaged; some partially destroyed and others reduced to ruins. The rich cultural heritage of this two thousand-year-old city was ravaged. More than one hundred churches, monasteries and noble palaces were severely or totally destroyed. Even today, controversy arises over this devastating bombardment by the Allied Forces. Was such devastation necessary to demonstrate the consequences of siding with Mussolini's regimen?[1]

One of the palaces affected by the bombing was located on Via Ruggiero Settimo. The Palazzo Guccia was a large noble palace, not yet a century old. The Palazzo Guccia will play an important role in our story and, consequently, in the history of the international organization of modern mathematics.

The words of Edmund Landau will introduce us to the owner of the palace and the main character of our story, and to his work, the main object of our story. In his capacity as representative of the Göttingen Mathematical Society, Landau delivered a speech in 1914. The occasion of the speech will be revealed in short.

---

[1] A. Chirco, *Memoria del 9 maggio 1943*, Fondazione Salvare Palermo, 2003, Palermo.

Ladies and gentlemen,
Dear members of the Circolo Matematico di Palermo,
Dear M. Guccia,

I am sorry for not being able to make use of your beautiful Italian language: my German is not sufficiently known here, despite the great friendship that unites both of our countries, so my only solution is to speak in French. Despite this, what I am going to say comes from the depths of my heart, and it is for saying this that I came to Palermo.

We are celebrating the jubilee of a society which has only a minority of its members from the city where it is situated, but which reunites almost a thousand mathematicians from all countries of the world and, among these, the greatest and most illustrious scholars from Italy, Germany, England, France, the United States, Hungary, and from all nations where our science is cultivated. It is the only permanent international organization that we have; we thus consider Palermo as the center of the mathematical world. This is not because Palermo is the seat of a society whose sessions we can hardly attend. This is not because of the pleasure and the honor that we get from relating with the well-known mathematician and charming man who is M. Guccia. The reason lays mainly in the journal, the *Rendiconti*, which the Circolo Matematico publishes under the direction of its founder, M. Guccia, who has dedicated himself to its oversight for the last thirty years. Being himself aware of all chapters of the mathematical sciences, no one other than he could be devoted to such a task. The *Rendiconti* is the best mathematical journal in the world. M. Guccia has managed to successfully gain as friends of his journal serious mathematicians from all countries. We would have to enter into details that do not interest this assembly in order to explain why it is preferred to publish the best of our results in the *Rendiconti* of Palermo rather than in any other journal of the world. I want to mention two details, pointing out, first of all, the so amiable and encouraging way in which M. Guccia treats the authors, and especially mention the great rapidity in which the memoirs are composed and printed. To illustrate, it has occurred to me that it took three days for me to arrive to Palermo, while it takes only eight days for me to receive the galley proofs after submitting a manuscript. And thanks to the perfection of the printing there was not even much to correct.

The greater part of the world is not interested in mathematics but I am convinced that in Palermo, even before this celebration, the Circolo Matematico was well known; whereas, it is certainly not so well known that in Germany there is a city called Göttingen (much smaller than your large and illustrious city), in which we do a lot of mathematics, and which has a close connection with the Circolo. I need to speak about this place, since I am commissioned to bring their congratulations.

First of all, from our Mathematical Society of Göttingen, the center of our scientific life, which you kindly invited to be part of this celebration, and that has given me the honor of representing it here.

I bring also the congratulations of four members of your Editorial Board living in Göttingen: Klein, Hilbert, Carathéodory, and myself, who, at the same time, are the only professors in pure mathematics in our university. Outside Palermo there is certainly no other city bringing a larger number of colleagues to the editorial board.

I do not think that M. Guccia intended to choose so many professors from the same university as collaborators: a university enjoying, since the times of Gauss, Dirichlet, Riemann, Minkowski and others, a great mathematical reputation. On the contrary, two of us were invited by M. Guccia to become members of the Editorial Board when we were in other places, and even before we had any official position in the education of our country. Perhaps the collaboration that links us to the Circolo has also helped us teach our students, and continue together the direction of the mathematical school of Göttingen. In any case— and it is one of the great and immortal merits of M. Guccia, and among those, one of the most remarkable—he has always judged mathematicians solely on their works, without looking at their age or official rank, and he has helped the beginners—as I was, ten years ago—to publish their research in such a major journal, and gain confidence in themselves. I came here to thank the Circolo Matematico, that is to say M. Guccia, since he has facilitated, for many mathematicians, their situation in science. I hope that he shall see, for a long time and in perfect health, the fruits of his work in laborious life, the gratitude of all of us who are proud to be his collaborators, and overall the progress of our dear mathematical science, to which he dedicated his life.[2]

In this intense speech, Landau was referring to the Sicilian mathematician Giovanni Battista Guccia (Fig. 1). Edmund Landau (Fig. 2)—author of mathematical gems such as the *Handbuch der Lehre von der Verteilung der Primzahlen*, and the *Grundlagen der Analysis*—was well known for his rigor and precision. Landau never strayed from fair, even blunt judgment for the sake of sentiment, making his praise of Guccia's character and his work managing the journal Rendiconti del Circolo Matematico di Palermo even more remarkable.

In 1914, Giovanni Battista Guccia was the leading figure of the Circolo Matematico di Palermo, the largest mathematical society in the world, and was the managing editor of the international mathematical journal with the largest print run worldwide. Guccia founded the Circolo Matematico di Palermo in 1884, and thirty years later, the society grew into a truly international society of around nine hundred members. Two thirds of its members were foreigners, among them, 140 were from Germany and 140 from the United States of America. The Rendiconti del Circolo Matematico di Palermo published yearly eight hundred pages of mathematical research and printed the outstanding figure of 1,200 copies of them. In the editorial board of the Rendiconti there were renowned mathematicians such as Borel, Carathéodory, Hadamard, Hilbert, Klein, Landau, Mittag-Leffler, Picard, Poincaré (until he passed away in 1912), and De la Vallée Poussin. Today the mathematical journal Rendiconti del Circolo Matematico di Palermo, so much praised by Landau, is known mainly for the many classical and influential papers published mainly at the beginning of the twentieth century. Much less well known is the mathematical society that supported its publication: the Circolo Matematico di Palermo.

Giovanni Battista Guccia was a prominent figure in the international mathematical arena of his time. A measure of such prominence was the celebration that was organized in Palermo on April 1914 to commemorate the 30th anniversary of the foundation of the Circolo: it was a worldwide celebration with some one hundred and twenty mathematical societies, academies and scientific institutions from twenty different countries represented. It was at the opening ceremony of the celebration when Landau delivered the speech quoted above. However, Giovanni Battista Guccia is not well known among mathematicians today. There are two reasons for this. The first and most important one regards his mathematical contributions. Guccia was a doctoral student of the nineteenth century Italian algebraic geometer Luigi Cremona. Algebraic Geometry changed deeply at the end of the nineteenth century as a result of the introduction of powerful tools from Algebra. Guccia, however, did not follow these changes. At the same time, Guccia started to devote

---

[2]*Discorso del Prof. Dr. Edmund Landau*, XXX anniversario della fondazione del Circolo Matematico di Palermo, Adunanza solenne del 14 aprile 1914, Suppl. Rend. Circ. Mat. Palermo, 9 (1914), pp. 16–17.

Fig. 1 Giovanni Battista Guccia (1855–1914). Courtesy of the Accademia Nazionale dei Lincei

Fig. 2 Edmund Landau. Courtesy of the Archives of the Mathematisches Forschungsinstitut Oberwolfach

much of his time and effort to running the society and the journal. As a consequence, his mathematical contributions lost the front line, and have not resisted the passage of time. The second reason is the decline, especially after World War II, of the Circolo Matematico di Palermo as a mathematical society and as a result, the decline of the influence of its journal, the Rendiconti del Circolo Matematico di Palermo.

Further, not much is known of Guccia's origins. The classical Treccani biographical dictionary of the *Enciclopedia Italiana di Scienze, Lettere ed Arti'* only mentions the fact that Guccia's family was wealthy, and that his father belonged to a minor branch of nobility. Unfortunately, the Palazzo Guccia does not exist any more; thus, no family archives exist for Giovanni Battista Guccia. However, we are fortunate to have the Archive of the Circolo Matematico di Palermo and of its Rendiconti. Their mathematical richness is an invaluable resource. Using the material of this archive, Aldo Brigaglia and Guido Masotto wrote a fundamental and pioneering study of the Circolo Matematico di Palermo in 1982.[3] This archive is also a privileged, though limited, source to learn about the person who dedicated his life both to the Circolo and the Rendiconti. In any case, the lack of biographical information inhibits our understanding of the deep personal involvement of Giovanni Battista Guccia in the creation and running of the Circolo Matematico di Palermo, and his willingness to make use of his personal fortune in the managing of the Rendiconti. A better understanding of the source and the extent of his wealth were required. For that, it was necessary to start tracing back his family origins.

---

[3] A. Brigaglia, G. Masotto, *Il Circolo Matematico di Palermo*, Edizioni Dedalo, 1982, Bari.

At this point, the interest in Giovanni Battista Guccia of the authors of this book merged. For Benedetto Bongiorno, Guccia was a familiar figure since he served as vice-president of the Circolo Matematico di Palermo from 1983 to 1997, helped organize the centennial of the foundation of the Circolo Matematico di Palermo in 1984, and created (together with Pasquale Vetro) the historical index of articles published in the Rendiconti from 1887 to 1977. Guillermo Curbera became interested in Guccia in 2006, when he organized for the International Congress of Mathematicians held in Madrid an exhibition on the visual history of the International Congress of Mathematicians from 1897 to 2006. A discussion in 2012 with the historian Orazio Cancila, from the University of Palermo, and with Salvatore Savoia, first General Secretary of the Sicilian Society for the "Storia Patria", led the authors to realize that the Guccia family was related to the Tomasi family, a noble Sicilian family whose wealth and decadence was masterly narrated in the novel *Il Gattopardo* (translated into English as *The Leopard*).[4] Indeed, Giovanni Battista Guccia was a nephew of Giulio Fabrizio Tomasi (known as the Prince of Lampedusa), who was the historical figure behind the main character in *The Leopard*. The relationship between these two families, the Guccias and the Tomasis, plays a key role in our story. That relation connected Guccia with the well-known astronomical interests of his uncle. This exposure to astronomy and science was a determining factor in young Giovanni Battista's own interest in science and in mathematics. The Tomasi family lived in the Palazzo Lampedusa, which was near the Palazzo Guccia, and was bombed during the bombardments of 1943 down to ruins.

We now know about Guccia's family, about its fortune, and about the influences that Giovanni Battista received from his family background. This provides a better understanding of his life and also of his activity as a mathematician, as founder of a mathematical society, and as editor of a mathematical journal. These details shed light on important episodes in the history of the Circolo and the Rendiconti. With this new information, this book is about Giovanni Battista Guccia, his circumstances, his projects and achievements, and his place in the development of an international community of mathematics.

The book is organized into seven chapters and an appendix consisting of eight parts.

The first chapter is devoted to the background of Giovanni Battista Guccia. We present it organized into four concentric circles: Sicily and Palermo, the importance of astronomy in early nineteenth century Palermo, the origins of the Guccia family, and the early influences on the young Giovanni Battista. Guccia was Sicilian-born and lived all his life in Palermo. These two facts over-influenced his life course. Thus, a short account of the history of Sicily is presented. Giuseppe Piazzi's discovery of the asteroid Ceres at the beginning of the nineteenth century had an immense impact on science and society. Piazzi's discovery was especially

---

[4]G. Tomasi di Lampedusa, *Il Gattopardo,* Feltrinelli, 1957, Milan. In English: *The Leopard*, Pantheon Books, 1960, New York.

significant in Palermo, since he discovered the asteroid in the Palermo Observatory. This is relevant to our story, since it influenced the Prince of Lampedusa and hence his young nephew Giovanni Battista. To understand this development, we present a brief summary of the situation of astronomical knowledge before Piazzi. With respect to Guccia's family origins, the first family group carrying the name Guccia is found in the early eighteenth century in Palermo. Details from this family genealogy provide insight into the origin of the family's fortune, way of life, and character, all of which greatly influenced Giovanni Battista in his youth. Finally, we end the chapter by describing how the practical and business-oriented atmosphere of the Guccia family interwove with the astronomical activity of Giovanni Battista's uncle, the Prince of Lampedusa, in the education of Giovanni Battista.

The second chapter deals with the school and university education of Giovanni Battista. He was a boy when the Kingdom of Italy was founded and a new school system was created. We briefly describe the school system of the Risorgimento and then follow the steps of young Giovanni Battista through his middle school and technical education. With respect to the existence of universities, it is important to note that Sicily and Naples had remained isolated from the rest of the Italian peninsula for centuries. The establishment of a university in Palermo, then, was decided exclusively between Sicily and Naples. We trace the political difficulties that stalled the creation of the University of Palermo until 1805. A short intermission allows us to make a presentation of mathematicians born in Sicily, from Archimedes to Francesco Maurolico and Niccolò Cento. Giovanni Battista's way through the University of Palermo was cut short due to the most important encounter of his life. It was at a meeting of the Società Italiana per il Progresso delle Scienze in 1875 that Guccia met Luigi Cremona, the renowned algebraic geometer. After meeting Cremona, Guccia moved to Rome and enrolled in the university, where he came into contact with the Italian mathematical academic world. Five years later he presented his thesis under the guidance of Cremona.

The third chapter is an intermission that is necessary for understanding the scientific context in which Guccia devised his professional projects. We review three important processes regarding the organization of mathematics as a scientific discipline that took place in the second half of the nineteenth century: the foundation of national mathematical societies, the creation of research mathematical journals, and the organization of mathematics at a level larger than nations. Since we will follow the process of the creation of a mathematical society, the Circolo Matematico di Palermo, it is most appropriate to review the creation of two prominent mathematical societies, the London Mathematical Society and the Société Mathématique de France. These societies were formed not long before the Circolo, and several of their mathematicians later became very close to Guccia and likely influenced his work at the Circolo. As we are going to follow the process of creation and management of a mathematical journal, the Rendiconti del Circolo Matematico di Palermo, it is also appropriate to review the creation and management of an early and important mathematical journal. We focus on Acta Mathematica, because of the similarities with the Rendiconti and also on account of the close relationship between Guccia and the founder of Acta Mathematica, Gösta Mittag-

Leffler. Regarding the international organization of mathematics, we will discuss the origins of the International Congresses of Mathematicians, and review the first congresses before the 1908 Rome Congress in which Guccia was deeply involved.

The fourth chapter narrates the process leading up to Guccia's conception of a mathematical society, as well as the founding of both the Circolo Matematico di Palermo and its journal, the Rendiconti del Circolo Matematico di Palermo. We follow Guccia's post-doctoral journey in the summer of 1880 through Paris, Reims and London. During that trip, he established professional relations and friendships that lasted for life, and determined his personal viewpoint of mathematics and its modern structuring. After devoting himself to research with his mentor Luigi Cremona for some time, he decided in 1884 to found a mathematical society. He was at that moment twenty-eight years old. Despite his youth, he was able to navigate the academic environment in Palermo without any professional relation with the university. His early success encouraged him to lead the society towards internationalization four years later, and to create the journal in order to garner its international participation.

The fifth chapter is devoted to the most ambitious of Guccia's goals: developing the Circolo into an international association of mathematicians. We start by reviewing the relation of Guccia with two important mathematicians: Vito Volterra and Henri Poincaré. Guccia's close relationships with these two mathematicians were a great influence on him and his projects for the Circolo. In the period from 1889 to 1908, several important events also occurred: he applied to a professorship in the University of Palermo through a somewhat turbulent process, and he had to face a rebellion inside the Circolo that threatened the nature of the society as Guccia had conceived of it. On both issues, he succeeded. The Circolo experienced a burst of success after the Heidelberg International Congress of 1904: its membership increased, and its fame expanded worldwide. These were the contour conditions for Guccia's visionary project: transforming the Circolo into the international association of mathematicians. The issue was discussed and decided at the Rome International Congress of 1908.

The sixth chapter reviews the period from 1908 to 1914 when the Circolo and the Rendiconti reached their maximum splendor. The society grew at a strong, steady rate, and the journal strengthened its international scientific prestige, playing an important role in the publication of first-line results. To illustrate the quality reached by the Rendiconti we comment on several mathematical papers published in the journal. The international nature and the internal organization of the society played a crucial role in such success, and we analyze these causes and effects in detail. Finally, the celebration in 1914 of the 30th anniversary of the foundation of the society was the highlight of the Circolo Matematico di Palermo, of the Rendiconti and of Guccia's projects.

The Epilogue narrates Guccia's death, which occurred as World War I began. The dramatics of the war deteriorated his already poor health; as a true internationalist, he felt great pain at the spectacle of Europe destroying itself. The death of Guccia and the aftermath of the war had serious consequences on the Circolo and the Rendiconti. However, the society's members were able to resist the pressure of

Mussolini's regime until World War II, when the bombing of Palermo literally destroyed the project. We finish this story with an evaluation of the work, the efforts and the contributions of Guccia to the main aim of his life: the development of an international community of mathematicians.

The Appendixes contain: the list of the mathematical works of Giovanni Battista Guccia, as listed by the Jahrbuch über die Fortschritte der Mathematik; the Statute of the Circolo Matematico di Palermo, in its three stages of development, 1884, 1888 and 1908; the list of members joining the Circolo Matematico di Palermo from its foundation in 1884 until 1914 (scanned pages of the Annuario Biografico del Circolo Matematico di Palermo); statistical data on the membership of the Circolo, in particular, the distribution of membership by country (scanned pages of the Annuario Biografico del Circolo Matematico di Palermo); the list of members of the Steering Council of the Circolo Matematico di Palermo from 1888 to 1914 (scanned pages of the Annuario Biografico del Circolo Matematico di Palermo); and the original text of the excerpts of most of the documents quoted in the book, i.e., speeches, letters, and other published materials.

The issue of images of Guccia deserves a comment. Up until now, there were only three images of Giovanni Battista Guccia known: a photograph taken at a photographic studio in Palermo, kept in the Archive of the Circolo Matematico di Palermo; a similar photograph that Guccia dedicated to Volterra on the occasion of the 1914 celebration honoring Guccia as founder of the Circolo Matematico di Palermo, kept in the Biblioteca dell'Accademia Nazionale dei Lincei in Rome (Fig. 1); and a portrait by the renowned Italian painter Ettore De Maria Bergler, kept in the Dipartimento di Matematica e Informatica of the Università degli Studi di Palermo (Fig. 7.7). They all show Guccia at an older age, and their similarity suggests that the painter used the photographs for the portrait. This was the situation until March 2016 when the authors of this book exploited the connection between the Guccia and the Tomasi families. They contacted 82-year old Cesare Crescimanno, great grandson of the Prince of Lampedusa, who was married to the sister of Giovanni Battista Guccia's father.[5] The Crescimanno family photo album contained a photograph of young Giovanni Battista (Fig. 2.1), probably at the time when he was studying at the university, two photographs of his sister Carolina, and a photograph of a painting of his parents, Giuseppe Maria (Fig. 1.4) and Chiara (Fig. 1.5). In the Epilogue, we explain the reason for the lack of images, and, in general, memories, of the Guccia family.

On October 29, 2014, a celebration of the centennial of the death of Giovanni Battista Guccia was organized at the University of Palermo. The chairman of the Dipartimento di Matematica e Informatica, Camillo Trapani, and the president of the Circolo Matematico di Palermo, Pasquale Vetro, presided over the celebration. The program of the event included the presentation of the portrait of Guccia after its restoration by the Dipartimento dei Beni Culturali (Fig. 7.7); a lecture on the mathematical works of Guccia by the president of the Unione Matematica Italiana,

---

[5] A. Vitello, Giuseppe Tomasi di Lampedusa, Sellerio, 2008, Palermo, Albero genealogico.

Ciro Ciliberto; a lecture on the history of the Circolo Matematico di Palermo by
Aldo Brigaglia; and the public presentation by Benedetto Bongiorno and Guillermo
Curbera of their findings on the origins of the Guccia family, the links of Giovanni
Battista Guccia and the Tomasi family, and their consequences for the formation of
the scientific interest of the young Giovanni Battista Guccia.

Palermo, Italy   Benedetto Bongiorno
Sevilla, Spain   Guillermo P. Curbera
April 2017

Acknowledgements

We have greatly profited from collaboration with many institutions. Special thanks to the Circolo Matematico di Palermo, via its president Pasquale Vetro, and the Dipartimento di Matematica e Informatica of the Università degli Studi di Palermo, via its chairman Camillo Trapani.

The writing of this book has required an intense recourse to archival material. We are pleased to acknowledge the access granted to us and the professional aid obtained from a number of archival institutions and archivists:

The Archive of the Circolo Matematico di Palermo; the Archive of the Mittag-Leffler Institute; the Center for History of Science of the Royal Swedish Academy of Sciences and the archivist Anne Miche de Malleray; the Archives Henri-Poincaré and the historian Scott Walter; the Archivio di Stato di Palermo; the Archivio di Stato di Roma and the archivist Giuliana Adorni; the Biblioteca dell'Accademia Nazionale dei Lincei e Corsiniana and the archivist Alessandro Romanello; the Osservatorio Astronomico di Palermo "Giuseppe Salvatore Vaiana" and the astronomer Ileana Chinnici; the Biblioteca Centrale della Regione Siciliana and its directors Francesco Vergara and Carlo Pastena; the Archivio Storico Diocesano di Palermo; the Archivio Notarile Distrettuale di Palermo; the Archivio Notarile Distrettuale di Temini-Imerese; the Archivio storico dell'Università degli Studi di Palermo; the Istituto Tecnico "Filippo Parlatore" di Palermo; the Archivio Storico Comunale di Palermo; and the Bibliothèque Nationale de France.

We have received help from many people, whom we acknowledge with pleasure:
Cesare Ajroldi, Juan Arias de Reyna, Sara Arias de Reyna, Aldo Brigaglia, Julio Bernués, Emilio Bujalance, Orazio Cancila, Gianluca Capaccio, Giovanna Cerami, Cinzia Cerroni, Ciro Ciliberto, Sonia Claesson, Cesare Crescimanno, Jaime Curbera, Aldo de Franchis, Melchiorre Di Carlo, Mario Di Liberto, Luisa Di Piazza, Bernhard Dierolf, Dieter Geyer, Livia Giacardi, Giovanni Fatta, José Ferreirós, Patrick Ion, Gioacchino Lanza Tomasi, Javier Lerín, Mario Martínez, Natividad Gómez, Washek Pfeffer, Werner J. Ricker, Amelia Rizzo, Emma Sallent Del Colombo, Salvatore Savoia, John Toland, Elisa Tomarchio, Maurizio Urbano Tortorici, and Steven Wepster.

Especially deserving of mention is the generous collaboration that we have received from Julia Curbera from New York, Eileen O'Brien from Canberra, Giuseppe Sole from Palermo, and Juan Luis Varona from Logroño.

Contents

List of Figures

Chapter 1
Background

This chapter is devoted to the background of Giovanni Battista Guccia. It is organized into four concentric circles: Sicily and Palermo, the importance of astronomy in early nineteenth century Palermo, the origins of the Guccia family, and the early influences on the young Giovanni Battista. Guccia was Sicilian-born and lived all his life in Palermo. These two facts over-influenced his life course. Thus, a short account of the history of Sicily is presented.[1] Giuseppe Piazzi's discovery of the asteroid Ceres at the beginning of the nineteenth century had an immense impact on science and society. Piazzi's discovery was especially significant in Palermo, since he discovered the asteroid in the Palermo Observatory. This is relevant to our story, since it influenced the Prince of Lampedusa and hence his young nephew Giovanni Battista. To understand this development, we present a brief summary of the situation of astronomical knowledge before Piazzi. With respect to Guccia's family origins, the first family group carrying the name Guccia is found in the early eighteenth century in Palermo. Details from this family genealogy provide insight into the origin of the family's fortune, way of life, and character, all of which greatly influenced Giovanni Battista in his youth. Finally, we end the chapter by describing how the practical and business-oriented atmosphere of the Guccia family interwove with the astronomical activity of Giovanni Battista's uncle, the Prince of Lampedusa, in the education of Giovanni Battista.

---

[1]We mostly follow the classical works by Denis Mack Smith FBA *Medieval Sicily: 800–1713* and *Modern Sicily after 1713,* Chatto & Windus, 1968, London. Other recent texts have also been used for particular facts, mostly F. Renda, *Storia della Sicilia dalle origini ai giorni nostri,* 3 vol., Sellerio, 2003, Palermo.

© Springer International Publishing AG, part of Springer Nature 2018
B. Bongiorno, G. P. Curbera, *Giovanni Battista Guccia,*
https://doi.org/10.1007/978-3-319-78667-4_1

Fig. 1.1 Map of Sicily. Courtesy of the Norman B. Leventhal Map Center at the Boston Public Library

1.1   Some History

The Mediterranean Sea is divided into its Western and Eastern parts by the island of Sicily. In the 1st century the Greek geographer Strabo described it in this way:

> Sicily is triangular in shape; and for this reason it was at first called "Trinacria," though later the name was changed to the more euphonious "Thrinacis." Its shape is defined by three capes: Pelorias, which with Caenys and Columna Rheginorum forms the strait; Pachynus, which lies out towards the east and is washed by the Sicilian Sea, thus facing towards the Peloponnesus and the sea-passage to Crete; and, third, Lilybaeum, the cape that is next to Libya, thus facing at the same time towards Libya and the winter sunset. As for the sides which are marked off by the three capes, two of them are moderately concave, whereas the third, the one that reaches from Lilybaeum to Pelorias, is convex; and this last is the longest, being one thousand seven hundred stadia in length, as Poseidonius states, though he adds twenty stadia more. Of the other two sides, the one from Lilybaeum to Pachynus is longer than the other, and the one next to the strait and Italy, from Pelorias to Pachynus, is the shortest, being about one thousand one hundred and thirty stadia long. And the distance round the island by sea, as declared by Poseidonius, is four thousand stadia (Fig. 1.1).[2]

In size, Sicily is almost a third of the island of Ireland. It is mostly hilly and mountainous, especially on the northern coast. On the Eastern coast, Mount Etna,

---

[2] Strabo, *Geographica,* Book VI, Chapter 2, Section 1.

the largest active volcano in Europe, stands 3,300 meters above sea level. A massive sulfur vein runs across the island, from Southwest to Northeast. Winters are mild and rainy, but summers are extremely hot and dry. The central and southwestern parts of the island are practically without any forests. Sicily is a prime example of anthropogenic deforestation, which began during Roman domination, when the turning of a naturally fertile island into an agricultural region led to a severe decline of rainfall and the drying of rivers.

Historian H.G. Koenigsberger wrote: "If Gibbon's famous aphorism, that history is the story of the crimes and follies of mankind, has any truth, then this will surely be found in the history of Sicily during the last thousand years."[3] Sicily's location in the center of the Mediterranean Sea invited all ancient civilizations passing by to stop, settle and leave their imprint. First was the visit of the Phoenicians around 900 BC. Later the Greeks arrived in 750 BC and established colonies, the most important being the coastal city of Syracuse in the southeastern part of the island. With the arrival of the Greeks, Sicily's destiny was joined to that of the South of Italy, creating the so-called Magna Graecia where the Pythagorean School flourished. The Greeks brought to the island profitable agriculture with olive groves and vineyards, as well as a vibrant culture and a rich religious tradition. While the Greek colony of Syracuse controlled much of Sicily, there were a few Carthaginian colonies in the west of the island. Greece and Carthage, two expanding naval and commercial empires in the Mediterranean, eventually clashed, and their conflict escalated into the Greek-Punic Wars. The wars lasted several hundred years, and finally ended when the Roman Republic annexed Sicily as the first Roman province outside of the Italian Peninsula. The Second Punic War between Rome and Carthage, from 218 to 202 BC, ended the Carthaginian presence in Sicily, as well as the life of Archimedes in the siege of Syracuse. For seven hundred years, Sicily remained a Roman province, and became an important part of the republic and later of the empire. Sicily served as Rome's granary, and this activity started and accelerated the island's deforestation process. Despite this long period of Roman domination—and some attempts under the first emperor Augustus to introduce Latin as the island's language—Sicily was never completely Romanized, and preserved its ties to Greek culture.

Christianity first appeared in Sicily in the third century AD. Before Emperor Constantine the Great issued the Edict of Milan in 313 establishing religious toleration for Christianity within the Roman Empire, a significant number of Sicilians had been martyred. Christianity spread rapidly in Sicily over the next two centuries. For Sicily, the fall of the Western Roman Empire in the year 476 with the deposition of the last emperor was followed by a sequence of invasions and changes of rulers: first the Vandals, then the Ostrogoths, and finally the Byzantines. For three centuries, Sicily was under the rule of the Greek-oriented Byzantine Empire based in Constantinople.

---

[3]H.G. Koenigsberger, *review of "A History of Sicily: Medieval Sicily 800–1713; Modern Sicily after 1713" by Denis Mack Smith.* The English Historical Review, 85, no. 336 (1970), pp. 560–562.

The end of Byzantine rule came via the military forces of the Muslim Arabs, who entered Sicily in 827. The Emirate of Sicily was subsequently founded in 831. The Arabs brought land reforms, improved irrigation systems, and the expansion of small farms to Sicily. New crops were introduced to the island, such as oranges, lemons, pistachios and sugar cane. As a result, agricultural productivity increased substantially. Palermo was the capital of the Emirate. Situated on a northern coastal plain known as the Conca d'Oro (so named for its abundant production of citrus fruits), Palermo had first been a Phoenician colony, then the Greek city of Panormo, and later a Carthaginian seaport. Under Arab domination it became a large and wealthy city, the second largest city in Europe,[4] smaller only than Cordoba, the capital of the Muslim Caliphate in Spain. The Muslim conquest was rather tumultuous, however, and it took over a century for them to gain control of the full island. Sicilians of Byzantine origin, especially in the eastern part of the island, continued to revolt even after the Muslim victory. During this period, Sicily was a religious and cultural mixture of Muslims, Greek Orthodox Christians and Jews. The Emirate lasted two centuries.[5]

The twelfth century is mythical in the memory of Sicilians.[6] During the previous century, Norman mercenaries from northern Europe started to participate in political quarrels between the lords in the south of mainland Italy. Eventually the Normans took over the whole area below the Papal States, which included conquering Sicily from the Arabs to form the first kingdom of Sicily. This Norman kingdom became a prosperous and tolerant state that appreciated the pre-existing culture of the island, and allowed each of its constituent communities to keep their cultural identity. The influence of Muslim culture remained strong, and the Norman kings even dressed like the previous Muslim emirs. Palermo remained the capital under the Normans, and became an important cultural center, attracting artists and scholars from all over the Mediterranean. As we will later see, Palermo became (along with Toledo in Spain) an important center for the translation of Greek philosophy and science into Arabic and Latin. Strong immigration from Northern and Central Italy during this period brought immigrants who practiced Catholicism and spoke Latin, leading Sicily to become more Romanized, with Latin becoming the dominant language, and Roman Catholicism the ascendant religion.

After a century, the Norman dynasty, which had been so close to the heart of Sicilians, extinguished. The next century was consumed by disputes among the German Hohenstaufen dynasty—the rulers of the Holy Roman Empire—, the Pope, and the French Angevins from the House of Anjou over the control of Sicily. The popular insurrection against the French ruling in the so-called Sicilian Vespers ended this turbulent thirteenth century, and started a less agitated period of two centuries of uneventful rule by the Spanish crown of Aragon. The political tranquility was counterbalanced by the appearance of the Black Death plague.

---

[4]O. Cancila, *Palermo*, Editori Laterza, 1988, Palermo.

[5]D. Mack Smith, *Medieval Sicily: 800–1713*, Chatto & Windus, 1968, London, section 1.

[6]*Ibid.*, sections 2 to 4.

In 1347, Sicily was one of the entry points of the disease into Europe. The disease was said to have halved the population of the island.

At the end of the fifteenth century, the Spanish crowns of Aragon and Castile joined, and as a result Sicily became one of the many possessions of the King of Spain, who named viceroys to rule the island.[7] Spaniards brought new crops from America (such as tomatoes, corn, tobacco, and also the prickly pear cactus), as well as the Inquisition, causing the expulsion of Jews from Sicily in 1492. A far away and small part of a large and widespread empire, Sicily remained largely forgotten and became somewhat ungovernable by the Spanish governors. Earthquakes in 1542 and 1693, a return of the Plague, and recurring pirate raids also had a devastating effect on the population. The end of the Habsburg Spanish Empire was declared in 1713 in the Treaty of Utrecht. As a result, control of Sicily was assigned to the House of Savoy;[8] then transferred to the Austrian Habsburgs;[9] and finally to a prince from the Spanish Bourbon family.[10] The King Don Carlos of Bourbon established his court in Naples, now united again with Sicily, and completely forgot about the island. Sicily was ruled by four different dynasties in three decades, but Spain's influence remained strong: Spaniards continued to own large parts of the island, and Spanish continued to be the language of official documents. However, a deep change was slowly taking place: by the mid-eighteenth century, Italian cultural influence was becoming stronger. Consequently, the Italian language became increasingly common (alongside the Sicilian dialect). A clear example of this Italianization of Sicily can be seen in the annual opening speeches in parliament: in 1738, the viceroy Bartolomeo Corsini delivered it in Spanish, while in 1741 the same viceroy used Italian.[11] Don Carlos was succeeded by his son Ferdinand (IV of Naples, III of Sicily and I of the Two Sicilies) who reigned for the next sixty-six years. Meanwhile, the links binding Sicily to Italy continued to develop.

Historian Denis Mack Smith is blunt when describing the Sicilian economy at the time: "The middle class, to the small extent that one existed, remained professional and bureaucratic, rather than commercial or industrial, while the local leaders of society lived off their rents and nourished a resolute prejudice against trade."[12] Moreover, "Sicilian landowners seemed to accept without question that debts need not be paid and that productive work was undignified."[13] Meanwhile, the French Revolution and Napoleon's First Empire shook the European continent. In 1799 Napoleon took over Naples, forcing King Ferdinand to retreat to Sicily. There, with the aid of British naval protection, Ferdinand resisted Napoleon's attacks and eventually regained Naples. However, the British backing had a price: the imposition

---

[7] *Ibid.*, section 10.

[8] D. Mack Smith, *Modern Sicily after 1713,* Chatto & Windus, 1968, London, section 24.

[9] *Ibid.*, section 25.

[10] *Ibid.*, section 27.

[11] *Ibid.*, p. 263.

[12] *Ibid.*, p. 373.

[13] *Ibid.*, pp. 373–374.

of a constitutional monarchy. Thus, a parliament with real powers was formed and feudalism was abolished. These reforms were short-lived, however, and the old system returned to the so-called Kingdom of the Two Sicilies, which united Sicily and Naples. Popular rebellions against the Bourbons occurred in 1820 and 1848, but were quashed, with some effort. Finally, in 1860 Giuseppe Garibaldi defeated the Sicilian armies and later took Naples, marking the end of the Kingdom of the Two Sicilies. In the same year, the newly founded Kingdom of Italy annexed Sicily and Naples.

The Unification of Italy, known as "Il Risorgimento", involved a centralized development effort that affected the new country unevenly. Northern Italy was already involved in industrial and agricultural modernization, which was largely aided by infrastructure projects sponsored by the new political regime. In the South, the situation was different. Despite the new regime's initial intentions to remain sensitive to Sicily's local culture, Piedmont officials proved to be rigorous and insensitive. Governing officials communicated poorly with Sicilians, and felt the cultural distance as a difference of civilizations. They believed that there were no conditions for a peaceful settlement, and declined the responsibility of acting as initiators and guarantors of a fair new order. As a consequence, they intended to govern the island in a reactionary way.[14] After centuries of backwardness and illiteracy, bureaucracy and corruption, inefficient agriculture and abandonment of lands, Sicilians felt that these changes were impositions from the government in the wealthy North, who used military occupation to enforce tax increases and indirectly impoverished the island by removing internal customs duties. The resentment grew even stronger as centralization brought the opposite of the desired—and promised—autonomy.[15]

Is there an explanation for Sicily's longstanding backwardness? Or, rather, an explanation that is more scientifically based than one blaming fate or the natural inclinations of Sicilians? A historically established fact is that Sicily flourished under devoted and effective rulers. This occurred during Greek colonization, when Syracuse became one of the most important commercial and cultural cities in the Mediterranean Sea, and during the Emirate rule, when Palermo blossomed into a city of several thousand inhabitants. Sicily flourished most under the Norman kings, whose fair and wise leadership generated economic wealth and social consensus among a diverse society of Byzantine, Arab, and Norman cultures.

After the fourteenth century, however, Sicily suffered as a forgotten part of larger political entities that left Sicily to be ruled by the local aristocracy. The Sicilian aristocracy originated from the series of conquests of the island. Every conqueror granted his leading warriors land to settle and cultivate. Landlords would administer justice, both civil and criminal, in their feudal estates. Marriages were carefully planned to keep estates intact; intermarriages between cousins or uncles and nieces

---

[14]F. Renda, *Storia della Sicilia dalle origini ai giorni nostri*, vol. 1, Sellerio, 2003, Palermo, pp. 186–187.

[15]D. Mack Smith, *Modern Sicily after 1713, op. cit.*, p. 445.

were favored in order to prevent the loss of wealth through the division of property. This expedient helped to maintain the predominance of the aristocracy. Until the seventeenth century, the aristocracy contributed to the development of agriculture and the foundation of towns. In the eighteenth century, however, there was a change. Many properties were neglected, and were left for hunting, grazing and wood supply. The nobility lacked any practical interest in agriculture, even though agriculture was the source of its wealth. An aristocrat remaining in his estates and devoted to agriculture could incur the disapproval of his peers. Many aristocrats resented any economic responsibilities and were indifferent towards earning money. They became absentee landlords attracted by the glamour of the court of the Viceroy and the social life of Palermo. However, life in a luxurious palace or villa in Palermo was expensive. Despite their power, most of the aristocrats were living on credit, and some of them were quite poor. Large debts accumulated from inefficient farming, inability to administer large estates, or through family lawsuits over inheritances that would carry on for years. In the nineteenth century, the nobility began to mortgage their estates and hand over their management to an emerging social class, the so-called *gabellotti*. The gabellotti in turn rented small properties to poor farmers, and hired armed groups, composed of thieves and bandits, to guard the land. The gabelloti gained extra revenue by extorting the farmers and shepherds with the help of these armed groups. The profits earned from the soil did not go back into the land; they went into the towns for the building of palaces and churches. Meanwhile, country roads were neglected and the farming community was abandoned to bandits and malaria.

One noble Sicilian family plays an important role in our story.[16] At the end of the sixteenth century, Mario Tomasi arrived in Sicily with the Spanish Viceroy Marcantonio Colonna, with whom he had fought in the Battle of Lepanto against the Ottomans. Legend traces the Tomasi family origins to a Byzantine prince named Thomas, known as "the leopard," whose sons fled from Constantinople in the mid-sixth century for Italy, where they were called "the Thomasij," that is, the sons of Thomas. Mario Tomasi entered the Sicilian aristocracy by marrying the daughter and only heir of the Baron of Montechiaro (near Licata, in the province of Agrigento) and Lord of Lampedusa, a small island—20 square kilometers— between Sicily and Africa, which was given to the Barony of Montechiaro by Alfonso the Magnanimous, king of Aragon, in the fifteenth century.

Mario Tomasi had twin grandsons, Carlo and Giulio, who were given a broad humanistic and ecclesiastic education. They founded the city of Palma, in the barony of Montechiaro, and Carlo was named the Duke of Palma by King Phillip IV of Spain. Some years later, after a mystical experience, Carlo gave up the title of Duke to his brother Giulio, and joined the Theatine religious order in Palermo. He later moved to the seat of the order in Rome. Carlo was a man of great wisdom, and published many theological studies. His brother Giulio transformed his palace into a Benedictine monastery, and built another palace for himself. He also built many

---

[16]A. Vitello, *op. cit.*, pp. 48–52.

churches, including the magnificent baroque cathedral of Palma. Giulio added to his noble titles that of Prince of Lampedusa.

The Tomasi brothers were not only devoted to religion. In addition to religion, the Tomasi brothers were also fond of scientific inquiry. For more than twenty years they hosted on their property the Sicilian astronomer Giovanni Battista Hodierna, a Roman Catholic priest. Hodierna was an enthusiastic follower of Galileo's physical theories at a time when adhering to the astronomical theories of Galileo was a dangerous option. Hodierna studied planets, satellites and nebulae with a telescope, and had some correspondence with the Dutch scientist and astronomer Christiaan Huygens. The Tomasi brothers also funded Hodierna's scientific publications.[17]

Giulio Tomasi married the Baroness of Falconeri and had four daughters and two sons. All four daughters became Benedictine nuns; one of them was declared Venerable a century later. In 1661, the Pope granted Giulio and his wife a divorce so that they could lead religious lives: Giulio became "Il Duca Santo", the Saintly Duke, and his wife "Suor Maria Seppellita", Sister Mary "the Buried". The eldest son, Giuseppe Maria, had a mystical vision at the age of fifteen similar to that of his uncle Carlo, gave up his aristocratic title and joined the Theatine order. He became a cardinal in 1712, was beatified in 1803, and was declared a saint in 1986. The second son, Ferdinando, was also very religious; he would join religious processions in the streets, wash the feet of the poor, and was very generous, especially in times of famine. He was known as "Il Principe Santo", the Saintly Prince. He died very young, leaving a one-year-old son, Giulio, who, following his father's footsteps, became very devoted and died young, leaving—after marrying his cousin—a one year old child named Ferdinando Maria.

The beginning of the eighteenth century and the influence of Ferdinando Maria brought change to the Tomasi family. They moved from Palma di Montechiaro to Palermo and became involved in the social and political life of the country. Ferdinando Maria participated in many activities, from politics and administration to arts and literature, becoming a patron of many artists. He was appointed Knight of Malta, Grandee of Spain of first class, Gentleman of the Chamber of Charles VI, among other honors. The Tomasi family flourished through the next two generations with Giuseppe Maria and Giulio Maria. They bought a magnificent palace, later called Palazzo Lampedusa. Giulio Maria tried to profit from the island of Lampedusa. In 260 years, the Tomasi family had obtained nothing from that inhabited and barren land. In 1800, a long-term lease for the enormous annual rent of 300,000 ducats was agreed. However, the rent was never paid.

The social upheaval caused by the French Revolution had severe economic consequences on the Tomasi family. In 1823, the duchy of Palma went under judicial administration. When the son of Giulio Maria, Giuseppe Tomasi, died in 1831 his widow, who was from a Neapolitan middle-class family, was able to avoid bankruptcy by selling the island of Lampedusa in a wisely maneuvered international

---

[17]I. Chinnici, *Giovan Battista Hodierna e l'astronomia,* Giornale di astronomia, 34 (2008), pp. 10–17.

and political auction among Britain, Russia and the Kingdom of the Two Sicilies. With the money from the sale of Lampedusa, the family was able to pay off all debts and buy the Villa Lampedusa, a sumptuous property on the outskirts of Palermo. The firstborn of Giuseppe was Giulio Fabrizio, born in 1815. We will encounter Giulio Fabrizio Tomasi again in what follows.[18]

Italian literature has produced several novels reflecting the world created by the Sicilian nobility. Two of them stand out, and the characters that inspired them are involved in our story. *La lunga vita di Marianna Ucria* (translated into English as *The Silent Duchess*) shows the splendor and the darkness of the Sicilian nobility in the eighteenth century through the eyes of a deaf and mute marquise. The character of the marquise is modeled on the historical character of Marianna Valguarnera, an ancestor of the novel's author Dacia Maraini.[19] The other novel, *The Leopard,* recounts the decadence of the Sicilian nobility through the story of Fabrizio Corbera, Prince of Salina. The historical character behind the fictional protagonist is Giulio Fabrizio Tomasi, Prince of Lampedusa, and great-grandfather of the author of *The Leopard*, Giuseppe Tomasi di Lampedusa.[20]

1.2 The Astronomical Observatory of Palermo

At the end of the fifteenth century the Aristotelian view of the Cosmos remained as a secure legacy of classical Antiquity. It was believed that the Cosmos was enclosed by the sphere of fixed stars, and divided into the superlunary and the sublunary worlds. In the superlunary world, the sun, moon, stars and the wandering stars, the planets, revolved in perfect and eternal harmony. The sublunary world contained the Earth fixed at the center of the Universe, taking in all that was variable and imperfect. The observable movement of the planets, however, did not fit well the superlunary model of uniform and circular movement. Planetary movements were instead described via a complex mechanism of rotating circles devised by Claudius Ptolemy in his treatise *Almagestum* (from the Arabic word meaning "the mightiest") in the second century. The five planets identified were those that could be seen with the naked eye, and which had accompanied humans since they started to stare with curiosity into the heavens: Mercury, Venus, Mars, Jupiter, and Saturn.

In 1543 Nicolaus Copernicus published *De revolutionibus orbium coelestium,* where he presented the heliocentric model of the Universe. This model located the Sun in the center of the Universe, and added the Earth as the sixth member to the list of planets. Later, in 1610, Galileo Galilei published his observation of Jupiter's four satellites in his work *Sidereus Nuncius,* and showed that the earth was

---

[18]A. Vitello, *op. cit.,* p. 375.

[19]D. Maraini, *La lunga vita di Marianna Ucría*, Rizzoli, 1990, Milan. In English: *The Silent Duchess,* The Feminist Press at CUYN, 2000, New York.

[20]G. Tomasi di Lampedusa, *op. cit.*

not the only celestial body with satellites. These findings shattered the Aristotelian view of the Cosmos and allowed the possibility of enlarging the list of planets. As early as 1596, Johannes Kepler, in his work *Mysterium Cosmographicum,* had used mathematical and theological arguments to speculate, as a mere intellectual exercise, on the existence of new planets:

> Between Jupiter and Mars I place a new planet, and also another between Venus and Mercury... Yet the interposition of a single planet was not sufficient for the huge gap between Jupiter and Mars.[21]

Interest in these astronomical questions was growing rapidly, and by the eighteenth century, thousands of professional and amateur astronomers began to observe the sky in search for new planets.

This quest was reinforced by a hypothesis built on the ideas of the Scottish astronomer David Gregory. In his 1702 work *Astronomiae physicae et geometricae elementa* Gregory compared the distance of the Earth to the Sun to the distance of other planets to the Sun. The German astronomer Johann Daniel Tietzt presented a more appealing version of Gregory's ideas, which in 1772 was included in the book *Anleitung zur Kenntniss des gestirnten Himmels* by the German astronomer Johann Elert Bode (who later became director of the Observatory of Berlin).

Bode explained:

> ...the astonishing relation which the known six planets observe in their distances from the Sun. Let the distance from the Sun to Saturn be taken as 100, then Mercury is separated by 4 such parts from the Sun. Venus is $4 + 3 = 7$. The Earth $4 + 6 = 10$. Mars $4 + 12 = 16$. Now comes the gap in this so orderly progression. After Mars there follows a space of $4 + 24 = 28$ parts, in which no planet has yet been seen. Can one believe that the Founder of the universe had left this space empty? Certainly not. From here we come to the distance of Jupiter by $4 + 48 = 52$ parts, and finally to that of Saturn by $4 + 96 = 100$ parts.[22]

This Pythagorean flavored presentation of the distances from the planets to the Sun became known as the Titius–Bode law. Excluding Mercury, the distances correspond to the values of the sequence $4 + 2^n \cdot 3$, taking $n = 0$ for Venus, $n = 1$ for the Earth, $n = 2$ for Mars, $n = 4$ for Jupiter, and $n = 5$ for Saturn. As a proper Pythagorean explanation, however, the Titius–Bode law left an intriguing question unsolved: Which was the planet corresponding to the case when $n = 3$? This was to be a planet whose distance to the Sun was $28/100$ of the distance from Saturn to the Sun, thus situated between Mars and Jupiter, two planets whose orbits were considered to be too far apart.

In 1781, German-born but English-based astronomer William Herschel, while still an amateur astronomer, was carrying out what he called a "second revision of the sky" when he discovered somewhat unexpectedly the seventh planet: Uranus. This discovery had a great impact, as it was the first truly new planet discovered

---

[21]G. Foderà Serio, A. Manara, P. Siclo, *Giuseppe Piazzi and the discovery of Ceres.* In: W.F. Bottke, A. Cellino, P. Paolicchi, R.P. Binzel (eds.), *Asteroids III,* The University of Arizona Press, 2002, Tucson, pp. 17–24, p. 18.

[22]*Ibid.*

since Antiquity. The position of this new planet was in accordance with the Titius–Bode Law, as the mean distance from Uranus to the Sun corresponded to $196/100$ the distance from Saturn to the Sun, and 196 was precisely $4 + 2^n \cdot 3$ for $n = 6$. This gave further reinforcement to the search for new planets, especially the search for the "missing planet."

The German astronomer Franz Xaver von Zach was convinced that a planet between Mars and Jupiter, at a distance to the Sun $28/100$ of the distance from Saturn to the Sun, should exist according to the Titius–Bode law. From his observatory at Seeberg in Saxony, he proposed organizing an alert network that would search the sky until the missing planet was found. The task was immense, so Zach divided the zodiacal region into 24 zones and assigned each zone to a team headed by prestigious astronomers from different observatories. The observatories of Celle, Lilienthal, and Bremen in Germany, and many others, joined the teams. Zach also considered collaborating with the Palermo Observatory, which was the southernmost observatory in Europe. The director, and creator, of the Observatory of Palermo was Giuseppe Piazzi.

On January 1, 1801, Piazzi observed that a celestial body with very dim luminosity in the constellation of Taurus was regularly changing its position. Was it a comet, or could it be a planet? For almost a month and a half Piazzi continued observing the new celestial body. He communicated his finding to the press—as was the custom at the time—and by letter to other astronomers. At that moment, Piazzi was not sure of what he had found:

> I have announced this star as a comet, but since it shows no nebulosity, and, moreover, since it has a slow and rather uniform motion, I surmise that it could be something better than a comet. However, I would not by any means advance publicly this conjecture.[23]

Piazzi's notice with all the precise observations appeared on September 1801 in the astronomical journal *Monatliche Correspondenz,* edited by Zach. By that time, it was difficult to see the celestial body due to its closeness to the Sun. That was the moment when Carl Friedrich Gauss came onto the scene. With the little observational data supplied by Piazzi's report, Gauss was able to quickly calculate the body's trajectories and predict its position. For this Gauss had to develop the method of least squares. On December 31, 1801 Zach found the celestial body where Gauss had predicted. It was a major achievement for science in general, and for Gauss in particular, given that he was only 24 years old at time. In 1809, in his work *Theoria Motus Corporum Coelestium,* Gauss wrote

> Nowhere in the annals of Astronomy do we meet with so great an opportunity, and a greater one could hardly be imagined, for showing most strikingly the value of this problem, than in this crisis and urgent necessity, when all hope of discovery in the heavens this planetary atom, among innumerable small stars after the lapse of nearly a year, rested solely upon a sufficient approximate knowledge of its orbit to be based upon these few observations.[24]

---

[23] *Ibid.,* p. 19.

[24] *Ibid.,* pp. 20–21.

After some controversy, the celestial body was eventually named Ceres after the Roman goddess of agriculture and protector of Sicily. Ceres was found in the position predicted by the Titius–Bode law for the missing planet. Piazzi published in 1802 the work *Della scoperta del nuovo pianeta Cerere Ferdinandea ottavo tra i primarj del nostro sistema solare*, where he presented Ceres as the eighth planet. The fate of Ceres as a planet was, however, short. In the following years, similar celestial bodies were discovered: Pallas in 1802, Juno in 1804, and Vesta in 1807. Fifty similar celestial bodies were discovered by 1857, and all of them were substantially smaller than the original planets. Because of this, they were reclassified as asteroids. The Titius–Bode law was discredited by these and later discoveries.

Let us turn our attention to Giuseppe Piazzi.[25] He was born in 1746 into a noble and wealthy family in a small town in Lombardy in the north of Italy, at that time part of Switzerland. Following the tradition established for the younger children of noble families, he joined a religious order, in his case, the Theatines. This was a religious congregation founded in Rome in 1524, in accordance with the movements in the sixteenth century for the reformation of the Catholic Church. The order first expanded throughout Italy, later in Spain, France, Portugal and Bavaria. By their vows, its members were subjected to a severe rule and committed to an edifying life. Joining the Theatines allowed Piazzi to follow classical studies and learn mathematics. He took holy orders and became a presbyter. At the request of his superiors in the order, Piazzi taught philosophy and mathematics in Genoa, Malta, Ravenna and Rome.

In 1781, Piazzi was appointed to the chair in mathematics of the Accademia dei Regi Studi of Palermo (the Royal Academy of Studies), a newly created institution that later became the University of Palermo in 1805. The creation of the Accademia was a major project in the attempt by the viceroys of the time to reform Sicily from its century-old backwardness. The plans included creating the Accademia in Palermo, and also a botanical garden, a theatre of anatomy, and an astronomical observatory. Sicilian leaders were determined to advance Sicily from the scientific peripheries of Europe by bringing the best teachers and scientists from overseas to these new institutions. The Viceroy Caracciolo approached the mathematician and astronomer Joseph-Louis Lagrange, who politely declined the invitation.

In 1787, Piazzi was appointed to the chair in astronomy of the Accademia. He was also charged with overseeing the construction of the proposed astronomical observatory. The observatory was to be built on top of the Santa Ninfa tower of the Palazzo dei Normanni in Palermo, the royal palace built by the Normans. For this task Piazzi traveled through the main astronomical centers of Europe, Paris and London, for three years. He gained knowledge of construction and management of contemporary observatories, as well as the esteem and friendship of some of the most important astronomers of the time, such as Lalande, Messier, Cassini, and Maskelyne, the director of the Greenwich Observatory. Piazzi was part of the group

---

[25]*Memoir of the life and writings of M. Piazzi, Director General of the Observatories of Naples and Palermo,* Edinburgh J. Science, 6 (1827), pp. 193–199.

of astronomers who determined the longitudinal difference between the Paris and Greenwich observatories, and he observed the solar eclipse of June 3, 1788 with Maskelyne. These relationships allowed Piazzi to acquire first class observational instruments for his observatory. In less than two years, he was able to obtain a complete set of astronomical instruments built by Jesse Ramsden, the renowned designer of mathematical, astronomical, surveying and navigational instruments. In particular, he acquired the beautiful telescope known as the Palermo Circle. This was a novel instrument with a diameter of five feet, equipped with a circular scale for measurements and considered to be one of the most accurate instruments at the time. This was the instrument that Piazzi used for first observing Ceres.

Piazzi was not a scientist; he had not published scientific papers in any area, nor had he shown any interest in astronomy before going to Palermo. However, with his high-quality instruments at hand, he established a thorough scientific program to accurately measure stellar positions. This meticulous work resulted in the publication of two catalogues of stars. The first one appeared in 1803 and contained 6,748 stars. The Académie des Sciences in Paris awarded Piazzi a prize for this catalogue, and he was elected member of the Royal Society in London. The second catalogue was published in 1814. It contained 7,646 stars and also received the prize of the Académie des Sciences. The discovery of Ceres was an unexpected outcome of the observational method that Piazzi had used. Another astronomical achievement of Piazzi was the discovery in 1804 of the star that he named "Flying Star" because of its large angular motion. This star, currently known as star 61 of the constellation Cygni, came to international attention in 1838 when Friedrich Wilhelm Bessel, who was at that time director of the Königsberg Observatory, measured— quite accurately, indeed—its distance to the Earth and its stellar parallax. It was the first measurement of the distance from a star to the Earth, apart from that of the Sun. Piazzi left Palermo in 1817 for Naples, where he was commissioned to finalize the seemingly unending construction of the Observatory of Capodimonte. The observatory opened in 1820 with Piazzi as director. In 1826, Piazzi died of cholera.[26]

The discovery of Ceres and the work of Piazzi placed Palermo in the first line of European observational astronomy. It had been a long time since Sicily had such a position in science. The loss of Piazzi, along with the political instability in Sicily prior to the Unification of Italy in 1861, were impediments to the successful consolidation of the observatory and the advancement of its scientific activity to its full potential. However, the observatory remained a fine scientific institution in Europe, with strong links to other scientific bodies within Italy, and highly respected in the city of Palermo.

---

[26]C.J. Cunningham, *Piazzi, Giuseppe*. In: Th. Hockey, V. Trimble, Th.R. Williams, K. Bracher, R. Jarrell, J.D. Marché, F.J. Ragep (eds.), *Biographical Encyclopedia of Astronomers,* Springer, 2007, Berlin, pp. 902–903.

1.3   The Fortune of the Guccia Family

Giovanni Battista Guccia was born in 1855 into a Palermitan family whose particular character and history had an important imprint on his personality, and, as we will see, on his scientific career. We will briefly review the genealogy of the family. In the course of almost two hundred years, the family prospered and mixed with the nobility of Sicily.[27]

Several members of the family carried the name Giovanni Battista Guccia. This causes complications that will require some care in the tracing of the genealogy.[28] To aid the reader in this task, we will use a custom regarding family names that was common in Sicily during the Spanish rule: each person appearing in this account will carry his or her given name, together with two last names, first the father's last name and second the mother's last name, with an "e" (standing for "and") between them.

In 1713, Sicily was assigned to the House of Savoy. When King Victor Amadeus II arrived in Palermo, he was impressed by the size of the city compared to his capital Turin. He asked about the number of inhabitants in Palermo, but to his surprise this figure was not known. The city had previously been exempt from taxation, and thus had no need to determine its population count. The King ordered a census to be done. The document thus produced, known as the "numerazione delle anime", that is, the counting of the souls, is a detailed account of the people of Palermo, by neighborhood, street and house. There we find the first Guccia family: Gaspare Guccia with his wife Elisabetta and their six children. The family was not short of means: they lived in a comfortable house with four servants.[29] In 1726, Gaspare paid for a daily and perpetual mass for his soul at the Monastery of Saint Rosalia in Palermo, where two of his daughters were nuns and one of them was the abbess.[30]

---

[27]The complete disappearance of the family archives presents a fundamental difficulty for tracing the genealogy of the Guccia family. The alternative sources available are the registries of births, marriages and deaths of the city of Palermo. They are stored in the Archivio di Stato for the period 1820–1865, and in the Comune di Palermo for the period from 1865 and on. Before 1820, the registries are located in the different parishes of the city. Some of these documents were lost in the bombardments of Palermo in World War II. Other important sources are the notarial acts and the court cases involving the family that are kept in the Archivio Notarile Distrettuale di Palermo, in the Archivio Notarile Distrettuale di Termini Imerese, and in the Archivio di Stato di Palermo. The section on "Congregazioni religiose soppresse" of the Archivio di Stato holds the archives of the monasteries. Since many women from the Guccia family were nuns, in these files there is a wealth of documents on testaments, mortgages and accounting of the Guccia family.

[28]The confusion generated by the existence of several persons carrying the name Giovanni Battista Guccia has caused many misunderstandings in books on urban development, architecture and the history of Palermo.

[29]A. Lo Faso di Serradifalco, *Palermo 1713: numerationi delle anime di tutte le Parochie della Città di Palermo fatte nel mese di 9mbre 1713*, 3 vol., Ila-Palma, 2004, Palermo, Cap. IX La Parrochia di Sant'Antonio Abate, Strada delli Panneri, p. 413.

[30]Folder Guccia, Suora Concetta Eucaristica in seculo Elisabetta. Section "Ordini religiosi soppressi", Monastero del Santissimo Salvatore. Archivio di Stato di Palermo, section "La Gancia".

The first-born of the family was named Giovanni Maria.[31] Giovanni Maria Guccia married Rosalia Bonomolo, whose brother Rocco was a baron with some estates. In 1751, when Rocco Bonomolo bequeathed his state to his sister and brother-in-law, Giovanni Maria acquired the title of baron.[32] Their eldest son, Giovanni Battista Guccia e Bonomolo, was born in 1733,[33] and became the patriarch of the Guccia family.

The precise nature of Giovanni Battista Guccia e Bonomolo's profession is not known. In legal documents, he was addressed as a doctor in civil and canon law.[34] Since at that time there was no university in Palermo, he must have pursued his degree at the Università degli Studi di Catania in the Eastern part of Sicily, although this fact has not been confirmed in the university archive in Catania. In later documents, he exhibited a sound and thorough knowledge of agricultural production, in particular regarding orchards.[35] By 1792, he was wealthy enough to acquire a property in a public auction held by the Senate of Palermo (the governing body of noblemen) in order to build a palace.[36] The property was not in the noble part of the center of the city—like Via Toledo, nowadays Corso Vittorio Emanuele, or Via Nuova, nowadays Via Maqueda—as he did not belong to the old aristocratic elite. It was near the Porta d'Ossuna, in the upper part of the city near the Palazzo dei Normanni and the Cathedral. The property consisted of a military rampart, known as Bastione della Balata, which was part of the old city wall, which was reinforced at the beginning of the sixteenth century to defend Palermo from the Ottoman Armada.

A year later, Guccia e Bonomolo built an opulent and Baroque palace, the Palazzo Guccia del Bastione. Situated on top of the rampart, several meters above the street level, the palace occupied a third of the surface of the rampart, some 1,300 square meters. The entrance level hosted the palace service rooms. The four stories above had one hundred rooms of different sizes and uses, as well as noble staircases, patios, and balconies. There were private areas, as well as rooms for receptions and balls. A twenty-meter-high tower allowed an impressive view of the city. What remains of the interior decorations shows rooms with walls and ceilings decorated with beautiful paintings in allegorical style. The palace grounds included

---

[31] Section "Parrocchia Santa Margherita alla Conceria". Archivio Storico Diocesano di Palermo.

[32] Folder Guccia, Suora Concetta Eucaristica in seculo Elisabetta. Section "Ordini religiosi soppressi", Monastero del Santissimo Salvatore. Archivio di Stato di Palermo, section "La Gancia".

[33] Section "Parrocchia Santa Margherita alla Conceria". Archivio Storico Diocesano di Palermo.

[34] Ibid.

[35] Folder Guccia, Suora Concetta Eucaristica in seculo Elisabetta. Section "Ordini religiosi soppressi", Monastero del Santissimo Salvatore. Archivio di Stato di Palermo, section "La Gancia".

[36] Notar Giuseppe Maria Cavarretta, October 26, 1792. Section "Notai". Archivio di Stato di Palermo, section "La Gancia".

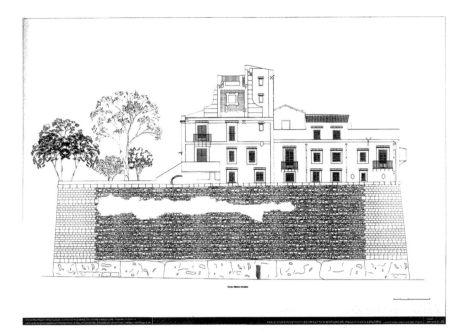

Fig. 1.2 Drawing of the Palazzo Guccia del Bastione della Balata. Courtesy of Giovanni Fatta (tutor of Gianluca Capaccio and Elisa Tomarchio, authors of the drawing)

a magnificent 2,200 square meter garden. Rather than a luxurious ornamental piece, the garden was home to a bountiful orchard that produced the excellent fruits of the Conca d'Oro: oranges, lemons, mandarins, grapefruits, olives, pears, figs, and medlars. What now remains of the palace is a high stonewall bordering Via Guccia, the Vicolo Guccia, and Piazza Porta Guccia, crowned by a chaotic cluster of ruined apartments, windows and terraces, and accompanied by an overgrown forest of old fruit trees (Figs. 1.2 and 1.3).[37]

The influence and power of the Sicilian nobility class was quite strong: it possessed feudal lordship over 280 out of 360 villages of the island.[38] A noble title meant social recognition and pride. Not having a title equated to social insignificance. At the end of the eighteenth century in Sicily, according to the figures given by Denis Mack Smith, there were 142 princes, 788 marquises, and about 1,500 barons and dukes. Most of the nobility was concentrated in Palermo. The access

---

[37] A. Rizzo, *Palermo, città antica e sottosuolo antropizzato: ipotesi progettuale nell'area del Bastione del Papireto,* tesi di laurea, Faculty of Architecture, 1998, University of Palermo; G. Capaccio, *Progetto di restauro del Palazzo Guccia a Palermo,* tesi di laurea, Faculty of Engeneering, 2010, University of Palermo; E.M. Tomarchio, *Analisi Storico-Costruttiva del Palazzo Guccia a Palermo,* tesi di laurea, Faculty of Engeneering, 2010, University of Palermo.

[38] D. Mack Smith, *op. cit.,* p. 284.

Fig. 1.3   The Palazzo Guccia del Bastione della Balata nowadays

to this elite group from other social classes was closed. However, the financial necessities of the Monarchy at the end of the eighteenth century placed heavy taxation on title transmissions. This, in turn, forced many insolvent aristocratic families to sell their titles and rights to wealthy landowners who were well aware of the value of a noble title to their business and trade opportunities.

Guccia e Bonomolo became baron upon the death of his father in 1776. At that time, the title baron had lost part of its social value. In order to enhance his nobiliary situation, Guccia e Bonomolo acquired the feudal estate of Ganzaria in 1793. Ganzaria was located in the territory of Cammarata, southeast of Palermo, in the province of Agrigento.[39] This estate belonged to Marianna Valguarnera (the fifth Princess of Valguarnera, tenth Countess of Assoro, and seventh Princess of Gangi), the historical figure upon which the novel *The Silent Duchess* is based. This acquisition allowed Guccia e Bonomolo to use the title of Baron of Ganzaria and

---

[39]Notar Francesco Salvatore Cirafici: January 12, 1793. Section "Notai". Archivio di Stato di Palermo, section "La Gancia"; and Section "Conservatoria di registro. Serie Investitura", vol. 1181, foglio 108 retro. Archivio di Stato di Palermo, section "La Catena".

to have a coat of arms.[40] The official description of the coat of arms of the Guccia family was:

> In blue, a stormy country, lies a bivalve shell, topped by eight drops, juxtaposed with two stars of the seven rays, all in silver.[41]

The legal process was completed in 1797. Thus, at the age of 63, Guccia e Bonomolo joined the Sicilian nobility. Some years later, he was able to enhance his aristocratic position. The opportunity arose when the Bourbon King Ferdinand, forced again by severe financial needs, sold many noble titles, just before the abolition of feudalism in 1812. Guccia e Bonomolo bought the title associated to his estate, becoming *Marchese di Ganzaria* (in English, Marquis of Ganzaria) when he was 76 years old.[42]

In 1761, at the age of 27, Giovanni Battista Guccia e Bonomolo had married Agatha Agati.[43] They had seven surviving children: Luigi, Ignazia, Rosalia, Carlo, Elisabetta, Giovanni Maria and Maria Stella. Three of the daughters became nuns; among them was Elisabetta. Among the descendants of this family was Giovanni Battista Guccia e Di Maria, son of Carlo. He graduated in law from the newly founded Università degli Studi di Palermo, and later became vice president of the Grand Court of Audit and Gentlemen of the Chamber under the Bourbon Dynasty. He married his cousin, Carolina Cipponeri e Guccia, daughter of Rosalia. This marriage produced several children, one of which was Chiara Guccia e Cipponeri, born in 1822 and future mother of the mathematician Giovanni Battista Guccia.

While this line of Guccias was approaching its third generation, in February 1814 Agatha Agati died, leaving her husband Giovanni Battista Guccia e Bonomolo as an 80-year-old rich and aristocratic widower. Five months later, in July 1814, he married the 29-year-old Concetta Vetrano.[44] In the following eight years, the new Guccia e Vetrano family welcomed four children: Maria Stella, born in 1815; Francesca, born in 1818; Giuseppe Maria, born in 1820; and Francesco, born in 1822.[45]

Over the course of his life, the marquis accumulated a fortune. This was due to his careful spending, his strong work ethic, and his ability to take advantage of

---

[40]F. San Martino De Spucches, *La storia dei feudi e dei titoli nobiliari di Sicilia dalla loro origine ai nostri giorni,* Boccone del povero, 1924, Palermo, vol. 4, quadro 414, pp. 24–26.

[41]A. Mango di Casalgerardo, *Il nobiliario di Sicilia,* A. Reber Editore, 1912, Palermo.

[42]Section "Protonotario del Regno", vol. 997, foglio 87 retro; section "Conservatoria di registro. Serie Mercedes", vol. 664, foglio 8 retro. Archivio di Stato di Palermo, section "La Catena".

[43]Section "Parrocchia Santa Margherita alla Conceria". Archivio Storico Diocesano di Palermo.

[44]*Ibid.*

[45]Registry. Parrocchia San Ippolito, Palermo.

opportunities. The fortune consisted of feudal estates throughout Sicily (including Ganzaria, in Cammarata, in the province of Agrigento, southeast of Palermo; of Ballata e Rifalsafi, in Castronovo, in the province of Palermo; and Malluta, in Cerda, east of Palermo), large productive properties (Santa Maria di Burgitabus, in Cerda; Pirato, in Caccamo; Amari and San Martino, in Ciaculli; Maggio and Montaperto, in Monreale; and also in Castronovo), and several houses and warehouses, most of them under rent. He was also a tax collector for the King in the area of Castronovo, and collected a share of the market-place taxes in Palermo. The majority of his wealth, his financial jewel, however, lay in the property rights to two natural springs in the city of Palermo, which he bought from the Crown in 1802 and 1812. This allowed him to own the water supply for the palaces and wealthy properties in Via Toledo and Via Maqueda, the main streets of Palermo. This was a true fountain of current and future revenue. The figures of this business, as seen in the legal deed and family documents, were truly enormous.[46]

The Sicilian nobility was ranked according to seniority. The oldest and most prestigious families date back to Norman times, followed by the Spaniards of Aragon and Castile, and finally by the nobility from the Habsburg era. The Guccia family was part of the newer nobility created by the Bourbons who ruled during the eighteenth and nineteenth centuries. The Guccias owed their title and their wealth, gained via economic concessions and administrative posts, to the Bourbon dynasty.

In 1834, the old Marquis of Ganzaria died. It was a sad, though long-expected event. He was 100 years old, and had enjoyed a long and fruitful life. The old marquis took careful steps before his death to support the management of his great estate into the future. Ten years before his death in 1824, at the age of 90, he had divided his estate among his large and diverse family. He did this by calling the notary to the Palazzo Guccia del Bastione and delivering his decisions from his bed, surrounded by all his family (as certified by the notary).[47] First he took care of his daughters who were nuns in convents. They were each given a generous annual lifelong pension. One of them, Elisabetta, had previously obtained a pontifical exemption to leave the convent and take care of her father,[48] and thus required a larger pension. Next came the three remaining Guccia e Agati children (from his first marriage), each of them around 70 years old: Luigi, Carlo, and Rosalia, the latter represented by her daughter Carolina (the other son, Giovanni Maria, had died in 1817). The feudal estates and all other properties were divided evenly among them. The adjudication of the Palazzo Guccia del Bastione was more difficult, and the marquis knew that the palace was impossible to divide amongst his heirs. To resolve

---

[46]Folder Guccia, Suora Concetta Eucaristica in seculo Elisabetta. Section "Ordini religiosi soppressi", Monastero del Santissimo Salvatore. Archivio di Stato di Palermo, section "La Gancia".

[47]Notar Diego Lo Bianco e Lumia, October 9, 1824; October 13, 1824; November 2, 1825. Archivio Notarile Distrettuale di Termini Imerese.

[48]*Ibid.*, November 2, 1825.

this, the marquis established an auction for Carlo and Carolina that would give the palace to the heir that offered the largest sum in annual rent to the other. Lastly came the most delicate issue of all: the legacy for the Guccia e Vetrano family, which included the marquis' second wife Concetta and the three surviving minors, Maria Stella, Giuseppe Maria and Francesco. They each obtained large annual pensions, and for Giuseppe Maria and Francesco a cash deposit equivalent to twenty years of a sizable pension. Concetta and her children were also granted the right to remain in the palace until the children married. The daughter nun Elisabetta was entrusted with ensuring that the terms of the division were followed.

This detailed explanation will later allow us to properly gauge the wealth of the family of the mathematician. The final example of the preciseness and care that the old marquis devoted to his affairs was the division of the largest source of revenue: the water rights. These were distributed among Luigi, Carlo, and Carolina, and were to be administered by Giovanni Battista Guccia e Di Maria, Carolina's husband, and son of Carlo. The revenue from the water rights would first have to pay all the pensions for the other family members. The rules for dealing with the lumps sums of cash generated were extremely precise. Some points deserve being quoted:

> The money thus obtained should be placed into an iron safe, in an institution chosen by decision and at the risk of the proprietaries, with the care of having three different keys for it [corresponding to different locks]; one key held by Elisabetta [the nun], another one by the administrator [Giovanni Battista Guccia e Di Maria, grandson], and the third one by the person hired as cashier of the society.[49]

The will of the marquis shows some of his appreciation for art. There is a special clause regarding a Nativity and five other paintings,[50] all done in wax, by Anna Fortino, an artist very much celebrated at the time. She was a student of the painter Rosalia Novelli, daughter and a student of Pietro Novelli, Sicily's most important painter of the seventeenth century.[51]

Economic and social life in Palermo was constantly entangled in lawsuits, which could continue for years in appeal courts. This judicial hyper-activity is easily seen in the Giornale del Regno delle Due Sicilie, the newspaper of Sicily during the 1850s. About two of the six pages of each issue were devoted to judicial announcements. The legacy of the old marquis was no exception to this: thirty years after his death, the auction regarding the Palazzo Guccia del Bastione was still being resolved in court.[52]

---

[49]*Ibid.*, March 2, 1829, Art 14.

[50]*Ibid.*, March 2, 1829, Art 19.

[51]A. Gallo, *Elegio storico di Pietro Novelli da Monreale in Sicilia, pittore, architetto ed incisore*, Stamperia reale, 1830, Palermo.

[52]Notar Gaspare Spinoso, April 4, 1865; notar Girolamo Guarnaschelli, August 13, 1872. Archivio Notarile Distrettuale di Palermo.

Abridged version of the genealogical tree of the Guccia family

Giovanni Battista Guccia e Bonomolo (1733–1834)
I Marquis of Ganzaria
First marriage in 1761 with Agatha Agati e Bonanno (1742–1814)

Luigi Guccia e Agati (1762–1831)
    Married with Marianna Russo

    Pietro Guccia e Russo (1806–1832)

    Agata Guccia e Russo (1808–1861)

Rosalia Guccia e Agati (1763–1828)
    Married with Giuseppe Cipponeri

    Carolina Cipponeri e Guccia (1793–1845)
        Married with Giovanni Battista Guccia e Di Maria (*)

Ignazia Guccia e Agati (1764–1848), nun

Carlo Guccia e Agati (1765–1845)
II Marquis of Ganzaria
    First marriage with Chiara Di Maria

    Giovanni Battista Guccia e Di Maria (1798–1871)
    III Marquis of Ganzaria
        Married with Carolina Cipponeri e Guccia (*)

        Rosalia Guccia e Cipponeri (1821–1888)

        Chiara Guccia e Cipponeri (1822–1900)
        Mother of the mathematician
            Married in 1841 with Giuseppe Maria Guccia e Vetrano (**)

        Salvatore Guccia e Cipponeri (1827–1898)
        IV Marquis of Ganzaria
            Married with Elisabetta Cannizzaro e Di Benedetto

            Giovanni Battista Guccia e Cannizzaro (1850–1922)
            V Marquis of Ganzaria

Elisabetta Guccia e Agati (1768–1850), nun

Giovanni Maria Guccia e Agati (1777–1817)

Maria Stella Guccia e Agati (1781–1842), nun

Abridged version of the genealogical tree of the Guccia family

Giovanni Battista Guccia e Bonomolo (1733–1834)
I Marquis of Ganzaria
Second marriage on 1814 with Concetta Vetrano e Agnetta (1785–1865)

Maria Stella Guccia e Vetrano (1815–1886)
    Married with Giulio Fabrizio Tomasi di Lampedusa

    Giovanni Battista Tomasi e Guccia
        Married with Carolina Guccia e Guccia (***)

Francesca Guccia e Vetrano (1818–?)

Giuseppe Maria Guccia e Vetrano (1820–1900)
Father of the mathematician
    Married in 1841 with Chiara Guccia e Cipponeri (**)

    Concetta Maria Guccia e Guccia (1842–1901), nun

    Carolina Guccia e Guccia (1844–1903)
        Married with Giovanni Battista Tomasi e Guccia (***)

    Maria Stella Guccia e Guccia (1849–1908)
        Married with Salvatore Jacona e Notarbartolo

    Giovanni Battista Guccia e Guccia (21/10/1855–29/10/1914)
    Noble of Marquis of Ganzaria
    Mathematician

    Francesco Guccia e Guccia (1859–1924)
        Married with Adriana Fatta (1870–1943)

        Chiara Guccia e Fatta (1893–1939)

        Carmela Guccia e Fatta (1896–1955)

Francesco Guccia e Vetrano (1822–?)

The two lines of the Guccia family joined in 1841 when Giuseppe Maria Guccia e Vetrano (Fig. 1.4), son of the old marquis and his second wife Concetta, married his relative Chiara Guccia e Cipponeri (Fig. 1.5), great granddaughter of the old marquis from his first wife Agatha, and daughter of Giovanni Battista Guccia e Di Maria and Carolina Cipponeri e Guccia. The starting financial situation of the new marriage was quite ample, as the couple joined wealth from several sources. Giuseppe Maria, being one of the late sons of the old marquis, had received the large pension assigned to the Guccia e Vetrano family, as well as a large cash deposit. The marriage also involved properties from the Guccia e Agati family: via Rosalia,

Fig. 1.4 Giuseppe Maria Guccia e Vetrano, father of Giovanni Battista. Courtesy of Cesare Crescimanno

the mother of Chiara's mother Carolina; via Carlo, the father of Chiara's father Giovanni Battista Guccia e Di Maria; and via the oldest son of the marquis, Luigi. Luigi and his son Pietro passed away before the marquis, and their estates were inherited by his daughter Agatha. She was a nun but left the convent and accepted a generous pension and a cash deposit from her cousin Giovanni Battista Guccia e Di Maria in exchange for her inheritance. This flow of wealth is not easy to follow without the genealogic puzzle at hand but ultimately, the Guccia e Guccia family consolidated almost the full fortune that Giovanni Battista Guccia e Bonomolo, Marquis of Ganzaria had accumulated throughout his life.

Fig. 1.5 Chiara Guccia e Cipponeri, mother of Giovanni Battista. Courtesy of Cesare Cresci-manno

1.4 Youth and Influences

After they married, Giuseppe Maria and Chiara rented the first floor of the magnificent Palazzo Francavilla,[53] situated outside the gate of the old city known as Porta Maqueda (near where some years later the Teatro Massimo, the renowned opera theatre of Palermo and the largest one in Europe after the Paris Opera, was to be built). They had ten children, who carried the last names Guccia e Guccia. Three of them were named Giovanni Battista, two of whom died before the age of two. The third one, born in 1855, was Giovanni Battista Guccia e Guccia, the mathematician.[54] Two other children also died in their childhood. The other children

---

[53]L.M. Majorca Mortillaro, *Il palazzo Francavilla in Palermo*, Alberto Reber, 1905, Palermo, p. 50.

[54]In the municipal archive of births of Palermo he was registered under the name Guccia, Giovanni Battista. Since the name was too long, he was addressed within the family and friends as Giovanni, or more fondly as Giovannino. This is confirmed on his gravestone in the family tomb built in 1939.

Fig. 1.6 Palazzo Guccia around 1880

were Concetta, born in 1843, Carolina, born in 1844, Maria Stella, born in 1849, and Francesco, born in 1859. Concetta became a nun, and all the others married, except for the mathematician, who remained single all his life.

In 1850, the Guccia e Guccia family acquired a palace under construction.[55] The palace was located on Via Ruggiero Settimo, where it met the Piazza Marchese di Regalmici in a new area of the city that expanded towards the northwest. The palace was large: it comprised street numbers 2, 3 and 4 of Piazza Marchese di Regalmici, and 26, 28, 30, 32 and 34 of Via Ruggiero Settimo (Fig. 1.6). It had four stories, two of them very grand, with large windows and balconies. On the ground floor, there

---

In the cemetery of the Cappuccini in Palermo by his sister in law Adriana Fatta. It is also confirmed by a letter from Joseph Whitaker sent to him on April 19, 1910 and kept in the Archive of the Circolo Matematico di Palermo. Due to this custom, when Guccia registered in the Universities of Palermo and Rome, he shortened his name to Giovanni Guccia, as can be confirmed in the archives of both universities. Later, he used the abbreviation G.B. Guccia, which was used in all of his papers, in all volumes of the Rendiconti, in all letters kept in the Archive of the Circolo Matematico di Palermo, and in all articles published in newspapers. The first time that his name was shortened to Giovan Battista Guccia was in his obituary announcement published in the Giornale di Sicilia. This was probably due to an issue of typesetting space in the newspaper. In fact, a few days later, the Rendiconti announced his death writing his full name Giovanni Battista Guccia. The abbreviation Giovan Battista Guccia has also been used by A. Brigaglia and G. Masotto in their cited work, and by A. Barlotti, F. Bartolozzi and G. Zappa in *L'attività scientifica di Giovan Battista Guccia*, Suppl. Rend. Circ. Mat. Palermo, 67 (2001), pp. xi–xxvi.

[55]Notar Giovanni Pincitore, August 6, 1850. Archivio Notarile Distrettuale di Palermo.

Fig. 1.7 Maria Stella Guccia, aunt of Giovanni Battista and spouse of the Prince of Lampedusa. Courtesy of Cesare Crescimanno

was an inner patio, and commercial rooms opening to the street.[56] This was to be the family home, where the mathematician Giovanni Battista Guccia e Guccia lived all his life, and where the Circolo Matematico di Palermo was founded and had its seat. It was known in Palermo just as the Palazzo Guccia. Unfortunately, the building no longer exists, and in its place, there is now a bank building. The street that was on the back-side of the palace, however, is still named Via Giovanni Battista Guccia.

In the vicinity of the Palazzo Guccia was the Palazzo Lampedusa of the Tomasi family. We have already encountered the Tomasi family, a traditional noble Sicilian family. In 1837, the firstborn son Giulio Fabrizio Tomasi, Prince of Lampedusa, had married the young and wealthy Maria Stella Guccia e Vetrano (Fig. 1.7), aunt of the mathematician. They lived in the Palazzo Lampedusa, where they had twelve

---

[56]A. Chirco, M. Di Liberto, *Via Ruggero Settimo ieri e oggi,* Dario Flaccovio Editore, 2002, Palermo, pp. 83–84.

children. The eldest, Giuseppe, was born in 1838, and was the grandfather of Giuseppe Tomasi di Lampedusa, author of the novel *The Leopard*.

In 1856, when Giovanni Battista was one year old, Maria Stella's mother, Concetta Vetrano, also moved to the Palazzo Lampedusa as her youngest son Francesco had already died.[57] It then became common for the Guccia e Guccia family to get together with the Tomasi e Guccia family, visiting the grandmother and playing with the cousins. This closeness favored matrimony: Giovanni Tomasi e Guccia married his cousin Carolina Guccia e Guccia, sister of the mathematician.

The young Giovanni Battista possibly also grew a relationship with his uncle Giulio Fabrizio Tomasi. Known to the family as "the astronomer", Giulio Fabrizio Tomasi was born in Palermo in 1815 (Fig. 1.8). Nothing is known about his formal studies, but the Tomasi family had a good educational background. While many Sicilian aristocrats were poorly educated (and some even close to illiterate), the Tomasi family had shown respect for and interest in education for a long time. Giulio Fabrizio's father had been one of the few nobles who attended the boarding school for ancient nobility created by King Ferdinand in response to the request for a university in Palermo. According to Andrea Vitello, the unofficial biographer of the Tomasi family, Giulio Fabrizio had a private teacher, named Abbot Foschi.[58] However, no record of this abbot has been found.

What is certain is that Giulio became a well-known amateur astronomer. This interest in astronomy was not unusual; in fact, many Italian noblemen at the time were interested in astronomy. Guilio Fabrizio's family greatly influenced his interest in astronomy. As previously mentioned, his ancestors Carlo and Giulio Tomasi had hosted, for more than twenty years, the astronomer Giovanni Battista Hodierna on their property in Palma di Montechiaro, and had funded his scientific publications. It was likely that there still were books, documents, and instruments from Hodierna in the family palace in Palma that Giulio Fabrizio could have seen. We can say that there was some proximity to science in the family.

The Tomasi family connection to the Theatine order may have also influenced Giulio Fabrizio's interest in astronomy. Carlo Tomasi had been an important member of the order in Rome in the seventeenth century, and Giuseppe Maria Tomasi had been a Theatine cardinal in the early eighteenth century and beatified in 1803, just two years after the Theatine Giuseppe Piazzi discovered Ceres. The Tomasi family had a chapel at the Theatine church of Saint Giuseppe in Palermo, which still exists today. It is also possible that Guilio Fabrizio made contact with the Palermo Observatory through the Theatines.

---

[57]Which is deduced from Concetta Vetrano's personally handwritten will, dated October 10, 1856, eight years before her death on January 27, 1865, notar Francesco Anelli. Archivio Notarile Distrettuale di Palermo.

[58]A. Vitello, *op. cit.*, p. 375.

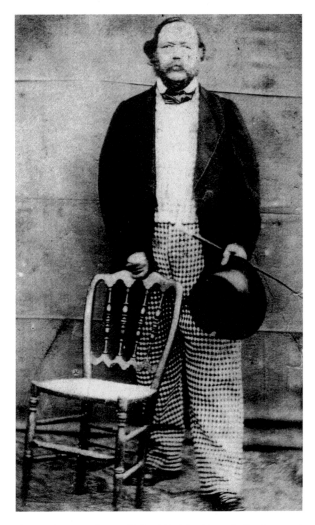

Fig. 1.8 The Prince of Lampedusa, Giulio Fabrizio Tomasi, Il Gattopardo. Courtesy of
Gioacchino Lanza Tomasi

Finally, Giulio Fabrizio observed Halley's Comet in 1835 when he was 20 years
old. The sighting was reported to be quite a celebrated event, and would quite likely
have made a strong impression on him.[59]

Early in his life Giulio Fabrizio managed to set up a true astronomical facility,
taking his interest in astronomy beyond a mere hobby. By 1852, at the age of
37, Giulio Fabrizio Tomasi had built a private observatory in his summer estate

---

[59]Personal communication by Ileana Chinnici, Osservatorio Astronomico di Palermo.

Fig. 1.9   Osservatorio ai Colli of the Prince of Lampedusa. Courtesy of Gioacchino Lanza Tomasi

in the outskirts of Palermo known as Villa Lampedusa in San Lorenzo ai Colli, or Villa ai Colli (Fig. 1.9). At that time, Giovanni Battista Guccia e Guccia had not yet been born. The observatory had two stories, the upper one being a rotating dome. It was equipped with telescopes, astronomical clocks and pendulums, marine and portable chronometers, barometers and hygrometers, globes and stars atlases. Giulio Fabrizio had several types of telescopes: a Gregorian telescope, a refractor telescope (built by the firm Merz und Sohn from Munich), an equatorial telescope (built by the firm Lerebours et Secretan from Paris, Fig. 1.10), and an azimuthal telescope (by the firm Worthington from London).[60] It is not known how and when most of these instruments were acquired, but an Italian nobleman, also an amateur astronomer, donated one of the telescopes to Giulio Fabrizio. The director of the Palermo Observatory at the time, Gaetano Cacciatore, could have been the person who helped Giulio Fabrizio equip his observatory. Gaetano Cacciatore was the son

---

[60]I. Chinnici, *Old astronomical instruments on a movie set: the case of "The Leopard"*, Bull. Scientific Instrument Soc., 109 (2011), 9–13.

Fig. 1.10 Lerebours et Secretan telescope of the Prince of Lampedusa. Courtesy of the Astronomical Observatory of Palermo

of Niccolò Cacciatore, an astronomer who was a student and collaborator of Piazzi and the first director of the observatory after him.

Giulio Fabrizio started publishing his astronomical observations in 1853 in the newspaper Giornale Officiale di Sicilia. On August 26, the newspaper reported that

> The Prince of Lampedusa, who in his pleasant Villa ai Colli has built a small observatory equipped with a beautiful telescope [. . .], has seen the comet. . . [61]

In the following years, he regularly sent letters to the editor of the local newspaper, which appeared in the section Notizie Varie, announcing his observations concerning sighting of comets, the positioning of stars, and the conjunction of planets. These letters appeared mostly in the period 1853–1859. In 1857, confusion in the measurements combined with the rush of the press led to the erroneous

---

[61]Giornale Officiale di Sicilia, August 26, 1853.

announcement of the discovery of a new comet, which was immediately named by the press as comet Tomasi. The error in measurement was easily explained as a result of normal observational activity. However, from that time on, Giulio Fabrizio Tomasi reduced the number of observations he published. The last letter he sent was in 1881, when he was 66 years old.

A highlight of Giulio Fabrizio's astronomical observations was the solar eclipse that occurred on December 22, 1870. The eclipse was visible along the southern coast of Sicily, near the town of Agrigento. He went there to observe it, carrying the necessary instruments, and aided by Father Saverio Pirrone, his collaborator in the astronomical observations and the family priest.[62] His observations were incorporated in the official report of the Palermo Observatory, which was edited by Gaetano Cacciatore. The handwritten notes with the measurements taken during the expedition are kept in the Archive of the Palermo Observatory.[63]

According to Gaetano Cacciatore, the observatory had first-class observational instruments, and some of them were even better than those in use at the time at the Palermo Observatory. As a result, it is a fair to assume that Guilio Fabrizio's observations were of a very high quality. Indeed, in 1883, the renowned astronomer Pietro Tacchini conducted an official survey of the Italian observatories as part of their reform. In his report, he considered that the only private Italian observatory worth mentioning was that of the Prince of Lampedusa.[64] Modern evaluations highlight that Giulio Fabrizio Tomasi was very skilled at measurements and at calculations, pointing out that he must have had specialized training.[65] Unfortunately, it is not known how he acquired such astronomical training.

We do know that Giulio Fabrizio Tomasi had books on science. The library of the Palazzo Lanza Tomasi in Palermo today holds the few books that survived the bombardments in 1943 of the Palazzo Lampedusa. Those books show Giulio Fabrizio's amateur scientific curiosity. Indeed, there is *Cosmos. Essai d'une description physique du monde*, a translation into French of the great classic work by Alexander von Humboldt, and also *Il cielo. Nozione astronomiche* by Dionysius Lardner, a successful author who specialized in the popularization of science. Both books include almost one thousand pages with explanations of the planets and their movements, the sun and the stars, the comets and the nebulae, as well as practical recommendations on how to make astronomical observations.[66]

There are two mathematical books in what remains of Giulio Fabrizio's library. One is the *Dizionario delle matematiche pure et applicate*, a translation into Italian of a nine-volume work written by Alexander Sarrazin de Montferrier and first published in Paris in 1835–37. The book is a compilation of diverse materials.

---

[62] A. Vitello, *op. cit.*, pp. 377–380.

[63] G. Cacciatore, *Rapporti della Commissione italiana per l'osservazione dell'ecclisse del 22 dicembre 1870*, Osservatorio Astronomico di Palermo, 1872.

[64] P. Tacchini, *Eclissi totali di sole*, 1888, Rome, pp. 111–112.

[65] Personal communication by Ileana Chinnici, Osservatorio Astronomico di Palermo.

[66] Personal library of Gioacchino Lanza Tomasi, Palermo.

Indeed, we find, for example, entries such as "Fatio de Duillier", one page devoted to the life and works of the Swiss mathematician and his role in the Newton versus Leibniz controversy over the invention of calculus. There is also the entry "Fattoriale", which contains fifty pages of facts about infinite expansions and the Gamma function. Each volume has some 500 pages, the ninth volume consisting of 253 mathematical tables of different type. Giulio Fabrizio could have used the tables of trigonometric functions and logarithms for his astronomical measurements.

The other mathematical book is *Élèments d'Algébre*, written in 1817 by Louis Pierre Marie Bourdon, a professor at the École Polytechnique in Paris. The book presents equations and systems of equations, polynomials and rational functions, sequences, exponentials and logarithms, the usage of tables, numerical methods for solving of equations, complex numbers, finishing with some facts on infinite series and power expansions. A large number of universities and technical academies around the world used it as a textbook for many years. The book could very well have been used by Giulio Fabrizio, as it is highly pedagogical in its style and moderate in its mathematical ambitions.

It is more than plausible that Giulio Fabrizio obtained these books following the advice of the director of the Palermo Observatory. This hypothesis is reinforced by the fact that the library of the Palermo Observatory also holds copies of the books just described.

The "Osservatorio ai Colli del Principe di Lampedusa" did not survive long after Giulio Fabrizio Tomasi's death. The family dismantled the observatory and sold the instruments, some of which were bought by the Palermo Observatory, where they remained in use for many years. At the beginning of the twentieth century, the observatory was demolished. Three of the telescopes mentioned above, as well as the renowned beautiful telescope known as the Palermo Circle used by Giuseppe Piazzi for the discovery of Ceres, can be seen today in the Museum of the Observatory of Palermo, situated on top of the Palazzo dei Normanni in Palermo.[67]

As a boy, Giovanni Battista Guccia was likely exposed to Guilio Fabrizio's career in astronomy. Surely he was aware of the scientific, however amateur, activities of his uncle Giulio Fabrizio; of the existence of the observatory; and of his uncle's publications in the newspaper. It was a type of activity quite different from the common lifestyle of nobility in Palermo, and surely memorable for that reason alone.

This long and winding genealogical journey has allowed us to establish that the Guccia family was a prosperous one, which acquired noble rank, and which was able to maintain its fortune. The patriarch, Giovanni Battista Guccia e Bonomolo, Marquis of Ganzaria, was a hard-working man, accustomed to pursuing and achieving ambitious goals. Over the course of time, the Guccia family became related to the Tomasi family, a traditional noble Sicilian family. From this contact a connection with the world of science was established.

---

[67]Museo di astronomia, Osservatorio Astronomico di Palermo.

The fate of the Tomasi family contrasted with that of the Guccia family. Since the abolition of feudalism, the laws on primogeniture in 1812 and the effects of the French Revolution, many Sicilian aristocratic families experienced severe economic difficulties. The Tomasi family was no exception to this. At the beginning of the nineteenth century, however, the family sold the island of Lampedusa, which restored its wealth. Further, at the end of the summer of 1885, a cholera outbreak spread throughout Sicily. In fear, Giulio Fabrizio Tomasi moved with his family to Florence. A few days after his arrival, the Prince of Lampedusa contracted the disease and died.[68] At the time of his death, nine of his children were alive. Although he had arranged for weekly masses for his soul since his youth, he had not taken the same care about earthly matters, and thus died intestate. His death caused serious disagreements over the family's inheritance, which ended in a legal quarrel which pitted the sons of the Prince of Lampedusa and one of his grandsons (the father of the author of *The Leopard*) against the widow, Maria Stella Guccia e Vetrano, and his daughters. This controversy was combined with financial incompetence and lack of interest in working or earning money, which was common in the Tomasi family. The situation was exacerbated by the everlasting lawsuit that for more than fifty years prevented the partition of the estate. The consequence was the gradual ruin of the Tomasi family.

---

[68] A. Vitello, *op. cit.,* p. 408.

Chapter 2
The Formative Years

This chapter deals with the school and university education of Giovanni Battista. He was a boy when the Kingdom of Italy was founded and a new school system was created. We briefly describe the school system of the Risorgimento and then follow the steps of young Giovanni Battista through his middle school and technical education. With respect to the development of universities in Sicily, it is important to note that Sicily and Naples had remained isolated from the rest of the Italian peninsula for centuries. The establishment of a university in Palermo, then, was negotiated exclusively between Sicily and Naples. We trace the political difficulties that stalled the creation of the University of Palermo until 1805. A short intermission allows us to make a presentation of mathematicians born in Sicily, from Archimedes to Francesco Maurolico and Niccolò Cento. Giovanni Battista's way through the University of Palermo was cut short due to the most important encounter of his life. It was at a meeting of the Società Italiana per il Progresso delle Scienze in 1875 that Guccia met Luigi Cremona, the renowned algebraic geometer. After meeting Cremona, he moved to Rome and enrolled in the university, where he came into contact with the Italian mathematical academic world. Five years later, he presented his thesis under the guidance of Cremona.

2.1 The Educational System of the Risorgimento

Giovanni Battista was five years old when Sicily, and the city of Palermo in particular, was struck by the Expedition of the Thousand. The Expedition of the Thousand, also known as the Red Shirts, was an army of mostly Northern Italian volunteers. Led by Giuseppe Garibaldi, the Expedition of the Thousand ended the Bourbon Kingdom of the Two Sicilies, and opened the way for the Unification of Italy. The army seized Palermo from King Francesco II's army in three days. During

B. Bongiorno, G. P. Curbera, *Giovanni Battista Guccia*,
https://doi.org/10.1007/978-3-319-78667-4_2

the seize, the Garibaldines freed Palermo's jails, and the Bourbon troops bombed the city.[1]

One of the first and main political goals of the new Kingdom of Italy was that of enhancing public education. Until the nineteenth century, education in Sicily—elementary, secondary, and higher education—had been in the hands of the Catholic Church. Traditionally, education, even elementary education, was restricted to the higher and urban classes. The influence of the Catholic Church on the education system became even more pronounced after the 1818 Concordat between the Kingdom of the Two Sicilies and the Catholic Church, as the two entities agreed that bishops and parish priests would act as school inspectors of their dioceses. Around the middle of the nineteenth century, the Bourbon regime made a weak attempt to introduce compulsory elementary education. Nonetheless, illiteracy in the island was widespread, ranging up to almost ninety per cent.

After the annexation of Sicily by the Kingdom of Italy, the school system throughout the country changed radically as a result of the Casati law of 1860. The law was established by the Minister of Public Education of Sardinia, Count Gabrio Casati. A member of the committee that developed the law was the mathematician Francesco Brioschi, tutor of Guccia's later mentor Luigi Cremona. This law regulated Italian education, and was based on the modern principles of compulsion, gratuity, and equality of gender. This was already a significant change from the existing approach to education. Elementary education was entrusted to municipalities, secondary to the provinces, and higher education was set under the control of the State.

The backbone of the new school system was secondary school education, with professors who were civil servants, appointed by competition, and with retirement pensions. Secondary school was divided into two branches: classical and technical studies. Classical studies, which were aimed at further studies in the university, consisted of eight years: five years of gymnasium and, through an entrance exam, three years of high school. The second branch, technical studies, consisted of six years, organized into three years of technical school and three years of technical institute. They were designed as a practical and professional apprenticeship. The exception was the physical-mathematical section of the technical institute, which allowed access to the faculties of Science and Engineering in the university, while students from other sections of the technical institute had to go through an individual entrance exam to be admitted into the university.[2]

The effect of the Casati law on Italian education was undeniably beneficial to the state of education. There was, though, a weak point of the Casati law: the elementary school system. Financing for the elementary schools depended on individual towns and cities, and thus the quality of teachers and resources was

---

[1] A. Scirocco, *Garibaldi: battaglie, amori, ideali di un cittadino del mondo*, Laterza, 2001, Bari.

[2] For the effect of the Casati Law in teaching of mathematics in secondary school, see L. Giacardi (ed.), *Da Casati a Gentile. Momenti di storia dell'insegnamento secondario della matemática in Italia*, Agora Publishing, 2006, Lugano.

dependent on each town's wealth and interest in education. Under these conditions, the recruitment of elementary teachers became a serious problem. This resulted in having poorly qualified teachers, with modest salaries and low social prestige. Too often schoolteachers had to hold a second job to be able to make up a living (which at times was quite a humble one, as tailor or shoemaker).

In Sicily, the ending of the Bourbon Kingdom meant the end of the stifling ecclesiastical control over public education. This fact itself was a genuine moral and intellectual revolution. The Casati law marked a crucial moment in the history of the island: from the first moment of the Unification, the primary, secondary and upper schools were secular, public, inspired by liberal principles and ruled by a staff only subjected to the State laws. The public expense in education (which before the Unification was a third of that of Piedmont) doubled and even tripled in a short period of time. Starting from the academic course 1860–61, the number of schools, classrooms, teachers and staff multiplied. During that school year, there were 34 elementary schools for boys and one for girls in the province of Palermo, whereas in the 1868–69 school year there were 554 elementary schools for both boys and girls. Thus, compulsory education was brought to places where its existence and need was unknown. When viewed quantitatively, these achievements were undoubtedly impressive. However, they nevertheless fell short of the needs of the Sicilian community at the time. At the fall of the Bourbon regime, 89 percent of Sicilians were illiterate, and only a few of the remaining were able to speak and write Italian. Forty years after the Unification, illiteracy in Sicily had gone down to 71 percent, but this was still fifty percent higher than the national average, and four times higher than in the Piedmont region. The reasons behind Sicily's continued illiteracy are many and complex, but one reason may have been the fact that voting rights were contingent upon a minimum level of education.[3]

2.2 Giovanni Battista in the Technical Institute

As we have established, all personal records of the Guccia family were lost. Consequently, the information that we have regarding Giovanni Battista's school years is necessarily indirect. Because of this, understanding and following his school and university education becomes particularly difficult.

According to Giovanni Battista's obituary, written by the mathematician Michele de Franchis, who was Guccia's successor in the Circolo Matematico di Palermo, Giovanni Battista received his primary education in a private school.[4] In 1865 he was ten years old, the age when secondary school started in the new system designed by the Casati law. We do not know whether he was sent to the gymnasium, to the

---

[3]F. Renda, *Storia della Sicilia dalle origini ai giorni nostri*, vol. 1, Sellerio, 2003, Palermo, pp. 251–253.
[4]M. De Franchis, *G.B. Guccia, Cenni biografici*, Rend. Circ. Mat. Palermo, 39 (1915), pp. i–x, p. i.

technical school, or whether he continued his education at the private school. It is possible that he spent two years in the gymnasium (or in a Catholic seminary, like his father did) and, disliking classical studies, chose not to continue. His early schooling may have also been affected by the onset of health problems that would afflict him throughout his entire life, and which obliged him to rest for two years. As we will later see, he was often affected by fever and even by bronchopneumonia, especially after strenuous efforts, such as, for example, long travel by train. We do know that on October of 1870, when he was about to turn fifteen years old, he was sent to follow technical education in the technical institute of Palermo. He started two years later than the educational regulations prescribed.

Let us take a brief look at the technical institute where young Giovanni Battista did his secondary studies. Recall that this type of institution was a novelty brought by the Unification of Italy and the Casati law.

The first Technical Institute of Palermo was created by the Ministry of Agriculture, Industry and Commerce in 1862. The institute started with many difficulties. As there was no facility available, classes were delivered in the main floor of an eighteenth century palace, Palazzo Comitini, in the center of the city, by Via Maqueda. Classes did not start until December that year, and the schooling period was shortened by one third. At first, the Technical Institute comprised four lines of study or sections: commercial and administrative studies, agronomy and agriculture, mechanics and construction, and mercantile marine studies. The full education system of Italy was still developing, especially the areas of industrial and professional education. Thus, the academic course 1864–65 started with changes: students could choose between the school of commerce and administration, the school of agronomy and surveying, the school of mechanics and construction, and a new school of tanning and leather finishing. These new technical studies reflected the strength in Palermo of the tanning industry. Merchant marine studies were returned to a specialized nautical college.

The year 1868 was another year of changes for the Technical Institute. The Institute acquired a temporary site at the Calasanzio Royal College (where a boarding school for nobles was situated before), and assumed the name "State Institute for Industry and Professions". The following year, the institute gained a renewed importance, as it was declared the seat for license exams in foreign languages. In 1872, a new section for physics and mathematics was established, and the industrial section opened in 1874. By 1876, the range of studies in the institute included: commerce, agronomy, land surveying, industry, and physics-mathematics. In 1882, following the request of the board of professors, the name was changed to Technical Institute Filippo Parlatore, after a distinguished naturalist and botanist from Palermo. There was high demand for the Institute, and it was very successful: by 1885 it was the fourth largest technical institute in Italy, only after those of Genoa, Naples and Milan. In 1887, it was moved to its permanent site, a new building where the old monastery of Montevergini used to be situated.[5]

Our source of information regarding Giovanni Battista in school are the records of the Technical Institute, precisely, the first book of official proceedings of the Institute.[6]

The first school year for Giovanni Battista ran from October 1870 to July 1871. There were twenty-seven students in the first course. That year, all first-year students from the different sections of the institute were instructed together. On July 1871, the board of professors of the Institute met to decide which students were admitted for the final examination. Of the twenty-seven students in the class, five received a grade of 9 or 10 (10 being the maximum); seventeen had a grade of 7 or 8; four had 5 or 6; and one failed with a grade of 3. Giovanni Battista passed, but with a low grade of 6. While many students were exempt from paying final exam fees, Giovanni Battista paid for the exam in full, as he came from a wealthy family. Some students in the section on mechanics and construction were excluded by the board of professors from the final examination, as they had attended the classes without being official students of the institute.

The next school year, 1871–72, Giovanni Battista enrolled in the mechanics and construction section, where there were twenty students. After the evaluations in June 1872, two students got a 9, five had a grade of 8, ten got a 7, and three a grade of 6. This time, Giovanni Battista did better than he did the previous year, both in absolute and relative terms, and achieved a grade of 8.

The last school year for him at the Institute was 1872–73. This was the year when the new section on physics and mathematics was introduced in the Institute, but it was only offered to the first-year students. This year, there were twenty-two students in Giovanni Battista's class. Looking at their final grades, it seems that the courses were becoming more difficult, and some students did not graduate. One student received a grade of 9, two an 8, sixteen had a grade of 6 or 7, and three had 5 or less. Giovanni Battista's grade was again a modest 6. In any case, he graduated and obtained his diploma in mechanics and construction from the Technical institute of Palermo in 1873 at the age of 18. His school record was quite modest.

Why did the Guccia e Guccia family send Giovanni Battista to the Technical Institute? Why did he not follow the more prestigious classical education?

The Guccia family, as far as we know, had gone through higher education. The family patriarch, Giovanni Battista Guccia e Bonomolo, according to his marriage certificate with his first wife Agatha Agati and to the baptismal certificates of his sons kept in the Archive of the Diocese of Palermo,[7] held the title of doctor in Civil and Canon law. His grandson Giovanni Battista Guccia e Di Maria, who was Giovanni Battista's grandfather through his mother's side, almost surely followed

---

[5] *150 anni di attività tra passato e futuro 1862–2012,* Istituto Tecnico Statale per Geometri "Filippo Parlatore", Annuario 2012–2013, Palermo.

[6] Records 1871–1873. Istituto Tecnico Filippo Parlatore, Palermo.

[7] Section "Parrocchia Santa Margherita alla Conceria". Archivio Storico Diocesano di Palermo.

classical education and graduated in Civil law, although his record is not found in the Archive of the University of Palermo, possibly due to incomplete cataloging. At the age of thirty-three he obtained the position of adviser of the Grand Court of Audit; some years later he became Vice-President of the court. At the time, the degree in law at the University of Palermo required three years of study: Natural and Canon law the first year, Roman law the second, and Civil law and political economy the third. In order to practice as a lawyer, it was necessary to complete an internship at a law firm and to pass a demanding examination that could last up to twelve hours. It seems that no other member of the Guccia family was interested in such a rigorous professional path. The father of Giovanni Battista, Giuseppe Maria Guccia e Vetrano, graduated from the Catholic seminary of the Congregation of the Oratory of Saint Filippo Neri.[8] His experience at the seminary affected him quite negatively, and caused him to withdraw from the religious vocation. He then married four years later and devoted his energy to his family, and to a number of profitable businesses that developed from his properties.

In the Guccia family, economic activities were constant, and there were many businesses to take care of daily. With time and effort, business activity expanded the family's estate. We have already mentioned that the patriarch Giovanni Battista Guccia e Bonomolo was able to significantly increase the family's properties. The new Guccia e Guccia family was also devoted to increasing its wealth. We have already commented that in 1850 Giovanni Battista's mother bought a palace on Via Ruggiero Settimo. In 1859, Giovanni Battista's father bought two properties: one in the area of San Lorenzo ai Colli, near the summer villa where the Prince of Lampedusa had his observatory; and another one, quite large, in the district of Ciaculli, on the outskirts of Palermo.[9]

The Guccia family members used funds and savings from other members of the family for these economic transactions, as well as for mortgages on properties and rents, legal services, and interventions by public notaries. In addition, there were the wills of relatives, in particular, of both of Giovanni Battista's grandmothers. Concetta Vetrano, mother of his father, died in 1865, and Carolina Cipponeri, mother of his mother, died in 1845. The latter's inheritance was disputed and finally divided between her legatees twenty-seven years later in 1872. These testaments were full of complicated, intertwined economic bequests, and at times, were followed by lengthy legal suits. This, for example, is what occurred with the property and subsequent division of the family's Palazzo Guccia del Bastione. These new properties and inherited monetary deposits also generated further economic transactions. The family also paid great attention to smaller business endeavors. The commercial rooms on the ground floor of the new Palazzo Guccia were soon leased, and the first

---

[8]G. Guccia, *Programma del saggio filosofico che darà il chierico Giuseppe Guccia novizio della Congregazione dell'Oratorio di S. Filippo Neri, nel Maggio 1837,* Roberti, 1837, Palermo. Biblioteca Centrale della Regione Siciliana, Palermo, MISC. A 523.

[9]Notar Giacinto Di Benedetto, March 27, 1859; notar Giuseppe Maria Terranova, April 5, 1859. Archivio Notarile Distrettuale di Palermo.

floor of the Palazzo was leased to a company crafting fine clothing which turned it into a large workroom with 150 young women sewing.[10]

The management of the water company (which, as we have seen, was the true fountain of the family wealth) required special and constant attention. In 1861, the "Amministrazione delle Acque di Casa Guccia" was temporarily put out to lease. This could have been due to the older age of Giovanni Battista Guccia e Di Maria, who had been managing the company, and the lack of a natural successor within the family. Years later, it was Giovanni Battista Guccia e Guccia, the mathematician, who took care of the daily running of the company. A critical situation occurred after the cholera outbreak of 1885. The Mayor of Palermo decided that the water supply of the city should be channeled by cast iron pipes. This required a quite expensive restructuring of the entire water supply system, and would have meant large expenses for the existing private water companies. In retaliation, the Guccia family and other private water companies started a legal process against the mayor's decree in 1886. We will later come back to this episode. What yielded from these changes, however, was the installation of many public water fountains throughout Palermo.

As seen, the family lived off the revenue of its businesses and properties. In fact, when Giovanni Battista enrolled in the University of Rome he listed his father's occupation as "Proprietary". The family businesses produced sufficient wealth to allow a pleasant life. The wealth was spent freely, but in an orderly manner. It was a tradition in the family to like sports, particularly horseback riding, in which young Giovanni Battista received special training and practiced regularly.[11] His father was one of the founders in 1862 of the first horse racing society of Palermo[12] (note the drawings in the background of the portrait of Giuseppe Maria Guccia, Fig. 1.4).

Back to the initial question: What could have been the reasons for young Giovanni Battista seeking a technical education? As we have just seen, this decision was surely not based on family tradition, nor on economic necessity.

Giovanni Battista was young at a time when technical and industrial developments were changing the world very rapidly. He had seen such jewels of technology as the telescopes of his uncle Giulio Fabrizio Tomasi. Giovanni Battista could have also witnessed the fleet of the industrialist Ignazio Florio, of more than one hundred steamboats at Palermo's harbor, which connected Palermo with Naples, Genoa, Malta and Tunis. Giovanni Battista's attraction to engineering and its technical creations can be traced throughout his life. He had one of the first cars in Palermo; in 1902, he had a model by the French builder *De Dion Buton*.[13] He installed modern facilities in the Palazzo Guccia. Indeed, in an invitation letter in 1905 he

---

[10]A. Chirco, M. Di Liberto, *op. cit.,* pp. 165–166.

[11]M. De Franchis, *op. cit.,* p. i.

[12]L.M. Majorca Mortillaro, *op. cit.,* p. 50.

[13]R. La Duca, E. Perricone, *Saluti da Palermo 1890–1940, Cinquant'anni di vita della città attraverso la cartolina illustrata,* Dario Flaccovio Editore, 2007, Palermo, pp. 101–102.

informed Mittag-Leffler of the installation of electrical heating in his home.[14] We also know that by 1913 there was an elevator in the palace.[15] The technical and industrial developments could have attracted young Giovanni Battista and induce him to choose a technical education.

2.3 University and Mathematics in Sicily

The first university to be founded in the world was the University of Bologna in 1088. When the first universities were founded in Sicily at the end of the Middle Ages, the island was under the rule of the Crown of Aragon. In accordance with the practice of other Italian states, political power and knowledge were kept separate: the Republic of Venice founded its university in Padua in 1222; the Republic of Florence in Pisa in 1343; and the Duchy of Milan in Pavia in 1361. Thus, King Alfonso the Magnanimous founded a Studium Generale in Catania while the capital remained in Palermo in 1434. Ten years later the Papal bull sanctioning the university in Catania arrived. The studies available were the usual ones at the time: Theology, Canon Law, Civil Law, and Medicine. Importantly, the Studium Generale in Catania was able to confer degrees.

By the middle of the sixteenth century the Jesuits established education in Palermo, with the permission of the Emperor, Charles V, and the blessings from the local nobility, represented by the Senate. The site they used was an old abbey, renamed Casa Professa, and the teachings they offered were many and diverse: Latin, Dialectics, Physics, Metaphysics, Philosophy and Theology. Degrees that could be granted—with the Pope's permission—included only Philosophy and Theology. Soon the patronage of the Senate of Palermo allowed the Jesuits to expand into a new and magnificent building, known as the Collegio Massimo (situated in today's Corso Vittorio Emanuele, where the Biblioteca Centrale della Regione Siciliana is located). Teachings were also expanded to Grammar, Logic and Mathematics. Other studies in Palermo such as Law and Medicine remained privately taught by skillful practitioners.

The main problem for the success in Palermo of this educational institution was that of obtaining a degree after these studies, since degrees could not be granted in Palermo. The only option was taking the courses and exams in the Università degli Studi di Catania, which was rather demanding, as Palermo and Catania are some three hundred kilometers apart. As a consequence, it was not an uncommon situation in Palermo, especially in the eighteenth century, to find lawyers and public

---

[14]Letter from Guccia to Mittag-Leffler, December 28, 1905, Folder Mittag-Leffler, Archive of the Circolo Matematico di Palermo.

[15]*Discorso del Prof. Ing. Michele Albeggiani*, XXX anniversario della fondazione del Circolo Matematico di Palermo, Adunanza solenne del 14 aprile 1914, Suppl. Rend. Circ. Mat. Palermo, 9 (1914), p. 15.

officials that held no valid degree. However, sometimes degrees could be obtained without attending the courses or taking the exams. Instead, a student could obtain valid degrees if two witnesses testified that he had attended the necessary courses. This was a common practice among other European universities at the time.

For a short period of time, just over a century, there was a university in Messina, which was the third most important city in Sicily, some two hundred kilometers east of Palermo, and three kilometers from continental Italy. With the teachings of the physiologist Marcello Malpighi (at the same time as the mathematician Alfonso Borelli) Messina became, for a short period, the center for modern science in Sicily. However, this did not last long: the insurgency of the city against Spain in the years 1674–1678 and the provincial rivalries within Sicily, especially between Catania and Palermo, caused the Università degli Studi di Messina to be shut down.[16] In the turmoil a proposal was ushered: establishing a university in Palermo. The lack of a university was felt as an offense to Palermo, while its creation was seen as a threat in Catania. The Spanish Viceroy at once rejected the proposal, however, and the case for a university for Palermo was lost for a long time to come.

In 1767 the Jesuits were expelled from Sicily, at the same time as they were also expelled from Naples, Spain, Portugal and France. This left the Collegio Massimo unoccupied. After two years, it reopened as a higher educational institution under the auspices—and the hopes—of the Senate of Palermo. However, the institution encountered various organizational and administrative problems. Studies led to no degree, since students still had to go through examination at the Università degli Studi di Catania. Another official request for a university in Palermo was presented in 1777, this time to the King in Naples, and was again immediately rejected. Instead, a Boarding School for Nobles was established. A committee was appointed to be in charge of the Boarding School: it consisted of the Bishop of Catania, the Inquisitor General of Sicily, the President of Court of Justice, two princes and a duke. Admission was highly selective, and followed the social ranking rules in Sicily at the time. Consequently, boarding school students were only accepted from families with more than one hundred years of nobility. The French Revolution was nowhere envisioned; even less in Sicily. In 1779, forty-seven students were admitted to the Boarding School, among them, Giuseppe Tomasi di Lampedusa, father of Giulio Fabrizio Tomasi.

Meanwhile, the Collegio Massimo had evolved into the Royal Academy of Studies, with chairs in Theology, Philosophy, Law, Medicine and Science. Laboratories in Anatomy and in Chemistry, and the Botanical Garden (which still exist today) were also added. Later came the chair in Astronomy that was entrusted to Giuseppe Piazzi. The subsequent story we know: the building of the Observatory of Palermo, and the discovery of Ceres in 1801. However, the granting of degrees was still in the hands of the Università degli Studi di Catania.

---

[16]For more on the Università degli Studi di Messina, see C. Dollo (ed), *Filosofia e scienze nella Sicilia dei secoli XVI e XVII,* Catania 1996, vol. I.

A rapid sequence of events, including the French Revolution and the ensuing fear, better to say panic, within the absolute monarchies; the Jesuits' return to Sicily; and the repossession of the Collegio Massimo in Palermo, resulted in the transformation in 1805 of the Royal Academy of Studies into the Università degli Studi di Palermo. The Theatines were instrumental in finding a building for the new university. In return, they were granted the offices of Rector and Librarian until 1860. The new university initially offered studies in four branches of knowledge: Theology, Philosophy, Medicine, and Law, and was now authorized to grant degrees.

The students, with few exceptions, still came from families from the aristocracy and from the ruling classes. For the lower classes, higher education remained almost inaccessible, unless they resorted to the seminaries to take holy orders.

The Faculty of Philosophy included several courses in mathematics. The courses and their first professors were: Elements of Algebra and Geometry, taught by the priest Giovan Battista Cancilla; Sublime Mathematics, taught by the architect Domenico Marabitti; Mixed Sublime Mathematics, taught by the priest Giuseppe Dalmasse; Mathematical Physics, taught by the priest Diego Muzio; and Astronomy, taught, obviously, by Giuseppe Piazzi.[17]

Years later, twenty-six year old Pietro Blaserna won the chair in Physics in 1863. He later moved to the University of Rome, where he was Rector from 1874 to 1877. As we will see, it was during this period that Giovanni Battista Guccia moved to the University of Rome. We will encounter Pietro Blaserna again, because he presided over the IV International Congress of Mathematicians, which was held in Rome in 1908 and in which Giovanni Battista Guccia and the Circolo Matematico di Palermo played an important role.

Mathematics has long been cultivated in Sicily. It is mandatory to begin with Archimedes, known as "the greatest mathematician of Antiquity".[18] Despite being well known, his close connection to the history of Sicily justifies a succinct review. Archimedes was born and lived in the Greek city of Syracuse in the third century BC. In Alexandria, he mastered the scientific and mathematical knowledge of antiquity, and of Greece in particular. His works are still cornerstones of scientific knowledge and masterpieces of creativity, rigor and style. Only a few of his works have reached us, among them *Quadrature of the parabola* (where the area of a segment of a parabola cut off by a chord is calculated), *On the sphere and cylinder* (where the volume and surface of a sphere are calculated), *Measurement of a circle* (where a procedure is devised to estimate the number π), and *The Sand Reckoner* (where a bound on the number of grains of sand required to fill the Universe is calculated). The works of Archimedes traversed the Mediterranean Sea from Alexandria to Constantinople, to Sicily during the Norman Rule, and later on to Rome, Venice, and the rest of Europe. Throughout this journey, Archimedes'

---

[17]O. Cancila, *Storia dell'Università di Palermo dalle origini al 1860,* Editori Laterza, 2006, Palermo.

[18]T.L. Heath, *The method of Archimedes. A supplement to the Works of Archimedes,* Cambridge University Press, 1897, Cambridge; p. 10.

manuscripts survived, but were also copied, collected, translated, and hosted by the main libraries of Europe. It is highly advisable to learn of these remarkable events through one of their discoverers, the Danish professor of classical philology John Ludwig Heiberg, who in 1906 found the manuscript of *The Method*, the work where Archimedes unveils his practical, even mechanical, procedures for formulating accurate mathematical conjectures. Archimedes was killed in 212 BC, in the conquering of Syracuse by the Roman Consul Marcellus during the Second Punic War (Syracuse had sided with the Carthaginians against the Roman Republic). Plutarch and other classical historians have narrated Archimedes' death as the price of devoting a life to science. According to Cicero, the tomb of Archimedes carried a sculptured drawing, illustrating what he himself considered his most significant mathematical proof: a cylinder circumscribing a sphere, which he used to calculate the volume of the sphere.[19]

Besides this high peak of science, there were other contributors and cultivators of mathematics who came from Sicily. They include: Francesco Maurolico, Giuseppe Moletti, Giuseppe Scala, Giovanni Alfonso Borelli, Giovanni Battista Hodierna (whom we have already met when speaking about the Tomasi family), Benedetto Maria Castrone, and Niccolò Cento.[20]

Francesco Maurolico was born in Messina in 1494, where he died in 1575. He was the son of a Byzantine physician who took refuge in Sicily during the Turkish invasion. He became a Benedictine monk, and later the abbot of the cathedral of Messina. A good example of a Renaissance humanist, he was interested in mathematics, astronomy, mechanics, optics, architecture, and history. He built optical and astronomical instruments, and studied methods for the correct measurement of the Earth. This allowed him to draw accurate geographical maps, which aided the Christian fleet leaving the port of Messina for the Battle of Lepanto. In mathematics, he had an instrumental role in recovering the knowledge of antiquity: he studied and translated ancient classical Greek works by Euclid, Archimedes, Apollonius, and other Greek mathematicians. Maurolico worked on many mathematical topics, among them Geometry, Number Theory and Algebra (he is credited with the first rigorous use of mathematical induction). He taught mathematics at the Università degli Studi di Messina. He left many printed publications (*Opuscula Mathematica, Arithmeticorum libri duo*, and others) and a large number of unpublished manuscripts.[21]

Two Sicilians, Giuseppe Moletti and Giuseppe Scala, participated in the committee of five learned scholars appointed by Pope Gregory XIII, and presided over by the astronomer Christopher Clavius, to reform the calendar. The reform of the

---

[19]Cicero, *Tusculan Disputations*, Book V, Sections 64–66.

[20]A. Brigaglia, *Le scienze matematiche in Sicilia dal riformismo settecentesco all'unità nazionale.* In: L. Pepe (ed.), *Europa matematica e Risorgimento italiano*, Ed. CLUEB, 2012, Bologna, pp. 307–330. See also G.E. Ortolani, *Biografia degli uomini illustri della Sicilia*, 4 vol., Arnaldo Forni Editore, Napoli, 1817.

[21]A. Brigaglia, *Maurolico e le matematiche del secolo XVI*, In: C. Dollo (ed), *Filosofia e scienze nella Sicilia dei secoli XVI e XVII,* Catania 1996, vol. I, pp. 15–27.

existing Julian calendar was a command of the Council of Trent (1545–1563) to adjust the date of the celebration of Easter to the original time of the year in which it had started to be celebrated, after the First Council of Nicaea, back in 325. The inaccurate count of days in the tropical year had caused a displacement over time of ten days from the initial setting. The final decision adopted was to reduce the number of leap years in a period of four centuries from one hundred to ninety-seven, by determining that only one of the centurial years should be a leap year. The reform was set out in 1582, and led to the creation of the Gregorian calendar. Moletti was born in Messina and was a student of Maurolico. He taught mathematics in the University of Padua and was in contact with Galileo Galilei, who took over the chair in mathematics in Padua after Moletti's death.

The birthplace of Giovanni Alfonso Borelli is disputed between Messina and Naples. In any case, he was born in 1608. He studied mathematics in Rome, where he became acquainted with Evangelista Torricelli and Bonaventura Cavalieri. In 1640, he was appointed professor of mathematics in Messina. There, Borelli became interested in the unpublished manuscripts left by Maurolico on Greek mathematics. In 1658, he wrote *Euclides restitutus,* where he intended to present Euclid's Elements in a more transparent way. A few years later, he found in the Laurenziana Library in Florence an Arabic manuscript containing a lost work of Apollonius: books 5, 6 and 7 of the *Conics,* which he published in Latin. He was also interested in astronomy, in particular, the orbits of planets. His work in this field was well considered by Newton, who had a copy of Borelli's work and referred to it on several occasions.

Benedetto Maria Castrone was a Dominican monk, born in Palermo around 1668. He traveled across Italy, France and Germany, deepening his mathematical studies at various universities. In 1705, he published in Venice the *Episagogicon geometricum,* which was designed as an introduction to Euclid's Elements, and included a discussion of axioms and postulates.

An interesting Sicilian figure in mathematics from the eighteenth century was Niccolò Cento, born in Palermo in 1719. His teacher was Melchiorre Spedalièri, a follower of the Cartesian method, who taught geometry at the Collegio Massimo in Palermo. Cento was professor of mathematics at the Academy of Palermo. He had carefully read the works of Newton and Leibniz. It seems that he intended to deduce consequences of this new knowledge emerging from the Scientific Revolution. Unfortunately, Cento did not publish his manuscripts on physics and philosophy.[22]

---

[22] A. Brigaglia, P. Nastasi, *Due matematici siciliani della prima metà del '700: Girolamo Settimo e Niccolò Cento,* Società di Storia patria per la Sicilia Orientale, 1981, Catania.

2.4   Giovanni Battista at the University of Palermo

Giovanni Battista obtained his Diploma from the Technical Institute on September 17, 1873. He could have continued on to the University of Palermo, but he did not. The reasons for this decision are unknown. He could have felt unprepared for the entrance examinations required at the university, and thus decided to spend one year preparing for them. He also may have suffered from health problems, as we commented before. In any case, Giovanni Battista (Fig. 2.1) remained out of school during the academic year 1873–74 with some dedication unknown to us.

At the time, the academic course started in November and ran until mid-June. Entrance exams for the university consisted in a five-hour written exam on the Italian language, a test on translation between Italian and Latin, and an oral exam on Geometry, Trigonometry, and Algebra. Giovanni Battista took the entrance exam on November 11, 1874, and passed it with quite a good grade, 27 points out of 30.

Fig. 2.1 Young Giovanni Battista Guccia. Courtesy of Cesare Crescimanno

He signed up in the Faculty of Mathematics, Physics and Natural Sciences, which at that time comprised the School for Applications to Engineering, the School of Pharmacy, and a degree course in Chemistry and Pharmacy.

There were thirty regular courses taught in the Faculty, which covered the disciplines in all areas. Apart from the regular courses, there were two 'free' courses offered. Of the regular courses, eight were in mathematics:[23]

Differential and Integral Calculus, taught by Giuseppe Albeggiani
Analytic Geometry, taught by Filippo Maggiacomo
Complementary Algebra, taught by Antonino Giardina
Rational Mechanics, taught by Michele Zappulla
Descriptive Geometry and Design, taught by Giuseppe Patricolo
Theoretical Geodesics, taught by Francesco Caldarera
Astronomy related to Geodesics, taught by Gaetano Cacciatore
Exercises in Astronomy, taught by G. Cacciatore aided by Pietro Tacchini

Of the two free courses, one was on Higher Analysis and Geometry, and was taught by Giuseppe Albeggiani.

Some of these professors we have already encountered while looking into the observatory of Giulio Fabrizio Tomasi; namely, the astronomers Gaetano Cacciatore and Pietro Tacchini. Cacciatore was the professor of astronomy at the University of Palermo. He was the son of Nicolò Cacciatore, one of Piazzi's collaborators, and he succeeded his father as director of the Palermo Observatory. He held this post for more than thirty years. Tacchini was a respected astronomer who directed several observatories in Italy and participated in many enterprises throughout Italy in astrophysics, seismology and meteorology. He founded several scientific societies and was member of the Royal Society.[24] We will encounter the mathematicians Albeggiani and Caldarera, and the astronomer Cacciatore, again when we discuss the creation of the Circolo Matematico di Palermo.

Giovanni Battista's choice to study within the School for Applications to Engineering followed his secondary studies at the Technical Institute. In that academic year, it also was not possible to study mathematics in the University of Palermo. It was in the next academic year of 1875–76 that the bachelor degree (licenza) in mathematics and physics started. The master degree (laurea) in mathematics in the University of Palermo was not established until ten years later in the academic year 1885–86.

The studies for a master degree in Engineering consisted of five years. In the first year, there were five courses: Analytic Geometry, Complementary Algebra, General Chemistry, General Design, and Physics. The daily schedule for the students was mild: there were classes from Monday to Saturday, three hours of lectures every

---

[23] Annuario Accademico 1874–75, Università degli Studi di Palermo.

[24] L. Gariboldi, *Tacchini, Pietro.* In: Th. Hockey, V. Trimble, Th.R. Williams, K. Bracher, R. Jarrell, J.D. Marché, F.J. Ragep (eds.), *Biographical Encyclopedia of Astronomers,* Springer, 2007, Berlin, pp. 1119–1121.

day, starting at 9 or 10 am. Classes were held in the building that previously was the Monastery della Martorana, on Via Maqueda.

Giovanni Battista only chose four courses, leaving aside the course in General Design, possibly because he was thinking of studying mathematics in the next years. He took the exams at the end of the academic year and on July 1875 he passed General Chemistry and Analytic Geometry, both with 27 points out of 30. On November that year he passed Complementary Algebra with 24 points out of 30. He dropped the Physics course.[25]

2.5   The Palermo Meeting of the Società Italiana per il Progresso delle Scienze

At the beginning of the nineteenth century there was a strong development in science—an explosion, one could say. The reasons for this were many and diverse, but two of them can be singled out. On the one hand, it was the economic growth deriving from the expansion of the Industrial Revolution throughout Europe which required scientific developments. On the other hand, higher education in France was expanded and restructured as a result of the French Revolution, the abolishment of Medieval universities, and the creation of the French *grandes écoles*. A new model of a research-oriented university emerged from the foundation of the University of Berlin by Wilhelm von Humbold in Prussia. The explosion in science was measurable in terms of production of results and publication of articles and books, and also in the number and diversity of practitioners of science, due mainly to the increase in the number of universities and positions for professors.

Communication within the scientific community had to be adapted to these new times and expectations. Established scientific institutions, such as the Académie des Sciences in Paris, the Royal Society, and the Berlin Academy, already provided a platform for dialogue among scientists. However, many young scientists viewed these old academies and scientific societies with skepticism and disillusionment due to their elitist and conservative attitude.

The first broad and open association of scientists to be created was the Gesellschaft Deutscher Naturforscher und Ärzte (the Society of German Natural Scientists and Physicians). Lorenz Oken, natural philosopher and physician, founded it in Leipzig in 1822. The Society aimed to become a forum for oral presentation of research results, favoring discussions and comments by the colleagues attending, enhancing personal contacts, and promoting a friendlier style in scientific publications. The Society met annually. Following this model, the British Association for the Advancement of Science was founded in 1831 in York (chosen for being "the most central city in the three kingdoms"). Charles Babbage

---

[25] Serie Scienze—Carriera scolastica degli studenti 1873–1875, 861/3369, no. 36. Archivio storico, Università degli Studi di Palermo.

was one of its founding members. The objectives of the association complemented those of the German society:

> To give a stronger motivation and more systematic direction to scientific inquiry, to obtain a greater degree of national attention to the objects of science, and a removal of those disadvantages which impede its progress, and to promote the intercourse of the cultivators of science with one another, and with foreign philosophers.[26]

In a similar vein, the American Association for the Advancement of Science was founded in Philadelphia in 1848. Some years later, its constitution was approved. The constitution declared the Association's objectives, which were similar to those of the other European societies, but included several differences that are worth mentioning:

> The objects of the Association are, by periodical and migratory meetings, to promote intercourse between those who are cultivating science in different parts of the United States; to give a stronger and more general impulse, and a more systematic direction to scientific research in our country; and to procure for the labors of scientific men, increased facilities and a wider usefulness.[27]

In this context of progressive organization of science, the first general meeting of Italian scientists was held in Pisa in 1839. In each of the following seven years there was a meeting in a different city: Turin, Florence, Padua, Lucca, Milan, Naples, and Genoa. At that time, Italy was divided into a puzzle of seven different states: the Kingdom of Piedmont-Sardinia, the Kingdom of Lombardy-Venetia, the Kingdom of the Two Sicilies, the Papal States, the Grand Duchy of Tuscany, the Duchy of Modena, and the Duchy of Parma. This complicated political situation, along with the Austrian Empire's political control over some of these states, prevented the creation of a 'national' scientific association, and placed constant impediments on the meetings. In fact, the Austrian police abruptly interrupted the ninth meeting of Italian scientists, held in Venice in 1847. This political intervention, together with the popular uprisings that spread throughout Europe in 1848, stopped the series of congresses.

The Unification of Italy in 1860 allowed the meetings of scientists to resume, the aims of which matched perfectly with the aims of the newly created state. In September 1861 there was an extraordinary meeting in Florence, intended as a celebration of the long-awaited unification of the country. At the following congress in Siena in 1862, scientists proposed the creation of the Società Italiana per il Progresso delle Scienze (the Italian Society for the Advancement of Science) in order to establish regular meetings of the Italian scientific community. The next congress had to wait more than ten years before it was held in Rome in 1873, where a Standing Committee for the society was elected. The physicist Pietro Blaserna was elected member of this committee.

---

[26]O.J.R. Howarth, *The British Association for the Advancement of Science: a retrospect 1831–1921*, 1922, London, pp. 16–17.

[27]Constitution of the American Association for the Advancement of Science, 1856.

The XII congress of the Società Italiana per il Progresso delle Scienze was organized and held in Palermo in 1875, from August 29 to September 7. For a city like Palermo, the meeting was a very important event, with many visitors from the peninsula and the presence of important scholars and politicians. Even the heir to the throne attended the meeting. The venue of the congress was the building of the former Collegio Massimo of the Jesuits. The congress was opened with a flowery speech by Count Terenzio Mamiani della Rovere, the president of the congress and a senator of the Kingdom. In his speech, he inevitably mentioned the discovery of Ceres by Giuseppe Piazzi, one of the scientific highlights of Palermo.[28]

The congress program was broad, as it comprised all branches of scientific knowledge. It was organized into ten sections (*classi*, in Italian), which were the following:

    I. Mathematics, Astronomy, Physics and Meteorology
   II. Engineering
  III. Chemistry and Mineralogy
  IV. Zoology, Comparative Anatomy, Botany and Geology
   V. Anatomy, Physiology and Medicine
  VI. Geography, Anthropology, Ethnography and Linguistics
 VII. Philology, History and Archeology
VIII. Statistics, Economy and Political Science
  IX. Legal Science
   X. Philosophy and Pedagogy

The congress was also large in terms of participation: according to the list printed in the proceedings, there were 788 participants, as well as some foreign guests. This figure is high, but it was roughly half of the attendance of the congresses held before the Unification of Italy. Many learned academies and societies sent their delegates. One of the peculiarities of these congresses was the large participation of educated public, alongside the professional scientists.

Approving the Regulations of the society was the first task of the meeting. Article 4 specified who was entitled to be member of the society. It is of interest to understand these criteria

The Italians who have already attended one of past congresses.
The members of the Academies or of other institutions that publish their proceedings.
The directors of higher education or scientific establishments.
The professors in service or emeritus.
All those who hold a degree or diploma from any University or from other higher institutions of the State.
The officers of the army or the navy.
Those who are proposed by two members of the society.

---

[28]*Atti del duodecimo congresso degli scienziati italiani tenuto in Palermo nel settembre del 1875*, Rome, 1879, p. xxiii.

These Regulations show a declared intention of exhibiting an open and liberal attitude directed at attracting the urban classes to approach science. Article 6 declared that women could be members of the Society, with the same rights and obligations as other members. It was decided that the Society would hold a meeting every three years. Despite these intentions, the following meeting was held thirty-two years later in 1907. The Standing Committee would consist of ten members, each elected by one of the different sections of the congress. The committee members would then elect one among them as President of the Society.

Pietro Blaserna presided over the first section on mathematics, astronomy and physics. It met every day from August 30 to September 7, for two or three hours a day. The first day was devoted to organization of the section: the members agreed to treat meteorology and matters of the astronomical observatory separately. In the second session, the first decision was to send a telegram to the mathematician Enrico Betti, who could not participate in the congress. Attendance was high, at least during the first sessions with some 50 participants. Most of the time was devoted to presentations on the electrostatic induction and other matters related to electricity; this topic created lively discussions among the participants. There were some isolated lectures on other topics (aberration of light; the Tiber River; the perception of colors). Gaetano Cacciatore and Pietro Tacchini, from the Astronomical Observatory of Palermo, gave speeches on the reorganization of meteorological studies in Italy, and on the situation of the astronomical observatories in Italy, respectively. The only presentations on mathematics were by Francesco Caldarera, from the University of Palermo. He introduced a theory on the development of functions "with very small variables" and presented an unpublished work entitled *Introduction to the study of Higher Geometry*.[29]

The geometer Luigi Cremona, from the University of Rome, was elected member of the Standing Committee representing Section I on mathematics, astronomy, physics and meteorology.

At the closing of the congress, the president of the Italian Geographical Society lectured on "Italy at the international geographical congress of Paris." Prince Humberto, the heir to the throne and future king from 1878–1900, attended the closing.

Among the participants of the congress, there were two of special interest for us: young Giovanni Battista Guccia and his uncle Giulio Fabrizio Tomasi.[30] The Prince of Lampedusa was likely referred to the Society (Fig. 2.2) by Gaetano Cacciatore and Pietro Tacchini. What is certain is that Tacchini mentioned the observatory of the Prince of Lampedusa in his Villa ai Colli, which he held in high regard.[31]

What we don't know is how Giovanni Battista Guccia got to be interested in the congress. He was not yet twenty years old and was in the first year of his studies in

---

[29] *Ibid.*, pp. 1–23.

[30] *Ibid.*, Giovanni Guccia, member 377 on page xvi-g, and Giulio di Lampedusa Tomasi, member 742 on page xvi-m.

[31] *Ibid.*, p. 2.

Fig. 2.2 Certificate of membership of Giulio Fabrizio Tomasi, Prince of Lampedusa, to the Società Italiana per il Progresso delle Scienze for the five-year period 1875–1879. Courtesy of Gioacchino Lanza Tomasi

engineering. He could have been involved in the congress by recommendation of his professors at the university, or by suggestion of his uncle, the Prince of Lampedusa. Another connection of Giovanni Battista with the congress was through Stanislao Cannizzaro, a renowned chemist, who was Rector of the University of Palermo from 1866 to 1868. One of Giovanni Battista's aunts was a sister of Cannizzaro. Cannizzaro was also a member of the Standing Committee of society, and his pupil Emanuele Paternò had been Giovanni Battista's teacher of General Chemistry (which he passed quite well, with 27 points out of 30).

Giovanni Battista's attendance at this congress turned out to be a crucial moment in his life.

2.6   Italian Science and Mathematics of the Risorgimento

In the autumn of 1858, three young Italian geometers set off together on a scientific journey. Their goal was to visit the universities of France and Germany, to establish relations with the most outstanding scholars, to learn about their scientific ideas and aspirations, and, at the same time, to spread their own work.

This trip, undertaken by Betti, Brioschi and Casorati, marked a memorable date. Italy was about to become a nation. From that moment on, Italy joined the mainstream of major scientific works and, through an ever-increasing number of scientists, started to add its contribution to the common work in the field.

Today, when so many mathematicians meet to inaugurate a fruitful exchange of ideas, I would like to recall those memories.

It would be impossible to understand and to follow the progress of Analysis in Italy, in the second half of the nineteenth century, without thoroughly knowing the work pursued with patience and energy for many years by these three geometers whose names I have just mentioned, together with the subsequent efforts of their best students.

It is to their teachings, their work, the tireless devotion with which they encouraged students and young scientists to do research, the influence they had in the organization of advanced studies, and the relationships they established between our homeland and countries abroad, that we owe the birth of a young school of Analysts in Italy.[32]

This is the beginning of a well-known lecture that Vito Volterra delivered at the second International Congress of Mathematicians held in Paris in 1900. It was a plenary lecture entitled *Betti, Brioschi, Casorati, three Italian Analysts, three ways to consider the questions of Analysis*, and it was delivered the day of the inauguration of the Congress. There were only four plenary lectures in that international meeting (as well as Hilbert's famous lecture *On the future problems of Mathematics*). Volterra decided not to speak on mathematical results of his own, but to pay tribute to the generation of his masters.

Volterra's praise goes beyond Analysis, and even beyond mathematics. It can be applied to a whole generation of Italian scientists, who shared, despite the differences in their scientific subjects, the dream of a united nation, linked in its scientific development to the rest of Europe. This was the cultural and scientific foundation of the Italian Risorgimento.

In the period between Napoleon's defeat and the Unification of Italy, the various Italian states did not encourage scientific research. The majority of Italian scientists often found considerable difficulties developing their research because of the lack of adequate facilities, laboratories and funding. Funding for science mostly depended on the protection and generosity of the state's prince. Further difficulties were due to the failure of communication between the Italian states and with Europe. In some of these states, there were "cultural exchanges", but most were of a bilateral and provincial character. The general social atmosphere did not favor interest in research, and lack of basic education, technological backwardness, and scarcity of modern machinery in industry also impeded the development of scientific research. Moreover, some parts of Italy were still dominated by a narrow obscurantism (the Papal States being the most obvious example).

A remarkable feature in Italian history was the great civil and political commitment of Italian scientists during the long social and political process that ended in the creation of the unified state. During the Risorgimento, scientists played a prominent

---

[32] V. Volterra, *Betti, Brioschi, Casorati, trois analystes italiens et trois manières d'envisager les questions d'analyse*. In: E. Duporcq (ed.), *Compte rendu du Deuxième Congrès Internationale des Mathématiciens*, Gauthier-Villar, 1902, Paris, pp. 43–57.

role, both in their research and in their involvement in the field of politics. Indeed, many Italian scientists were engaged in the front lines of the Wars of Independence and of the revolutionary movements. For them, science and politics represented two complementary aspects of a larger renovation project of Italian culture. The development of an appreciation for science was also an important element of the process that should lead to national unity, contributing, they thought, to spreading the ideas of freedom and tolerance.

Once unity was achieved, scientists also participated in the effort to equip the newborn nation with the necessary educational, formative, technical and administrative institutions. For this task, many important scientists contributed to the united Italy by participating in the construction of a modern university and new technical institutes, and many of them held important administrative and political positions, even as ministers and senators.

A few names stand out from the large list of "scientists of the Risorgimento." The chemists Stanislao Cannizzaro (who was awarded the Copley medal by the Royal Society, and was Vice-President of the Senate), Raffaele Piria (who was a senator), and Francesco Selmi (who was General Director of the Ministry of Public Education); the physicists Carlo Matteucci (who was awarded the Copley medal by the Royal Society, and was a senator and Minister of Public Education), Silvestro Gherardi (who was a member of Parliament, and Minister of Public Education), Macedonio Melloni, and Fabrizio Mossotti (who was a senator); the historian Michele Amari (who was Minister of Public Education); and the mathematicians Eugenio Beltrami (who was a senator), Enrico Betti (who was a senator), Francesco Brioschi (who was Secretary General of the Minister of Public Education, and a senator), Luigi Cremona (of whom we will speak soon), Angelo Genocchi (who was a senator), and Giuseppe Veronese (who also was a senator).[33]

We see that there was a generation of Italian mathematicians that, in the second half of the nineteenth century, shared these two different and intense activities, and produced first class mathematics while building the new Kingdom of Italy. Their scientific work established the foundations on which the Italian school of mathematics flourished in all areas of mathematical research in the late nineteenth and early twentieth centuries. We will briefly review how this group came to be.

The Renaissance produced in Italy not only beautiful art but also deep mathematical activity. The list of names to be highlighted speaks by itself of the quality of this group of scientists and of the level and diversity of their research: Luca Pacioli, Scipione del Ferro, Niccolò Tartaglia, Girolamo Cardano, Rafael Bombelli, Galileo Galilei, Bonaventura Cavalieri, Evangelista Torricelli and many others. The eighteenth century was moderate in the number of important mathematicians, the most renowned being Jacopo and Vincenzo Riccati, Maria Gaetana Agnesi, Lorenzo

---

[33]For the role played by mathematicians, see U. Bottazzini, *Mathematics in Unified Italy.* In: *Social History of nineteenth Century Mathematics,* H. Mehrtens, H. Bos, I. Schneider (eds.), Birkhäuser, 1981, Basel, pp. 165–178; and also U. Bottazzini, P. Nastasi, *La patria ci vuole eroi. Matematici e vita politica nell'Italia del Risorgimento*, Zanichelli, 2013, Bologna.

Mascheroni, Paolo Ruffini. Apart is the case of the Italian-French mathematician Joseph-Louis Lagrange.

Achieving national unity dominated the history of Italy during the first half of nineteenth century Italy. Poor connections between the different states, political prosecution, and emigration abroad to avoid imprisonment, all interfered with the development of science. This was, as we have seen, the historical context where the meetings of Italian scientists commenced.

During that turbulent period, three mathematicians from the University of Pavia can be accounted for setting the first steps for the new course for mathematics: Vincenzo Brunacci (1768–1818) and his students Antonio Bordoni (1788–1860) and Ottaviano Fabrizio Mossotti (1791–1863). Brunacci got to know the scientific atmosphere of the Napoleonic Paris at the beginning of the century, and Bordoni became acquainted with the work of Liouville and the ideas of Gauss. Their students included Francesco Brioschi, who graduated in 1845; Enrico Betti, who graduated in 1846; Luigi Cremona, who graduated in 1853; Felice Casorati, who graduated in 1856; and Eugenio Beltrami, who was expelled from the university in 1856 for political reasons. They are largely responsible for the big leap in Italian mathematics.

This group of mathematicians had diverse mathematical interests and worked in different directions but they all shared a common project, which on the scientific side included two main aims: linking Italian mathematics with the stream of mathematics in the rest of Europe, and adequately advertising Italian mathematics abroad. The first goal required awareness of works by the leading European mathematicians, not only by studying their work but also by personally contacting them. These were the impulses behind Betti, Brioschi and Casorati's famous trip throughout Europe in 1858. They visited Paris, where they met with Bertrand and Hermite; Berlin, where they met with Kronecker, Kummer and Weierstrass; and Göttingen, where they met with Dedekind, Dirichlet and Riemann.

The second goal pointed at publishing the results produced by Italian mathematicians so that they could reach an international audience. This was not so simple at the time, due to the scarcity of mathematical journals. In 1850 Barnaba Tortolini founded the Annali di scienze matematiche e fisiche in Rome, where Italian mathematicians started to publish their results. In 1858 Betti, Brioschi and Angelo Genocchi joined the editorial committee and the journal changed focus, by exclusively considering papers in mathematics, and also changed its name (to the current one), Annali di matematica pura ed applicata. The goal of having a well-known and respected journal as a platform to advertise Italian mathematics was achieved. Another mathematical journal was founded in 1860 by Giuseppe Battaglini in Naples, the *Giornale di Matematiche.*

Soon these efforts started to yield results and a considerable number of excellent young Italian mathematicians came out of the universities, forming a new generation of researchers. Between 1864 and 1882, a number of outstanding mathematicians graduated from the University of Pisa: Ulisse Dini graduated in 1864; Ernesto Padova graduated in 1866, and later was the advisor of Tullio Levi-Civita; Eugenio Bertini graduated in 1867; Giulio Ascoli graduated in 1868; Cesare Arzelà graduated in 1869; Salvatore Pincherle graduated in 1874; Gregorio Ricci-Curbastro graduated in 1875; Luigi Bianchi graduated in 1877; Carlo Somigliana graduated in 1881; and Vito Volterra graduated in 1882. The list of first class mathematicians graduating from Italian universities and doing excellent and influential mathematics continues and it is certainly longer.[34]

2.7 Luigi Cremona

Luigi Cremona is a perfect example of an Italian scientist of the Risorgimento (Fig. 2.3). He was a first-class mathematician, and is unanimously considered the man who laid the foundations of the prestigious Italian school of Algebraic Geometry. He was internationally recognized, and had good international contacts. He also devoted much effort to the Unification of Italy, and he contributed to building the new country's academic, administrative, and political institutions once the Unification was achieved. In the words of Jeremy Gray, "Luigi Cremona is one of those mathematicians whose life involved politics as much as science, and whose influence was as much political as scientific."[35]

Luigi Cremona was born in Pavia in 1830. His father, Gaudenzio Cremona, came from a wealthy family that had squandered their fortune. Gaudenzio married young, had three children and continued living luxuriously, despite working as a modest accountant. When he was almost sixty years old and a widower for many years, he married a young woman of humble origins in her twenties. Among the children of the new couple were the eldest Luigi and youngest Tranquillo. The death of Gaudenzio left the family in a dire economic situation. They were saved, however, by the generous aid of their stepbrothers, and Luigi and his brothers were able to continue their studies. Tranquillo Cremona became one of the most important painters of Italian artistic movement known as the "Scapigliatura".

At the time of Cremona's birth, Pavia was part of the Lombardo-Venetian Kingdom that was under the control of the Austrian Empire. His youth was determined by his participation in the nationalistic uprisings of 1848 against the Austrian domination. Through his stepsister, Luigi became acquainted with the wealthy and influential Cairoli family, one of the most important in the history of the Risorgimento. Four of the Cairoli brothers died in the different battles that eventually led to the Unification of Italy. Influenced by his lifelong friend Benedetto, the oldest of the Cairoli brothers, seventeen-year-old Luigi, left his home on April of 1848, without a word to his mother, to defend Venice with a group of Neapolitan

---

[34]See A. Durand, *Matematici in politica. Nuovi scenari di ricerca?*, Lettera Matematica, 98 (2016), pp. 35–38, and also A. Durand, *Matematici parlamentari in Italia: uno sguardo alla politicizzazione di un'élite (1848–1915)*. In: L. Pepe (ed.), *Europa matematica e Risorgimento italiano*, Ed. CLUEB, 2012, Bologna, pp. 125–136.

[35]J. Gray, *Worlds Out of Nothing. A course in the History of Geometry in the 19th Century*, Springer, 2007, London, p. 241.

Fig. 2.3 Luigi Cremona. Courtesy of the Archives of the Mathematisches Forschungsinstitut Oberwolfach

volunteer students. Sixteen months later, on August 1849, after the defeat of the Italian forces in Venice, Luigi walked back to Pavia to find that in the meantime his mother had died. He suffered violent typhus that kept him between life and death.[36] For the courage shown in the defense and in the bombing of Treviso, he was appointed corporal and shortly after promoted to sergeant. A personal outcome of

---

[36]G. Veronese, *Commemorazione del socio Luigi Cremona*, Rendiconti della Reale Accademia dei Lincei, 12 (1903), pp. 664–678.

the episode was his friendship with Niccolò Ferrari, one of the main collaborators of Giuseppe Mazzini, the ideologist of Italian independence. Years later, Luigi married the sister of his comrade, Elisa Ferrari.

Cremona's record of military activity against Austria made it difficult for him to enter the official educational system. As a result, he first became a private tutor in Pavia, and slowly entered the secondary school system. In 1859, the newly Italian government of Lombardy appointed him as gymnasium teacher in Milan.[37] During those years, Cremona had started to work in mathematics. Inspired by the research of his teachers Antonio Bordoni and Francesco Brioschi, he published several papers. In 1860, under the influence of Brioschi, the Minister of Public Education Terenzio Mamiani della Rovere created a chair of Higher Geometry in the University of Bologna to which Cremona was appointed professor. It was during his time in Bologna that Cremona became acquainted with a large number of Italian and European mathematicians.

Brioschi had founded in 1863 a Polytechnic school in Milan; on his recommendation, Cremona was transferred there in 1866. Six years later, Cremona was appointed director of the newly established Faculty of Engineering in the University of Rome, where he obtained in 1877 his definitive academic position.

The Unification of Italy, for which he had fought so intensively, caused Cremona to constantly struggle to combine his scientific research with his participation in the construction of the new Italian state by holding administrative and political posts. The English mathematician and good friend of Cremona T.A. Hirst recalled his first meeting with Cremona in 1859, "our conversation was first of all political and then mathematical."[38] He eventually succumbed to his political dream. In 1879, he was appointed senator, which brought an end to his research activity. Cremona was esteemed for his moral and intellectual qualities and, being also man of action, was often called upon to resolve sensitive and thorny issues. In 1880, the Minister of Education appointed him as Royal Commissioner for the reorganization of the Library Vittorio Emanuele in Rome. Despite great efforts and funding, the library ran into a series of scandals: there were accusations of large subtractions of books, unfavorable exchange rates, incompleteness and disorder in the inventory and catalog, and wrongdoings of some librarians. Cremona addressed the situation by closing the library, investigating the allegations, and informing the Minister. Some two years later, Cremona's appointment as director came to an end, the previous problems were solved, and the library reopened under the management of a new director.

Cremona had clear and strong views on higher education. His central idea was that of a selective university, which could access only the most deserving; this implied only few universities with highly qualified faculties. He advertised his ideas

---

[37] U. Bottazzini, L. Rossi, *Cremona, Luigi,* Dizionario Bibliografico degli Italiani, vol. 30, Treccani, 1984, Roma.

[38] A. Millán Gasca, *Mathematicians and the Nation in the Second Half of the Nineteenth Century as Reflected in the Luigi Cremona Correspondence,* Science in Context 24 (2011), pp. 43–72, p. 54.

often throughout his parliamentary life. In 1885, the Senate entrusted Cremona with the reorganization of university teaching. This project aimed to raise the level and quality of university studies in Italy. The project presented by Cremona proposed large institutions created by merging smaller ones; unifying chairs under a single denomination, such as mathematics, classical philology; assigning teaching on a yearly basis so that professors would be regularly preparing their lectures; and creating a new class of associate professors in order to develop a career for young professors. Cremona had great hopes for this project, and had placed all his efforts into its writing and its approval. After lengthy discussions, the Senate approved Cremona's project, but the Chamber of Deputies failed to discuss it due to the fall of the government. The project was never recovered.[39]

Cremona held the posts of President of the High Council of Public Education, Minister of Public Education (only for one month) and Vice-President of the Senate. He died in 1903, while still devoting his full energy to his political commitments. The Senate cancelled its session in mourning for his death and the Government took care of his funeral.

Cremona's moral character was described by Enrico D'Ovidio in his Obituary: "An upright, inflexible man, he possessed a strong sense of duty, and looked for the same in others. It was possible on occasions to differ from him in his judgment of men or things, but never to doubt the honesty of that judgment. On intimate acquaintance with him one recognized a man capable of impetuous admiration, of scornful contempt; qualities which revealed a nature rich in sensibility, eager for good, intolerant of all meanness."[40]

Cremona found his scientific inspiration not so much in the research of his teachers but in the ideas of Michel Chasles. Indeed, in 1860 he wrote to the French geometer: "My eyes opened to the sun of pure Geometry and all my gratitude is to you: it is your *Aperçu*, your treatise on Higher Geometry, that I will always bless!"[41] Cremona was referring to Chasles' influential work "Aperçu historique sur l'origine et le développement des méthodes en géométrie".

Cremona's most important contributions were on Projective Geometry and birational transformations. When Cremona started working on Projective Geometry, many theorems were stated without any proof, and there were few links between the various results of the theory. He improved the existing techniques, merging in an original way geometric intuition and algebraic results. In this way, he was able to fully rebuild the general theories of curves and algebraic surfaces, and also deepen the study of twisted curves of third order, the study of cubic surfaces, and the study

---

[39]For an extended discussion, see C. Ciliberto, C. Pedrini, *L'Autonomia dell'Università da Cremona a Oggi*, Lettera Matematica Pristem, 61 (2007), pp. 30–37.

[40]E. D'Ovidio, *Cenno necrologico*, Atti della Reale Accademia delle Scienze di Torino, 38 (1903). Cited by: E. Bertini, *Della vita e delle opere di Luigi Cremona, Opere matematiche di Luigi Cremona*, Milano 1917, t. III, pp. V–XXII, p. XX. English translation: *Obituary Note: Life and Works of Luigi Cremona*, Proc. London Math. Soc. 1904, s 2, vol. 1, pp. v–xviii, p. xvii.

[41]G. Veronese, *Commemorazione del Socio Luigi Cremona*, Rendiconti della Reale Accademia dei Lincei, (5), 12, 1903, pp. 664–678.

of some surfaces of fourth and fifth order. In 1866, the Royal Academy of Berlin awarded him the Steiner Prize [42] (together with Rudolf Sturm) for a study on cubic surfaces.

In 1862, he published *Introduzione ad una teoria geometrica delle curve piane.*[43] This work granted him fame throughout Europe, and it was soon translated into German. Regarding this paper, Max Noether wrote, "with his methods and his ideas Cremona has re-established the relations between pure geometry and the whole analytical-geometric development that were stated by Plücker, Hesse and Clebsch, and with Salmon and Cayley."[44]

In 1863 and 1865, Cremona published, in two installments, the paper *Sulle trasformazioni geometriche delle figure piane* where he conducted a complete study of birational transformations in the plane and gave a general method to characterize them. A measure of the impact of the paper among the geometers of the time was the fact that already in 1864 T.A. Hirst had presented a summary of the results of the first paper at the meeting of the British Association for the Advancement of Science. The fundamental contributions by Cremona to the study of birational transformations caused them to be known as Cremona transformations. These transformations had an instrumental role in the origins of Algebraic Geometry, as introduced in 1877 by Eugenio Bertini, a student of Cremona. He determined that two figures could be identified if it was possible to map one to the other by a Cremona transformation.[45] The study of Algebraic Geometry was developed by the powerful school of Italian geometers; in particular, by Guido Castelnuovo, Federico Enriques, Francesco Severi and their pupils.[46] Castelnuovo was a student of Giuseppe Veronese, who in turn was a student of Cremona. A student of Castelnuovo was Oscar Zariski, from whom a large dynasty of first class mathematicians derives, such as Michael Artin, Heisuke Hironaka, David Mumford and many others. The theory of Cremona transformations was also employed to resolve the singularities of a curve.[47]

As an evaluation of the scientific contributions of Cremona, Castelnuovo wrote, "Cremona [...] closed an era to open a new one."[48] He closed the golden era of Projective Geometry to open the era of Algebraic Geometry. Castelnuovo explained,

---

[42]J. Gray, *A history of prizes in mathematics.* In: J. Carlson, A. Jaffe, A. Wiles (eds.), *The Millennium Prize Problems*, American Mathematical Society, 2006, Providence, pp. 3–27, p. 16.

[43]L. Cremona, *Introduzione ad una teoria geometrica delle curve piane*, Memorie dell'Accademia delle Scienze di Bologna, Series I, 12 (1862) pp. 305–436.

[44]U. Bottazzini, L. Rossi, *Cremona, Luigi*, Dizionario Bibliografico degli Italiani, vol. 30, Treccani, 1984, Roma.

[45]L. Godeaux, *Les transformations birationelles du plan*, Mémorial des sciences mathématiques, vol. 122, Gautier-Villars, 1953, Paris, p. 2.

[46]J. Gray, *Algebraic Geometry in the late Nineteenth Century.* In: D.E. Rowe, J. McCleary, *The History of Modern Mathematics, Vol. 1: Ideas and Their Reception*, Academic Press, 1989, Boston, pp. 374, 376–379.

[47]J. Gray, 1989, *op. cit.*, p. 371.

[48]G. Castelnuovo, *Luigi Cremona nel centenario della nascita*, Rendiconti della Reale Accademia dei Lincei, 12 (1930) pp. 613–618.

"But it is not only for his researches that Cremona contributed to the development of mathematics in Italy [...] it is also for the impetus he gave to the branch of science that he liked, creating a flourishing school, which continues the traditions he left and boasts to have had him as a teacher".[49]

In 1874, the Royal Academy of Berlin awarded Cremona the Steiner prize, for the second time, this time for his overall scientific production. In 1879, he was elected a corresponding member of the Royal Society.

His comrade's reminiscences indicate that he had a strong and demanding, yet appealing, personality. Bertini recalled, "He did not save severe reprimands to his pupils for some impatience in their career, for a stopover or slowness in their work, or for some carelessness in the preparation of their publications, but, at the same time, he excited and encouraged them strongly and with dignity, giving them tips and valid affectionate aids."[50] Beppo Levi recalled the words delivered by Cremona during his first lecture at the University of Bologna: "Young students [...] without an unshakable constancy in the effort, you will fail to lead the possess of a science. On the contrary, if you have this noble purpose, I assure you that Science will appear to you as beautiful and admirable, and you will love it so strongly that, from then on, the intense studies will appear you as a sweet need of your life."[51]

2.8 Guccia in Rome

In his first year studying at the University of Palermo, Guccia was exposed to true mathematics. Measured by the grades that he obtained (27 points out 30 for Analytic Geometry, and 24 out of 30 for Complementary Algebra), it seems that mathematics was a pleasant discovery for him which he mastered and enjoyed. This, along with the possible influence and help of his uncle Giulio Fabrizio Tomasi, encouraged him to attend the meeting of the Società Italiana per il Progresso delle Scienze.

Later letters give evidence that Guccia personally met with the most significant mathematician attending the congress: Luigi Cremona. Their relationship began at the congress, and later developed into a close friendship. The figure of Luigi Cremona was of vital importance for the life of Giovanni Battista Guccia.

Guccia decided to move to Rome and continue his studies there in mathematics rather than in engineering. First, he had to pass the last of the mathematics courses of the first year in the University of Palermo. So, on November 4, 1875, he took the exam in Complementary Algebra (which, as we have seen, he passed with 24 points

[49] Ibid.

[50] E. Bertini, *Della vita e delle opere di Luigi Cremona*, In: *Opere matematiche di Luigi Cremona*, Hoepli, 1917, Milano, t. III pp. V–XXII.

[51] B. Levi, *Luigi Cremona*, In: Commemorazione tenuta nell'Istituto Matematico "S. Pincherle" di Bologna nel I centenario della nascita, 1930. See: http://www.luigi-cremona.it/download/Scritti_biografici/Comm_Levi.pdf

out of 30). Just one week later, Guccia was in Rome, requesting admission to the
laurea course in Mathematics and Physics. His application was submitted past the
deadline for admission to the upcoming school year, and he was required to submit
a special application along with supplemental documents. The final acceptance by
the Rector of the University of Rome, Pietro Blaserna, came soon.

The laurea course in Mathematics and Physics at the University of Rome had the
following syllabus.

First year:

Chemistry, taught by Stanislao Cannizzaro
Experimental Physics, taught by Pietro Blaserna
Analytic Geometry, taught by Angelo Armenante
Algebraic Analysis, taught by Luigi Biolchini

Second year:

Descriptive Geometry, taught by Nicola Salvatore Dino
General Physics, taught by Pietro Blaserna
Infinitesimal Calculus, taught by Giuseppe Battaglini
Geometric Design, taught by Francesco Chizzoni

Third year:

Rational Mechanics, taught by Luigi Cremona
Astronomy, taught by Lorenzo Respighi
Analytic Geometry, taught by Nicola Salvatore Dino
Determinants, taught by Angelo Armenante
Descriptive Geometry, taught by Nicola Salvatore Dino

Fourth year:

Higher Mathematics I, taught by Giuseppe Battaglini
Higher Mathematics II, taught by Luigi Cremona
Colloquium on Analytic Functions
Colloquium on Projective Mechanics
Colloquium on Higher Geometry, taught by Nicola Salvatore Dino

Guccia moved to Rome and entered into the second course of studies in the
academic year 1875–76. He rented a room in Rome, as was common at the time
for students, and his address was at Mrs. Depinglau's, in the first floor of Foro
Traiano 29.[52] The next academic year, 1876–77, he completed the third course
and stayed in the same accommodation. For the last course, in the academic year
1877–78, it seems that he changed his accommodation in Rome for a better place,
since at the moment of his academic registration, he indicated as his official address
the Hotel del Quirinale. His official record shows that he attended the classes and
passed the exams. The grades are not recorded in the documents found in the Fondo

---

[52]Folder Guccia, Fondo Università, Archivio di Stato di Roma.

Università of the Archivio di Stato di Roma. These documents show that Guccia had a tendency to leave some of the exams for October, since he would spend the summer in Palermo. He often had to request special permission to enroll in the courses, as he would present his application after the closing of the official period. When the 1877–78 academic year ended in June 1878, Guccia went back home to Palermo, having completed all the courses in mathematics and physics.

However, Guccia's period in Rome had not yet ended, since he had to complete his thesis under the direction of Luigi Cremona. At that moment, his relationship with his advisor had clearly grown more personal: in one of his letters, he thanks Mrs. Cremona for having returned by parcel post a book by Victor Hugo that he had lent her.[53]

In November 1878, he returned to Rome after his vacation. Having no official obligations at the university, he again attended the regular lectures by Luigi Cremona, and followed a seminar on the book *Vorlesungen über Geometrie* by Alfred Clebsch. He read mathematical papers and books with interest, deepening his knowledge, and learning the latest developments in Geometry. He spoke regularly about mathematics with Luigi Cremona and his colleague Giuseppe Battaglini (who was also devoted to non-Euclidean geometries). During this period, he became acquainted and established an enduring friendship with some of the former students of Cremona and Battaglini, in particular, with Giuseppe Veronese, Alfredo Capelli, Riccardo de Paolis and Giovanni Frattini.

This period of intense mathematical work and personal maturation lasted two years. Guccia was halfway through his thesis in March of 1879 when Luigi Cremona was appointed as senator of the Kingdom of Italy. The State required Cremona's experience and knowledge of science, education, and the school system; he had to halt his mathematical activity for the time being.

By the end of May 1880, Guccia was in Palermo and had completed the writing of his thesis. He wrote to Luigi Cremona to fix the date of the oral dissertation. Guccia suggested holding it in October, in order to avoid the unpleasant high temperatures of Rome during the summer.[54] Once this was agreed, Guccia, following the habits of Palermitan aristocracy, departed to spend the summer in some pleasant location in Switzerland with a milder climate. As he passed by Rome, he decided to stop and make a visit to Luigi Cremona in order to discuss certain points of his thesis and prepare the oral presentation. That visit completely changed the summer plans of Guccia, and, again, was a critical moment in his life and thus for our story. We will review that exciting summer of 1880 in Chapter 5, when we discuss the creation of the Circolo Matematico di Palermo. By the end of October, Guccia was in Rome, meeting with Luigi Cremona and preparing for the defense of his thesis.

The topic of Guccia's thesis was related to Cremona transformations. Before Cremona's work, the homographic and the quadratic (the inversion) transformations

---

[53]Letter from Guccia to Cremona, October 7, 1878. Folder Cremona, Archive of the Mazzini Institute of Genoa.

[54]*Ibid.*, June 4, 1880.

were the only known birational transformations.[55] In 1832, Ludwig Immanuel Magnus stated that birational transformations in the plane were only linear or quadratic. Many mathematicians, including Cremona, had used that statement, until 1863 when Cremona observed that in general the composition of two standard quadratic transformations is a transformation of degree 4, hence showing the existence of birational transformations in the plane that were not linear or quadratic.[56] This was the starting point of the so-called Cremona transformations.

To define such a transformation, consider on the projective plane $\mathbb{P}^2$ three homogeneous polynomials $f_i(\mathbf{x}) = 0$, for $i = 0, 1, 2$, and $\mathbf{x} = (x_0, x_1, x_2)$, which have no common divisor, are linearly independent, and have the same degree, say n. Define the set B of all points $\mathbf{x}$ of the plane lying on the three curves, that is, $f_i(\mathbf{x}) = 0$, for $i = 0, 1, 2$. The map ω from $\mathbb{P}^2 \setminus B$ to $\mathbb{P}^2$ defined by

$$\omega(\mathbf{x}) := (f_0(\mathbf{x}) : f_1(\mathbf{x}) : f_2(\mathbf{x}))$$

is called a Cremona transformation if it is generically invertible, i.e., invertible on an open subset. Given a Cremona transformation as above, the family of curves of the form

$$\lambda_0 f_0(\mathbf{x}) + \lambda_1 f_1(\mathbf{x}) + \lambda_2 f_2(\mathbf{x}) = 0,$$

with $\lambda_0, \lambda_1, \lambda_2$ scalars, is called the homoloidal net associated to the transformation ω. Each point lying on the three curves $f_0(\mathbf{x})$, $f_1(\mathbf{x})$, $f_2(\mathbf{x})$ is called a base point of the homoloidal net, and n—the degree of the polynomials—its order.[57]

The topic of Guccia's thesis was the study of algebraic surfaces in three-dimensional Euclidean space (Figs. 2.4 and 2.5). It was entitled *On a class of surfaces representable point by point on a plane* and was organized into six chapters:

1. Plane representation of a first order congruency
2. Generation of a surface representable point by point on a plane π
3. Plane sections of a surface F. Their images on spaces π', k, π
4. Representation on the plane π of singular straight lines of a surface F. Images of plane sections conducted through the straight lines and the singular points
5. A class of surfaces
6. The geometry of curves traced on the surface $F_{2mn-p-q+1}$

The surfaces considered in the thesis have a certain order $n+m+1$, with m and n integers (greater than or equal to one) and the property that they contain two straight-lines with multiplicities related to the order of the surface (m and n, respectively)

---

[55]L. Godeaux, *op. cit.*, p. 1.

[56]C. Ciliberto, E. Sallent Del Colombo, *Giovan Battista Guccia and the Circolo Matematico di Palermo*, Lettera Matematica, 3 (2015), pp. 177–188, p. 187.

[57]J. Gray, 1989, *op. cit.*, p. 373.

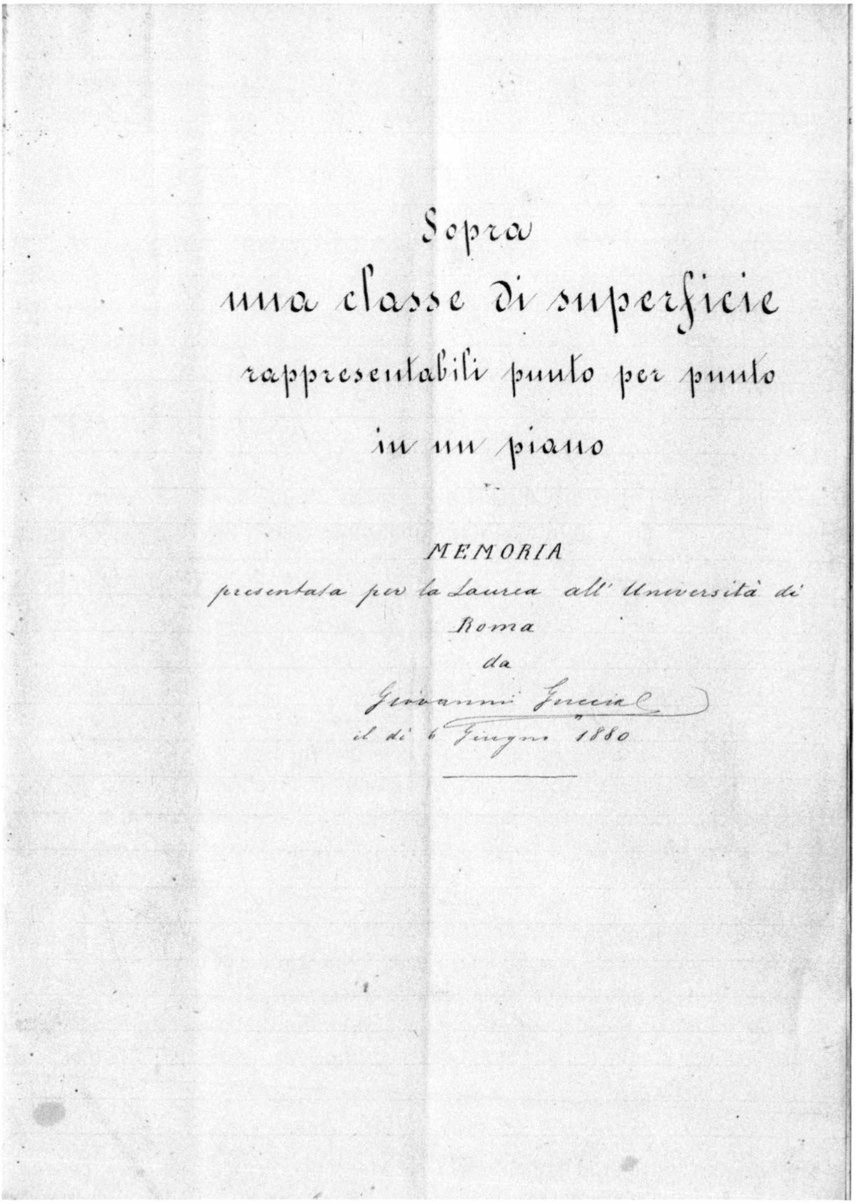

Sopra
una classe di superficie
rappresentabili punto per punto
in un piano

MEMORIA
presentata per la Laurea all' Università di
Roma
da
Giovanni Guccia
il dì 6 Giugno 1880

Fig. 2.4 Front cover of the thesis of Giovanni Battista Guccia, 1880. Courtesy of the Archivio di Stato di Roma

and which are skew. A surface of such type is rational, and thus it can be represented birationally on a plane. This type of surface was an extension of the case of surfaces of third order studied by Alfred Clebsch (in the papers *Sur les surfaces algébriques*,

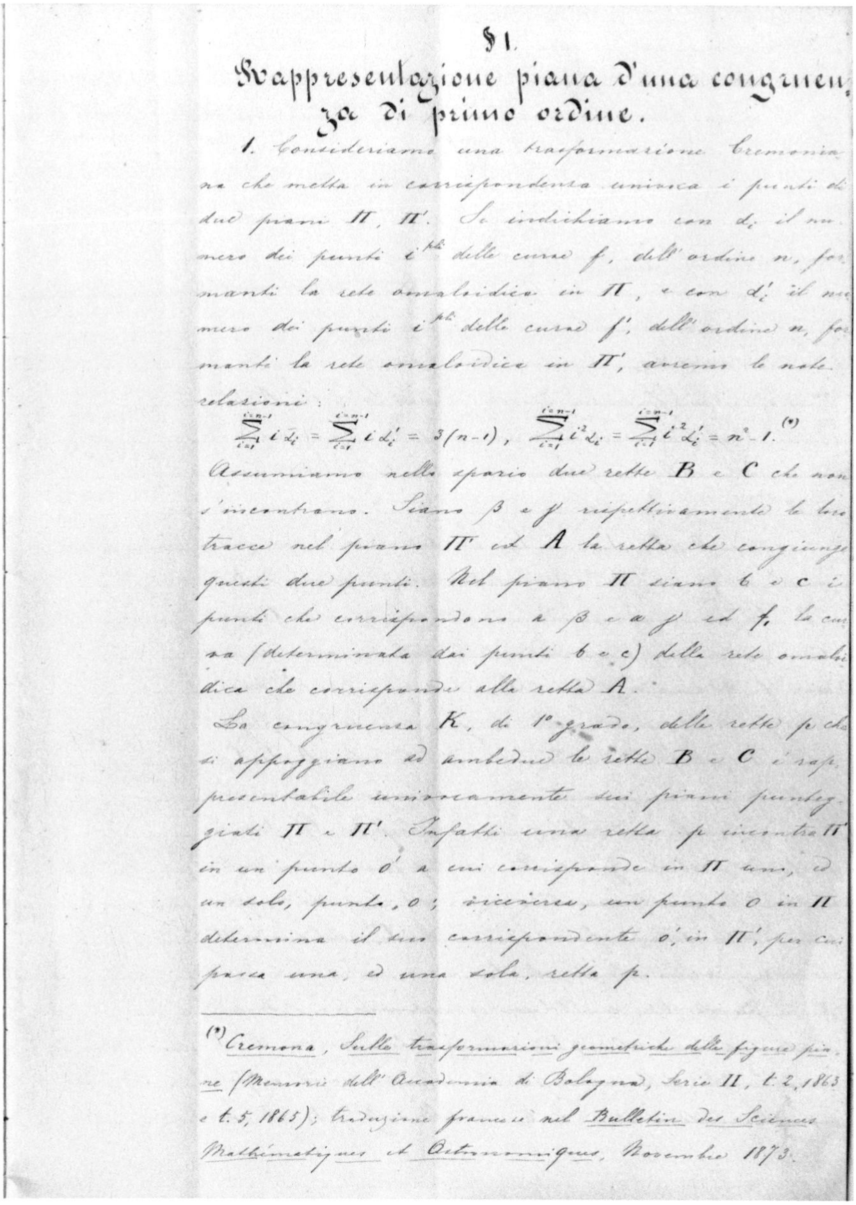

§1.
Rappresentazione piana d'una congruenza di primo ordine.

1. Consideriamo una trasformazione Cremoniana che metta in corrispondenza univoca i punti di due piani Π, Π'. Se indichiamo con α_i il numero dei punti i^{pli} delle curve f, dell'ordine n, formanti la rete omaloidica in Π, e con α'_i il numero dei punti i^{pli} delle curve f', dell'ordine n, formanti la rete omaloidica in Π', avremo le note relazioni:

$$\sum_{i=1}^{i=n-1} i\alpha_i = \sum_{i=1}^{i=n-1} i\alpha'_i = 3(n-1), \quad \sum_{i=1}^{i=n-1} i^2\alpha_i = \sum_{i=1}^{i=n-1} i^2\alpha'_i = n^2-1. \quad (*)$$

Assumiamo nello spazio due rette B e C che non s'incontrano. Siano β e γ rispettivamente le loro tracce nel piano Π' ed A la retta che congiunge questi due punti. Nel piano Π siano b e c i punti che corrispondono a β e a γ ed f_1 la curva (determinata dai punti b e c) della rete omaloidica che corrisponde alla retta A.

La congruenza K, di 1° grado, delle rette p che si appoggiano ad ambedue le rette B e C è rappresentabile univocamente sui piani punteggiati Π e Π'. Infatti una retta p incontra Π' in un punto o' a cui corrisponde in Π uno, ed un solo, punto, o; viceversa, un punto o in Π determina il suo corrispondente o' in Π', per cui passa una, ed una sola, retta p.

(*) Cremona, Sulle trasformazioni geometriche delle figure piane (Memorie dell'Accademia di Bologna, Serie II, t. 2, 1863 e t. 5, 1865); traduzione francese nel Bulletin des Sciences Mathématiques et Astronomiques, Novembre 1873.

Fig. 2.5 Page 1 of the thesis of Giovanni Battista Guccia, 1880. Courtesy of the Archivio di Stato di Roma

published in the Comptes Rendus de l'Académie des Sciences in 1868, and *Über die Abbildung algebraischer Flachen*, published in Mathematische Annalen in 1869), and also by Luigi Cremona in the paper *Sulle transformazioni razionali dello spazio*, published in the Rendiconti del Reale Istituto Lombardo di Scienze e Lettere in 1871.[58]

Part of Guccia's thesis was published under the title *Sur une classe de surfaces représentables, point par point, sur un plan*[59] (item 1, Appendix 1 on the mathematical works of Guccia).[60] The paper was reviewed in the Jahrbuch über die Fortschritte der Mathematik by Friedrich Wilhelm Meyer (who worked in algebraic and projective geometry and later collaborated with Felix Klein in the founding of the Encyclopädie der Mathematischen Wissenschaften).

The review explained that:

> The surfaces are of the order $n+m+1$ [in the projective 3-space] and contain two skew lines, with multiplicities n and m, respectively. A map is defined as follows: in the projective plane E take a homoloidal net N of order n. Then the curves of N containing a given point b in E form a pencil of curves f_b in the net; another point c in E gives another pencil of curves f_c in N. Take two such curves f_b and f_c. They define a linear pencil (f_b, f_c) [in the language of equations this is $t \cdot f_b + u \cdot f_c = 0$ with parameters t, u]. Now choose a projectivity π from (f_b, f_c) to a pencil (Φ, Ψ) of curves of order m with $\pi(f_b) = \Phi$ and $\pi(f_c) = \Psi$. This projectivity generates the image (S) of a plane section of the surface. The properties of the 1-dimensional object (S) allow several deductions with respect to the investigation of the singularities of the surface. Special cases of such surfaces are the general surface of order 3, and two surfaces of order 5 which have been investigated, respectively, by Clebsch and Cremona.[61]

On December 20, 1880, the oral presentation of his thesis took place. The copy of the thesis that is kept in the Archivio di Stato di Roma has fifty-nine handwritten pages; it contains some style corrections in pencil, probably done by Cremona.[62] Two days after the public defense of the thesis, Guccia requested the Rector of the University of Rome to issue his official diploma.[63] With his freshly issued diploma, Guccia returned to Palermo, just in time for Christmas. Giovanni Battista Guccia was the last doctoral student of Cremona.

---

[58] A. Barlotti, F. Bartolozzi, G. Zappa, *L'attività scientifica di Giovanni Battista Guccia*, Suppl. Rend. Circ. Mat. Palermo, 67 (2001), pp. xii–xiii.

[59] *Comptes-rendus de la 9ᵉ session Reims 1880*, Association française pour l'avancement des sciences, Paris, 1881, pp. 191–200.

[60] A. Barlotti, F. Bartolozzi, G. Zappa, *op. cit.*, pp. xi–xiii.

[61] F. Meyer, Jahrbuch über die Fortschritte der Mathematik, JFM 13.0652.03 (1881).

[62] Corrections done at the beginning of points 5, 6 and 12 of the thesis. Folder Guccia, Fondo Università, Archivio di Stato di Roma.

[63] Folder Guccia, Fondo Università, Archivio di Stato di Roma.

Chapter 3
The Scientific Context

This chapter is an intermission necessary for understanding the scientific context
in which Guccia devised his professional projects. We review three important
processes regarding the organization of mathematics as a scientific discipline that
took place in the second half of the nineteenth century: the foundation of national
mathematical societies; the creation of research mathematical journals; and the
creation of international mathematics organizations. Since we are going to follow
the process of the creation of a mathematical society, the Circolo Matematico di
Palermo, it is most appropriate to review the creation of two main mathematical
societies, the London Mathematical Society and the Société Mathématique de
France. These societies were formed not long before the Circolo, and several of
the mathematicians involved later became very close to Guccia and were likely
influences on his work and his viewpoints at the Circolo. As we are going to
follow the process of creation and management of a mathematical journal, the
Rendiconti del Circolo Matematico di Palermo, it is appropriate to review the
creation and management of an early and important mathematical journal. We
have focused on Acta Mathematica, because of the similarities with the Rendiconti
and also on account of the close relationship between Guccia and the founder of
Acta Mathematica, Gösta Mittag-Leffler. Regarding the international organization
of mathematics, we will discuss the origins of the International Congresses of
Mathematicians, and review the first Congresses before the 1908 Rome Congress
in which Guccia was deeply involved.

3.1 Two Mathematical Societies

While the London Mathematical Society was not the first one to be founded (it
was preceded by the Royal Dutch Mathematical Society in 1778 and the Moscow
Mathematical Society in 1864), there is no doubt that it was the first to be founded

© Springer International Publishing AG, part of Springer Nature 2018
B. Bongiorno, G. P. Curbera, *Giovanni Battista Guccia*,
https://doi.org/10.1007/978-3-319-78667-4_3

amongst the leading mathematical societies of the twentieth century. Not long after, the Société Mathématique de France (the French Mathematical Society) was founded following its model.

At the beginning of the nineteenth century, the established channels for the dissemination of scientific work were the learned societies, continuing with the model that was successfully developed throughout the eighteenth century. The most important learned societies were the Académie des Sciences in Paris, the Royal Society in London and the Royal Prussian Academy of Sciences, known as the Berlin Academy. As science developed rapidly during the nineteenth century, it also became increasingly specialized. One of the consequences of specialization was the rise in faculty positions at universities, which in turn led to an increase in the number of science practitioners. A need for more subject-oriented scientific societies arose, as the academies, with a strong elitist character, were not able to accommodate the new ways of addressing science.

In accordance with the practical character of British society, several mathematical societies consisting mostly of "studious artisans" had existed for a long time.[1] Despite Newton's legacy and despite being the largest producer of graduates in mathematics in Britain, the University of Cambridge was not the place where the mathematical society was to be founded.[2] London had a larger number of diverse practitioners of mathematics as well as a recently founded and active new institution, University College London. It was around this institution where the new mathematical society was born.

The London Mathematical Society was founded on January 16, 1865. The initiative came from two former students of University College London. The Society was created on the premises of University College London, and required the aid of the father of one of those students, Augustus De Morgan, who was a professor at University College London. The links of the new society with University College London were very strong. Indeed, twenty-six of the twenty-seven founding members of the London Mathematical Society were in one way or another associated with University College London. De Morgan was elected the first president and declared, "Our great aim is the cultivation of pure Mathematics and their most immediate applications."[3] A few months later, the most renowned English mathematicians of the time, Arthur Cayley and J.J. Sylvester, joined the Society and soon exhibited an active involvement in the development of the society. Not much later came George Salmon, James Clerk Maxwell and William Thomson (Lord Kelvin).

The Society grew rapidly, and in half a year it had doubled its membership. Two years after its foundation, the society had almost one hundred members. The

[1] A.C. Rice, *London Mathematical Society Historical Overview.* In: S. Oakes, A. Pears, A. Rice, *The Book of Presidents 1865–1965*, London Mathematical Society, 2005, London, pp. 1–16, p. 1.

[2] See J. Gray, *Nineteenth-century mathematical Europe(s).* In: C. Goldstein, J. Gray, J. Ritter (eds.), *L'Europe mathématique: Histoires, Mythes, Identités,* Maison des sciences de l'homme, 1996, Paris, pp. 347–360, pp. 352–354.

[3] A. De Morgan, *The London Mathematical Society. Speech of professor De Morgan at the first meeting of the society, January 16th, 1865,* Proc. London Math. Soc. 1 (1865), pp. 1–9, p. 1.

expansion transformed the society into the British national mathematical society. It is noticeable that at the start few members were professional mathematicians. Indeed, the society "brought together not only the leading mathematicians of the country but also others who were pursuing mathematical research in isolation, while earning a living in some profession."[4] In 1900, the Society had some 250 members; when it celebrated its centenary in 1965, it had 800 members.

While the growth in membership reflected the Society's success, it also caused significant difficulties, as publishing the society's proceedings became increasingly expensive. Quite soon, the Society had to significantly increase its membership fee. Despite the extra revenue from membership dues and expense cutbacks, the society went into a severe financial crisis. The Society was saved in 1874, however, by a donation of ten thousand pounds from Lord Rayleigh.

Not long after its foundation, the Society also founded a library. The Society subscribed to mathematical journals, and also obtained journals by exchanging its published proceedings. The first books came from the mathematical library of the astronomer Sir John Lubbock, which was donated by his son to the society. With time, an important part of the library was devoted to collected mathematical works of great mathematicians such as Brioschi, Euler, Gauss, Hermite, Kronecker, Lagrange, Newton and many others. As the library expanded, however, the society was in need of a place to properly house it. This was not easily achieved, and through the years the library occupied many different places.[5]

In 1884, nearly twenty years after its foundation, the society established an award for mathematical achievements. It was named the De Morgan Medal honoring the society's first president. The first medal was awarded to Arthur Cayley. From then on it was awarded every three years.

A key figure in the founding and first steps of the society was the geometer Thomas Archer Hirst (1830–1892, Fig. 3.1).[6] Hirst's efforts were instrumental in the success of the society in its beginnings. An Englishman, Hirst had studied in Germany, where he visited Berlin and Göttingen, attending lectures by Gauss, Dirichlet and Steiner. The influence of Steiner prompted Hirst to undertake research in geometry. Later he went to Paris, where he attended lectures by Liouville and Lamé. He spent a year in Italy and developed a friendship with Luigi Cremona. In 1883, he received the Royal Medal from the Royal Society for his work on Cremona's transformations. He held the post of professor at University College London; in fact, he succeeded Augustus De Morgan in 1871 in the chair of mathematics. Hirst was well known and respected, and had many friends and contacts not only in Britain but also across Europe. This was not so common for

[4]H. Davenport, *Looking back,* J. London Math. Soc., 41 (1966), pp. 1–10, p. 2.

[5]A. C. Rice, *The library of the London Mathematical Society*, Brit. Soc. Hist. Math. 9, Newsletter 27 (1994), pp. 37–39, p. 40.

[6]W.H. Brock, R.M. Macleod, *The life and journals of Thomas Archer Hirst, FRS (1830–1892),* Hist. Math. 1 (1974), 181–183.

Fig. 3.1 Thomas Archer Hirst. Courtesy of the London Mathematical Society

British mathematicians at the time. In the London Mathematical Society, he was Vice-President, Treasurer, and President from 1872 to 1874.

The London Mathematical Society was the first major mathematical society to be created, as De Morgan said in his inaugural speech: "If we look at what takes place around us, we shall find that we have no mathematical society to look to as our guide."[7] The London Mathematical Society attracted international attention, and served as a model for mathematicians in other countries who later formed their own mathematical societies. The French geometer Michel Chasles and the American mathematician Thomas Fiske recognized this fact. Hirst and Guccia met in London in the summer of 1880, and it is reasonable to suppose that the model of the London

---

[7] A. De Morgan, *op. cit.*, p. 1.

Mathematical Society influenced Guccia and so played a role in the creation of the Circolo Matematico di Palermo.

We end this brief overview of the founding of the London Mathematical Society with the words of De Morgan at the founding meeting of the society

> I must remind you that to societies such as the present the maxim "nascitur non fit" is applicable. The society should be what it will be: it will form itself. Its members will gradually adapt themselves to it and fall into their places. All similar societies take their growth from a few individuals, the most energetic, who give themselves up to the management and conduct of them. If these few members can catch the general tone of those who wish to form the society, the interest will be permanent, and the undertaking successful.[8]

The French Mathematical Society did not exist in 1870 when an official report by the French geometer Michel Chasles (1793–1880, Fig. 3.2) expressed a view that was extended among French mathematicians at the time regarding the existence of the Proceedings of the London Mathematical Society

> This fact, which we applaud, isn't it, for the cultivation of Mathematics, an element of future superiority that should worry us?[9]

There was no animosity in Chasles' observation, since he was captivated by the example of the London Mathematical Society (of which he had been in 1867 the first foreign member). Chasles intended to make a public statement regarding the necessity of a scientific society exclusively devoted to mathematics. His statement also reflected a general concern in France over the decline in science, in part a consequence of France's defeat in the 1870–71 Franco-Prussian war.

A forerunner of the mathematical society was the creation in 1870 by Gaston Darboux of the journal Bulletin des sciences mathématiques. The number of severe criticisms by the editors appearing in the first volumes of the journal on the stage of development of mathematics in France is certainly noticeable.[10]

In this atmosphere of necessity, the Société Mathématique de France was founded on November 6, 1872. Its internal regulations contrasted with the rigidity of the regulations of the existing academies: it allowed an unlimited number of members, the president was elected for one year and could not be immediately reelected, and the governing committee—consisting of eleven members—and the council—with sixteen additional members—were to be partially renovated yearly. The first president elected was Chasles, and Darboux was one of the vice-presidents. The average age of both the committee and the council was astonishingly low for the times: it was around forty. Among the council members there were two mathematicians that we will encounter later in our story: Victor Mannheim and

---

[8] *Ibid.*

[9] H. Gispert, *La France mathématique. La Société mathématique de France (1882–1914)*. Cahiers d'histoire & de philosophie des sciences, nouvelle série n. 34. Société francaise d'histoire des sciences et des techniques & Société mathématique de France, 1991, Paris, p. 14.

[10] *Ibid.*, p. 18.

Fig. 3.2 Michel Chasles. Courtesy of the Archives of the Mathematisches Forschungsinstitut Oberwolfach

George Halphen (who at the time was twenty-eight years old).[11] Liouville and Hermite joined the society the following year, but did not participate in its internal activity.

A few months later, on January 31, 1873, the first volume of the Bulletin de la Société Mathématique de France appeared. At that moment, according to the list published in that first volume, the society had 147 members, six of which were foreigners. One year later there were 178 French members; this number did not

---

[11] *Ibid.*, p. 20.

increase until 1900.[12] However, the number of foreign members, which had only increased to twelve in 1874, by the end of the century had gone up to 75.

Unfortunately, there are no records, archives, or personal collections available regarding the first period of the Society's history. The only information available comes from the published proceedings of the society.[13]

3.2 A Mathematical Journal

At the beginning of the nineteenth century, mathematical articles appeared in the academic journals of learned societies intermingled with articles from other fields of science, such as physics, astronomy and geography. The need for subject-oriented journals later emerged as a consequence of the increasing specialization of science. The first journal devoted solely to mathematics was the Annales de Mathématiques Pures et Appliquées, founded in 1810 by the French geometer Joseph Diaz Gergonne. The journal was mostly devoted to geometry, with publications by Chasles, Galois, Lamé, Poncelet, Steiner, among others. It was short-lived, however, and ceased publication in 1832. In 1826, the engineer and mathematician August Leopold Crelle founded the Journal für die reine und angewandte Mathematik (known as Crelle's Journal), which he edited for almost thirty years. The journal soon became quite respected and influential, mostly due to Crelle's fine ability to identify young and talented mathematicians. Ten years later, in 1836, Joseph Liouville at the age of twenty-seven founded the Journal de Mathématiques Pures et Appliquées. Crelle's Journal and the Journal de Liouville are both prestigious journals that still exist today.

In the second half of the nineteenth century, the Annali di Matematica Pura ed Applicata was founded in 1858 following the model of Liouville's journal. Some years later Francesco Brioschi and Luigi Cremona began managing the journal.[14] Other journals were soon founded. In general, these journals were mainly focused on a specific community of mathematicians, either national (Proceedings of the London Mathematical Society, 1866; Bulletin de la Société Mathématique de France, 1873) or attached to an academic institution (Matematicheskii Sbornik based at Moscow University, 1866; American Journal of Mathematics based at Johns Hopkins University, 1878).[15]

---

[12] *Ibid.*, p. 27.

[13] *Ibid.*, p. 13.

[14] H. Gispert, *Une comparaison des journaux français et italiens dans les années 1860–1875*. In: C. Goldstein, J. Gray, J. Ritter (eds.), *op. cit.*, pp. 389–406, p. 393.

[15] C. Gerini, N. Verdier (eds.), *L'émergence de la presse mathématique en Europe au 19ème siècle. Formes éditoriales et études de cas (France, Espagne, Italie et Portugal)*, College Publications, Collection «Cahiers de logique et d'épistémologie», vol. 19, Oxford, 2014.

Fig. 3.3 Gösta Mittag-Leffler holding an issue of Acta Mathematica. Courtesy of the Mittag-Leffler Institute. Photograph by Jonas Förare

Gösta Mittag-Leffler (1846–1927, Fig. 3.3) graduated from Uppsala University and later attended lectures by Charles Hermite in Paris and by Karl Weierstrass in Berlin. He spent several years in Helsingfors (the name of Helsinki in Swedish) as professor at the university and later moved to a new chair in mathematics in Stockholm's Högskola. He was thirty-five years old in 1881 when Sophus Lie suggested that he found a mathematical journal. Lie's idea was a journal with high international standards aimed at advertising and profiting from the recent blossoming of Scandinavian mathematics. Mittag-Leffler would be the editor.[16]

Mittag-Leffler did not hesitate and threw himself into the project. He kept Lie's idea of an editorial board consisting of the leading Scandinavian mathematicians,

---

[16]L.E. Turner, *Cultivating Mathematics in an International Space: Roles of Gösta Mittag-Leffler in the Development and Internationalization of Mathematics in Sweden and Beyond, 1880–1920,* Ph.D. thesis, 2011, Aarhus Universitet, chapter 3.

but changed the focus into publishing international mathematics. Articles in the new journal had to be written in French or German, exceptionally in English or Latin. The project for an international journal was well timed, given both the unfavorable international atmosphere after by the Franco-Prussian War in 1870–71 and the decline of mathematical publications that it caused. The international character of Crelle's Journal and Liouville's journal was seriously jeopardized, if not completely ceased.[17]

There were a number of difficulties in launching the project of the new journal, but Mittag-Leffler was a person of high practical and organizational skills and was able to address each one of the difficulties successfully. One of the difficulties regarded the journal's relationship with the German research community, which was the strongest in mathematical power at the time. After Crelle's death in 1855, C.W. Borchardt served as the editor of Crelle's Journal for twenty-five years. In 1881, Kronecker and Weierstrass took over the post. Mittag-Leffler was worried about the possible reaction of his former mentor to the new journal, which would be competing for quality papers with Crelle's Journal. The solution came via the intervention of the King of Sweden. Some German mathematicians (Kummer, Kronecker, and Weierstrass among them) had been awarded a Swedish royal distinction several years earlier. On this occasion, the King of Sweden sent a letter expressing his interest in the project of the new journal, and requesting their collaboration. The effect was very positive, and thus one of the problems was solved.

The creation of the journal also encountered financial difficulties. Before the journal could support itself, large financial contributions were needed. Mittag-Leffler looked for private donations to cover the expenses in the first years. He approached well-known Scandinavian sponsors of science and arts. The suggested contribution was of 1,500 crowns to be paid in three installments. The collection was a success, and 26,000 crowns were raised. King Oscar II of Sweden personally contributed 1,500 crowns. The King's generosity to the journal was due to the fact that he had studied mathematics in his youth himself, and was keenly interested in supporting mathematical research.[18] The governments of Denmark, Norway and Sweden also guaranteed an annual contribution of 1,000 crowns each. At an expected cost of 4,500 crowns per volume, the publication was ensured for the first years. In order to gauge these contributions, it is useful to note that Mittag-Leffler's annual salary at the time was some 7,000 crowns.[19]

Finally, it was important to launch the journal by publishing articles from top ranking mathematicians. To this end, Mittag-Leffler obtained invaluable support from Hermite (whose backing of the project also included a contribution of 720

---

[17] J.E. Barrow-Green, *Gösta Mittag-Leffler and the foundation and administration of Acta Mathematica.* In: K. Hunger Parshall, A.C. Rice, Adrian (eds.), *Mathematics unbound: the evolution of an International Mathematical Research Community, 1800–1945,* American Mathematical Society/London Mathematical Society, 2002, Providence, pp. 138–164, p. 141.

[18] *Ibid.,* p. 143.

[19] Y. Domar, *On the foundation of Acta Mathematica,* Acta Mathematica, 148 (1982), pp. 3–8, p. 4.

crowns). Hermite had commented on Henri Poincaré's brilliant talent to Mittag-Leffler. His idea was that:

> It was Abel, a Norwegian, who was responsible for the success of Crelle's Journal. Now
> [...] you, a Frenchman, might be as generous and make the success of our journal.[20]

Mittag-Leffler wrote to Poincaré in this way, and offered to publish Poincaré's large and breakthrough manuscript on Fuchsian groups in the first issue of the new journal, as well as four additional papers within the next year. Poincaré accepted the offer. This prompted other French authors, such as Hermite, Appell, Picard, Goursat, and Halphen, to participate in the launching of the new journal by sending their manuscripts.

Regarding the internal organization of the journal, Mittag-Leffler was fully in charge of its management. He arranged all the technical aspects: choosing the printing house, organizing the international distribution, and involving the young mathematician Gustav Eneström as proofreader. An instrumental role in all the preparations for creating the journal was played by the retired professor of mathematics Carl Malmsten, who was very close to Mittag-Leffler and had experience and contacts from the high political posts that he had held over the years.

Finally, before launching the journal, Mittag-Leffler sent information to academies, societies, journals and individual mathematicians, and embarked upon a promotional trip throughout Europe, making personal contact with mathematicians advertising the journal and collecting prospective submissions of manuscripts (Mittag-Leffler was certainly known to be charming and tactful at socializing). Mittag-Leffler won many German contributions to the journal in this way. A practical man, Mittag-Leffler organized so that the European journey was at the same time his honeymoon trip.

The last issue before publishing the first volume was choosing a name for the journal. The list of names that were considered was surprisingly wide, including Disquisitiones Mathematicae, Arkimedes, Analecta Mathematica, and Museum Mathematica. In the end the name chosen was Acta Mathematica Eruditorum. In a last-minute change, the genitive Eruditorum was dropped, and the journal was officially named *Acta Mathematica*.[21]

The journal very quickly attracted authors and readers from all around the world, and became a genuinely international journal. The success of the journal was reported just two years later in the British scientific journal Nature:

> The new Scandinavian journal, *Acta Mathematica,* has already gained such a reputation
> that the French government has decided to subscribe for 15 copies for the Facultés des
> Sciences.[22]

Mittag-Leffler was editor-in-chief of Acta Mathematica for forty-five years.

---

[20] J.E. Barrow-Green, *op. cit.,* p. 143.

[21] A. Stubhaug, *Gösta Mittag-Leffler: A man of conviction,* Springer, 2010, Berlin, p. 286.

[22] Nature, vol. 30, no. 763 (12 June 12 1884), p. 153.

3.3 The International Congress of Mathematicians

We end this chapter with a brief review of the origins and first steps of the International Congress of Mathematicians.[23] These congresses will have a determining importance for Guccia and the development of the Circolo Matematico di Palermo.

In 1893, an International Mathematical Congress was organized in Chicago in connection with the world's Columbian Exposition. One of the few Europeans participating in the Congress was Felix Klein, who attended as Imperial Commissioner of the Prussian Ministry of Culture. In a celebrated lecture entitled *The Present State of Mathematics*, Klein said,

> A distinction between the present and the earlier period lies evidently in this: that what was formerly begun by a single master-mind, we now must seek to accomplish by united efforts and cooperation. A movement in this direction was started in France some time since by the powerful influence of Poincaré. For similar purposes three years ago we founded in Germany a mathematical society, and I greet the young society in New York and its Bulletin as being in harmony with our aspirations. But our mathematicians must go further still. They must form international unions, and I trust that this present World's Congress at Chicago will be a step in that direction.[24]

Klein was reflecting the certainty among the major mathematicians in Europe of the importance of international cooperation for the future development of mathematics. Cantor, Hermite and Poincaré were among the most proactive leaders of this viewpoint.[25] As we later will see, Guccia was a true follower of this idea.

This positive international atmosphere was important to the beginnings of the International Congress of Mathematicians. Finally, in January 1897, 2,000 mathematicians received a letter of invitation to the first International Congress. The letters were signed by an international group of mathematicians coming from nine different countries: the Austro-Hungarian Empire, France, Great Britain, Germany, Italy, Russia, Sweden, Switzerland, and the United States of America. Luigi Cremona, Gösta Mittag-Leffler and Henri Poincaré were amongst them. The Congress took place in Zurich, "at the crossroad of the large railways from Paris to Vienna and from Berlin to Rome."[26] The promoting group of mathematicians decided to follow the model of the congresses "with itinerating venue" of the Swiss Society of Nature Scientists consisting of plenary sessions and specialized sections. There were four invited lectures in the plenary sessions, namely

---

[23] See G.P. Curbera, *Mathematicians of the World: Unite! The international Congress of Mathematicians. A human endeavor*, A.K. Peters, 2009, Wellesley.

[24] F. Klein, *The present state of mathematics*. In: E.H. Moore, O. Bolza, H. Maschke (eds.), *Mathematical Papers Read at the International Mathematical Congress held in Connection with the World's Columbian Exposition Chicago 1893*, American Mathematical Society, 1896, New York, pp. 133–135.

[25] J. W. Dauben, *Georg Cantor: His mathematics and philosophy of the infinite*, Princeton Univ. Press, 1979, Princeton, pp. 163–164.

[26] F. Rudio (ed.), *Verhandlungen des ersten Internationalen Mathematiker-Kongresses: in Zürich vom 9. bis 11. August 1897*, B.G. Teubner, 1898, Leipzig, p. 24.

"Sur les rapports de l'analyse pure et de la physique mathématique," by Henri Poincaré from Paris;

"Über die Entwickelung der allgemeinen Theorie der analytischen Funktionen in neuerer Zeit," by Adolf Hurwitz from Zurich;

"Logica matematica," by Giuseppe Peano from Turin;

"Zur Frage des höheren mathematischen Unterrichtes," by Felix Klein from Göttingen.

The rest of the lectures were organized into five specialized sections: Arithmetic and Algebra; Analysis and Function Theory; Geometry; Mechanics and Mathematical Physics; History and Bibliography. Thirty lectures altogether were scheduled in these sections. By way of example, let us mention that one of these lectures was "Sur la théorie des nombres premiers" by the young Belgian mathematician Charles de la Vallée Poussin, who one year before had proven the Prime Number Theorem (independently and at the same time as Jacques Hadamard). Unfortunately, the lack of time prevented De la Vallée Poussin from delivering his lecture.

This was the first occasion in which mathematicians from different countries met together, except for the Chicago 1893 meeting. The Congress was a success, both scientifically, as many of the most eminent mathematicians of the time attended the Congress, and in terms of attendance, as 208 mathematicians from sixteen countries participated in the Congress. The detailed attendance by country was the following:

Switzerland	60	U.S.A.	6	Great Britain	3
Germany	41	Sweden	6	Holland	3
France	23	Finland	4	Spain	1
Italy	20	Belgium	3	Greece	1
Austria-Hungary	17	Denmark	3	Portugal	1
Russia	12				

A delicate issue at the Congress involved balancing the role played by nationalities. The choice of the plenary lecturers reveals a careful balance: one Frenchman, one German, one Italian, and Adolf Hurwitz, who, although German, could be considered neutral as he was from the host institution in Zurich. This national equilibrium was also seen in the careful treatment of the language issue (the proceedings were edited in both languages, French and German).

Emile Picard, president of the Société Mathématique de France, summarized the general feeling after the Congress: "The success of our first meeting is a warrant for the future of the institution just founded."[27] It was unanimously decided that the next Congress would be held in 1900 in Paris. International cooperation in mathematics had officially started.

---

[27] *Ibid.*, p. 61.

The Paris Exposition Universelle of 1900 was a magnificent celebration of the beginning of the new century, and lasted for more than six months. Over two hundred scientific meetings were held in the city during that year. One was the second meeting of the International Congress of Mathematicians. Following the scheme adopted three years before in Zurich, there were four plenary lectures at the second Congress:

"L'historiographie des mathématiques," by Moritz Cantor from Heidelberg;
"Betti, Brioschi, Casorati—Trois analystes italiens et trois manières d'envisager les questions d'analyse," by Vito Volterra from Rome;
"Du rôle de l'intuition et de la logique en mathématiques," by Henri Poincaré from Paris;
"Une page de la vie de Weierstrass," by Gösta Mittag-Leffler from Stockholm.

The lectures of the sections took place in the amphitheaters of the Faculty of Science which were named after French mathematicians and astronomers: the Cauchy Amphitheater for the sections on Arithmetic and Algebra, and Geometry; the Le Verrier Amphitheater for the sections on Analysis, and on Mechanics; and the Chasles Amphitheater for the sections on Bibliography and History, and Teaching and Methods.

The Paris International Congress is mainly remembered for Hilbert's lecture, *Sur les problemes futurs des mathematiques*. In a long preamble, Hilbert discussed the nature of problems in mathematics and their role in the advance of the science. He expressed his faith that "In mathematics there is no ignorabimus," that is, there is no such valid statement as "we will not know." Hilbert then presented his renowned list of twenty-three problems, with the intention of illuminating the future of mathematics in the twentieth century.

The opening of Hilbert's lecture is now classical:

Who of us would not be glad to lift the veil behind which the future lies hidden; to cast a glance at the next advances of our science and at the secrets of its development during future centuries?[28]

The influence of Hilbert's list of problems in the development of twentieth century mathematics cannot be overestimated.[29]

The language used in science was a worrying issue at the time, and this concern was reflected in the Congress. It was explained at the Congress that, in the beginning of the nineteenth century, it was sufficient for a scholar to know three languages— Latin, English and French—; however, if the future required twenty or thirty languages to be mastered in order to follow the scientific developments, that would be a great danger for science. In response to this issue, the Congress recommended

---

[28]D. Hilbert, *Sur les problèmes futurs des mathématiques*. In: E. Duporcq (ed.), *Compte rendu du Deuxième Congrès Internationale des Mathématiciens*, Gauthier-Villar, 1902, Paris, pp. 58–114.

[29]Regarding Hilbert's problems, see J. Gray, *The Hilbert challenge: A perspective on twentieth century mathematics*, Oxford University Press, 2000, Oxford, and B.H. Yandell, *The Honors Class*, A K Peters, 2002, Natick.

"that the Academies and learned Societies from all countries study the proper means to remedy the harms coming from the increasing diversity of languages employed in the scientific literature."[30]

Regarding attendance, this Congress represented an improvement over the Zurich Congress, as the number rose to 250. The number of countries represented increased as well, jumping from sixteen to twenty-six. The largest national group was, naturally, the French with 95 participants, followed by the Germans with 26, Italians 23, North Americans 19, Russians 14, British 12, and Belgians 12. These figures show that the international Congresses were becoming a consolidated event within the mathematical community.

Charlotte Angas Scott, a British mathematician established in the United States, attended the Paris Congress and wrote a twenty-three-page report for the Bulletin of the American Mathematical Society. She reported bluntly on the presentation of papers in the Congress, which in her opinion were "usually shockingly bad" since the lecturer "instead of speaking to the audience, reads his paper to himself in a monotone that is sometimes hurried, sometimes hesitating, and frequently bored... so that he is often tedious and incomprehensible." The only lecturer that she valued positively was Mittag-Leffler, whose presentation Scott described as "admirable and engaging a style [...] It is not given to everyone to do it with this charm."[31]

The organization and the choice of venue of the next Congress were entrusted to the Deutsche Mathematiker-Vereinigung, the German Mathematical Society. The year was set for 1904.

At these Congresses, the feeling that mathematics was entering a new era based on international cooperation was very strong. Ferdinand Rudio, one of the Swiss organizers of the 1897 Zurich Congress, outlined some new initiatives that required international agreements. These initiatives included: unifying mathematical terminology and units; an international review journal for mathematics (the Jahrbuch über die Fortschritte der Mathematik, founded in 1868, and the Répertoire Bibliographique des Sciences Mathématiques, founded in 1885, were slow in reporting on the developments of a science producing many more results at a much faster speed than before). A general classification of mathematics that would help the bibliographic effort; an international directory of mathematicians, where one could find the address and field of specialty of all mathematicians in the world; and a biographical dictionary of current mathematicians (which would include portraits of the most important).[32] They were not thoughts from a visionary, but rather ideas shared by many mathematicians.

---

[30]E. Duporcq (ed.), *Compte rendu du Deuxième Congrès Internationale des Mathématiciens*, Gauthier-Villar, 1902, Paris, pp. 23–24.

[31]C.A. Scott, *Compte Rendu du Deuxieme Congrès International des mathematiciens tenu a Paris by E. Duporcq*, Bull. Amer. Math. Soc., 9 (1903), pp. 214–215.

[32]F. Rudio, *Über die Aufgaben und die Organisation internationaler mathematischer Kongresse*. In: F. Rudio *op. cit.*, pp. 31–37.

Despite some critical viewpoints that hold that "the universal brotherhood of mathematicians is an old-fashioned idea",[33] the words of Adolf Hurwitz in Zurich in 1897 remain for many mathematicians as a token of the spirit of the International Congress of Mathematicians:

> May the inspiring force of personal communication rise during these days, providing plenty of occasions for scientific discussions. May we together enjoy the relaxed and cheerful comradeship, enhanced by the feeling that here representatives of many different countries feel united by the most ideal interests in peace and friendship.[34]

---

[33] D.E. Rowe, *Review of "Mathematics unbound: The evolution of an international mathematical research community, 1800–1945" by K. Hunger Parshall, A.C. Rice (eds.).* Bull. Amer. Math. Soc. (N.S.), 40 (2003), pp. 535–542, p. 535.

[34] F. Rudio, *op. cit.,* p. 23.

Chapter 4
The Projects of Guccia: First Stage

This chapter narrates the process leading up to Guccia's conception of a mathematical society, as well as the founding of both the Circolo Matematico di Palermo and its journal, the Rendiconti del Circolo Matematico di Palermo. We follow Guccia's post-doctoral journey in the summer of 1880 through Paris, Reims and London. During that trip, he established professional relations and friendships that lasted for life, and determined his personal viewpoint of mathematics and its modern structuring. After devoting himself to research with his mentor Luigi Cremona for some time, he decided in 1884 to found a mathematical society. He was at that moment twenty-eight years old. Despite his youth, he was able to navigate the academic environment in Palermo without having any professional relation with the university. His early success encouraged him to lead the society towards internationalization four years later, and to create the journal in order to garner its international participation.

4.1 The Summer of 1880: Paris, Reims, London

We go back in time to the early summer of 1880. Having completed his thesis, and looking to avoid the unpleasantly high temperatures and humidity of the summer in Palermo, Guccia planned some relaxing vacation travel in the Swiss Alps. On his way he stopped in Rome to pay a visit to Luigi Cremona. His visit was academic as well as out of courtesy to his mentor, as he wished to postpone the public defense of his thesis dissertation until after the summer.

The letters exchanged between Guccia and his mentor Luigi Cremona shed light upon many of the events that occurred in the following years. The Archive of the Circolo Matematico di Palermo has a large collection of copies of these letters. However, the most important ones are in the Archive of the Mazzini Institute of

© Springer International Publishing AG, part of Springer Nature 2018
B. Bongiorno, G. P. Curbera, *Giovanni Battista Guccia*,
https://doi.org/10.1007/978-3-319-78667-4_4

Genoa.[1] The importance of the content of these lively and clarifying letters is reflected in our occasional inclusion of large excerpts from them.

Cremona's praise of Guccia's progress and mathematical results led him to suggest a change in Guccia's vacation plans and introduce him to his mathematician friends Thomas Hirst and Victor Mannheim.[2] Hirst was very close to Cremona, "probably [his] most affectionate friend and their correspondence covers family and personal affairs as well as mathematical problems."[3] Meeting Mannheim required travelling to Paris, and for Hirst it was necessary to go London. Guccia enthusiastically accepted this change of plan and promptly departed for Paris. This decision turned out to be of critical importance in the course of his life.

He arrived in Paris on June 13. The long train journey had affected his weak health: he did not feel well, and was diagnosed with bronchopneumonia. This obliged him to stay in bed, at the Hôtel de Bade.[4] This situation provoked a response typical of Guccia, one that would accompany him throughout his life: strict compliance with commitments, even at great cost to himself and his health. Despite his sickness, Guccia set out on an errand that Cremona had entrusted him with before leaving for Paris: he took care of delivering a parcel of documents to the General Inspector of Bridges and Roads and to the Director of the École des Ponts et Chaussees.

It took Guccia twenty days to recover from his illness. Once he did, he visited Mannheim. Victor Amédée Mannheim (1831–1906) followed the geometric tradition of Poncelet and Chasles, and was primarily devoted to projective geometry. He applied these studies to kinematic geometry and to the study of wave surfaces. In recognition of his contributions in 1872 he was awarded the Poncelet Prize by the Académie des Sciences. Most of his professional career was attached to the École Polytechnique, where he was a reputed and popular teacher. He was also a military man, attaining the rank of colonel in the engineering corps. His name is attached to the creation of the "Mannheim" slide rule, which is based on a double logarithmic scale. Mannheim received Guccia in quite a friendly manner. A few days after his arrival, he invited Guccia to a "déjeuner" (lunch), where he met many other French geometers.[5] A review of the geometrical works of Mannheim was later published by the Rendiconti del Circolo Matematico di Palermo in 1908.

---

[1] There are 43 letters exchanged with G.B. Guccia. See: C. Cerroni, *Il carteggio Cremona-Guccia (1878–1900)*, Mimesis, 2013, Milan. The Archive of the Mazzini Institute of Genoa contains over 6,000 documents, mainly consisting of Cremona's correspondence with scientific and institutional Italian interlocutors, coming from the legacy of Cremona's daugther Itala; see A. Brigaglia, S. Di Sieno, *The Luigi Cremona Archive of the Mazzini Institute of Genoa*, Hist. Math., 38 (2011), 96–110.

[2] The Archive of the Mazzini Institute of Genoa contains 86 letters exchanged with Hirst and 55 with Mannheim; see A. Brigaglia, S. Di Sieno, *op. cit.*, p. 105.

[3] A. Brigaglia, S. Di Sieno, *op. cit.*, p. 97.

[4] Letter from Guccia to Cremona, July 17, 1880, Folder Cremona, Archive of the Mazzini Institute of Genoa.

[5] *Ibid.*

In the following days, Guccia visited a mythical figure of nineteenth century Geometry: the eighty-seven-year-old Michel Chasles. A student of the École Polytechnique, Chasles had gained fame for his influential books. In 1837, he published the *Aperçu historique sur l'origine et le développement des méthodes en géométrie*; in 1852 the *Traité de géométrie* (where he introduced the cross ratio); and in 1865, he published the *Traité des sections coniques*. Chasles also became well known for solving the problem posed by Steiner on determining the number of conics that are tangent to five given conics (the precise solution being 3,264). In 1865, he was awarded the Copley Medal by the Royal Society. He is one of the seventy-two distinguished scientists who honored France and whose names are engraved on the first balcony of the Eiffel Tower (others are Cauchy, Fourier, Lagrange, Laplace, Legendre, Monge, Poisson, Poncelet, and Sturm). Guccia and Chasles enjoyed their first conversation, and Chasles gave the young and enthusiastic Sicilian a copy of an unpublished work of his from 1829. Guccia wrote to Cremona on July 17 that the visit to the old geometer was "a pleasure that I will always retain as a dearest memory."[6]

Guccia also got to know the renowned editor Jean-Albert Gauthier-Villars, and informed Cremona of the editor's interest in publishing a volume, in French, of three of Cremona's papers on plane curves, surfaces, and third order surfaces.[7] The project was never realized, however, due to the heavy load of public responsibilities that Cremona had assumed. Indeed, this was the time when Cremona was appointed as Royal Commissioner to address the scandals in the Vittorio Emanuele Library. Many years later, after Cremona's death, the scientific editor Ulrico Hoepli published between 1914 and 1917 under the auspices of the Royal Academy of Milan the complete works of Cremona.

Guccia's new friends invited him to two meetings of the Académie des Sciences, and to a meeting of the Société Mathématique de France, where Mannheim introduced him to every French mathematician present.[8] At these meetings, a popular topic of conversation was the upcoming meeting of the Association Française pour l'Avancement des Sciences (French Association for the Advancement of Science). The ninth meeting of the Association was to be held that August in Reims, and many mathematicians planned to participate. Guccia was strongly tempted to go along, but the fear of the hot climate in Reims made him decline the invitation and instead prepare an alternative trip to the Swiss Alps. In a letter to Cremona he wrote, "I have to fight for my health, which is very weak and requires a cold climate."[9] However, the discussions in the following days with other young mathematicians convinced him to change his plans and attend the meeting. He also thought about presenting a short communication based on the results of his thesis. This would be his first mathematical lecture, which he would have to deliver alone and in a foreign

---

[6] *Ibid.*

[7] *Ibid.*

[8] *Ibid.*

[9] *Ibid.*

country. He wrote to Cremona on July 22 to inform him of these new plans, to ask for his permission and to request his counseling.[10] Cremona was supportive, and wrote back with a (still nowadays conventional) recommendation:

> I suggest you to be short and concise, as much as possible [...] I know from experience that on such occasions only a few participants are interested in a given subject. Then, try to expose the substance of the topic without going into proofs or particular details.[11]

Finally, Guccia went to Reims and attended the meeting, which took place from August 12 to 19, 1880. The program consisted of one day devoted to the official opening and the reception banquet in the City Hall; five days devoted to plenary lectures and the sessions; and two days for excursions. Several afternoons ended with the so-called "industrial visits" to different businesses of the city. As we have already seen when looking at the Palermo meeting of the Società Italiana per il Progresso delle Scienze, these meetings were quite large and diverse (one of the plenary lectures was about wool, another one about sleepwalking). The scientific sessions were organized into four broad groups: Mathematical Sciences; Physical and Chemical Sciences; Natural Sciences; and Economic Sciences. Mathematics was grouped with Astronomy, Geodesy, and Mechanics in Sections 1 and 2 of the first group. Victor Mannheim was one of the delegates of these sections, although he was not able to attend the meeting. Over forty communications were presented (some participants presented two, three, or even four communications). Arthur Cayley, Gaston Darboux and James Joseph Sylvester were among those who presented communications. Sylvester was at Johns Hopkins University in the United States at the time, where he had founded the American Journal of Mathematics in 1878. In the session held on August 17 Guccia presented results from his thesis in the communication entitled *Sur une classe de surfaces représentables, point par point, sur un plan*. Afterwards, Sylvester spoke quite positively of the communication, and Darboux offered to publish the results in the Bulletin des Sciences Mathématiques, which he had founded in 1870. At a later meal held in the famous champagne cave Pommery, Darboux proposed a toast in honor of Guccia's advisor, Luigi Cremona.[12] Guccia's communication[13] was published in the proceedings of the meeting as a ten-page article[14] (item 1, Appendix 1 on the mathematical works of Guccia).

Guccia was enthusiastic about this experience. Recognizing him as a student of Cremona, many participants in the session instructed Guccia to give their regards to his mentor, in particular, Sylvester, Lucas, Darboux, Halphen, Catalan, Collignon,

---

[10] *Ibid.*, July 2, 1880.

[11] Letter from Cremona to Guccia, August 3, 1880. Folder Cremona, Archive of the Circolo Matematico di Palermo.

[12] Letter from Guccia to Cremona, September 5, 1880. Folder Cremona, Archive of the Mazzini Institute of Genoa.

[13] A. Barlotti, F. Bartolozzi, G. Zappa, *op. cit.*, pp. xii–xiii.

[14] *Comptes-rendus de la 9ᵉ session Reims 1880*, Association française pour l'avancement des sciences, Paris, 1881, pp. 191–200.

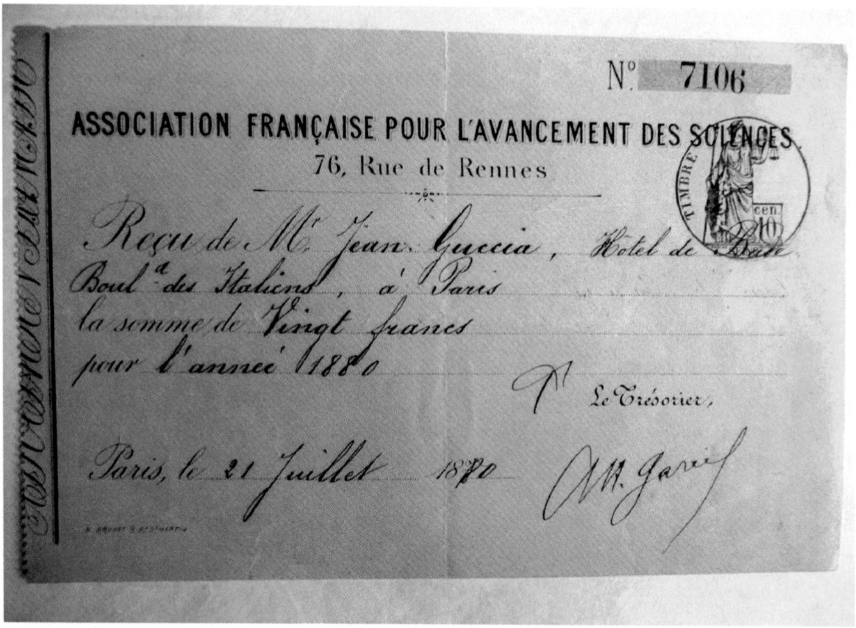
N° 7106
ASSOCIATION FRANÇAISE POUR L'AVANCEMENT DES SCIENCES
76, Rue de Rennes
Reçu de Mr Jean Guccia, Hôtel de Bade
Boul^d des Italiens, à Paris
la somme de Vingt francs
pour l'année 1880
Le Trésorier,
Paris, le 21 Juillet 1880

Fig. 4.1 Receipt for the annual association fee of Giovanni Battista Guccia to the Association Française pour l'Avancement des Sciences for 1880 (the year of the Reims meeting). Courtesy of the Circolo Matematico di Palermo

Schoute and Baehr (as Guccia carefully explained in a letter to Cremona).[15] Guccia appeared in the official list of members of the association as Jean Guccia, residing in Via Ruggiero Settimo 28, Palermo.[16] He kept the coated folder with documents concerning the participation in the meeting all his life; today it is in the Archive of the Circolo Matematico di Palermo (Fig. 4.1 shows one of those documents).[17]

Back in Paris, Guccia wrote to Cremona and explained that he had to decline an invitation for lunch with Chasles, because he had other commitments;[18] apparently, Guccia already had many friends in Paris. By mid-September, he left Paris and its active style of life to spend a week in the Northern city of Boulogne-Sur-Mer (where he found the beach too cold when compared with the Mediterranean Sea).[19]

[15]Letter from Guccia to Cremona, September 5, 1880. Folder Cremona, Archive of the Mazzini Institute of Genoa.

[16]*Comptes-rendus de la 9e session Reims 1880*, Association française pour l'avancement des sciences, Paris, 1881, p. lvii.

[17]Folder Guccia, Archive of the Circolo Matematico di Palermo.

[18]Letter from Guccia to Cremona, September 5, 1880. Folder Cremona, Archive of the Mazzini Institute of Genoa.

[19]*Ibid.*, October 10, 1880.

He arrived in London at the end of September of 1880 and immediately visited Hirst. As we have seen, Hirst was a major figure of the London Mathematical Society, and played a prominent role in its founding. Hirst took Guccia to the Royal Academy's Athenaeum Club and to the Royal Astronomical Society, where they visited the libraries. Most of the time they were together was spent discussing mathematical problems. Hirst communicated to Guccia that he was writing a paper on "Cremonian congruences" and expanded on the technical details. Hirst also shared his worries over a paper by Seigmann Kantor that had been published in the last issue of the Annali di Matematica Pura ed Applicata. Kantor's paper included some results that Hirst had communicated in 1865 by letter to Cremona. Luigi Cremona and Francesco Brioschi were the Editors of the Annali since 1867. Even though Hirst assured that he had no intention to raise a priority case, Guccia was very worried for the possible implications to Cremona. More than half of his letter to Cremona of October 10 was devoted to the details of the case.[20] He spent a few more days sightseeing in London. Then he went back to Paris, and finally to Rome at the end of October where he met his mentor and resumed preparations for the defense of the thesis.

The experiences of the summer and fall of 1880 were determining for Guccia. He had discovered a full new world that he could only have envisioned before: the atmosphere of research activity at its highest level and the internal life of an active mathematical community. This would shape his viewpoints on mathematical activity, and set his standards for his own work. He also made many friends, with whom he would share many of his projects in the future. Referring to this period of Guccia's life, his collaborator Michele de Franchis wrote: "The refinement of his manners, coming from the aristocratic environment in which he was raised, the personal affability and the brilliant wit made him pleasing to those who approached him."[21]

4.2   1881–1883: A Period of Maturation

When Michel Chasles died on December 18, 1880, his heirs decided to put up for auction all the books from his large library. The editor Ulrico Hoepli sent the catalog of the auction to Luigi Cremona, with the idea that some of the books could be of interest to the Vittorio Emanuele Library. After careful examination, Cremona prepared a detailed list of the books that he was interested in, indicating the maximum bid for each one of them.[22] He also requested the counseling of an expert librarian to carry out his instructions during the auction. Hoepli suggested that the right person for such assistance was the distinguished librarian Mr. Otto Lorenz

---

[20] *Ibid.*

[21] M. De Franchis, *op. cit.,* p. i.

[22] The full hand-written list is reproduced in C. Cerroni, *op. cit.*, pp. 241–274.

from Paris. Although Cremona trusted Hoepli, he decided that it would be better if the person representing the Library at the auction had a scientific background, who could make better judgments regarding the quality and the price of a book at auction. Cremona, realizing that Guccia was planning to stay in Paris for part of the summer, quickly sent a telegram to him at the Hotel Continental:

> Would you be willing to represent me in the Chasles book auction from July 4 to 18 buying for the Vittorio Emanuele Library? Tomorrow I write to you.[23]

That was June 25. The next day, he explained in a letter the full business and ended saying, "I rely entirely on you."[24] Guccia's positive and enthusiastic reply was immediate.

Cremona then sent Guccia the list of books to buy, detailing all the necessary financial issues, and giving him license to buy additional books if he believed they were interesting and well-priced. Cremona informed Mr. Lorenz that Guccia was going to be in charge of looking at the books and determining which ones should be bought. On the eve of the auction Cremona wrote to Guccia:

> I know that the Paris National Library is interested in buying everything that they lack. I think this will cause us little harm, since that Library is well stocked while we are missing everything, and also because we cannot take the luxury of hunting for rarities. We must content ourselves with that which is necessary and can be useful.[25]

On July 10 the first part of the auction was over. Cremona wrote to Guccia:

> You are highly meritorious to the studious in our country. Your victory over the Faculté des Sciences was a great pleasure for me [...]. Really bravo![26]

The letters of Cremona from July 16 and July 31 showed his deep appreciation for Guccia's success:

> Dearest Guccia,
> I thank you for your constant interest in sending me the information regarding the subsequent sessions of the auction. You do things excellently; and I am very happy that you are satisfied with Mr. Lorenz, to whom I am sending today a deposit for five thousand liras. I repeat that I am very satisfied with your work and I am very grateful to you.[27]

> The service that you have done to the Vittorio Emanuele Library is indeed great, and I cannot find words enough to express my deep gratitude. I also regret that, for my fault, you are left so many days in Paris enjoying an unbearable heat! [...] The boxes containing the books have started to arrive. Already, two arrived, and four more should be arriving. I have followed your advice in everything, including the one related to the ordering to Hermann for Cauchy. You did well in everything and I yield you ample testimony.[28]

---

[23]Letter from Cremona to Guccia, June 25, 1881. Folder Cremona, Archive of the Circolo Matematico di Palermo.

[24]*Ibid.*, June 26, 1881.

[25]*Ibid.*, June 26, 1881, bis.

[26]*Ibid.*, July 10, 1881.

[27]*Ibid.*, July 16, 1881.

[28]*Ibid.*, July 31, 1881.

The outcome of this event was extremely positive for Guccia. He had aided his mentor in an important and delicate situation, extended and deepened his contacts in the scientific environment of Paris, and discovered and mastered the world and business of the scientific book trade. He would profit from all of this experience in the future.

The beginning of 1882 was a difficult period for Luigi Cremona. First, he suffered a serious illness and, after recovering, his wife Elisa died. These sad events inevitably decreased his scientific exchange with Guccia.

During the summer of 1882, Guccia traveled through Northern Europe. Along his way, he met with several mathematicians, including Pieter H. Schoute at the University of Groningen, in Holland, who had attended the Reims meeting of the Association Française pour l'Avancement des Sciences; Leopold Kronecker at the University of Berlin; and Max Noether at the University of Erlangen. With them he discussed mathematical problems, in particular with Kronecker, as shown by some exchange of letters.[29]

However, his summer travels also had a different aim: a musical one. Guccia wanted to attend the first performance of Richard Wagner's opera "Parsifal" in Bayreuth on July 26. It seems that Guccia and Wagner had met in Palermo. Indeed, Guccia in a letter to Cremona, dated August 29, 1882, wrote, "I answer [...] from Bayreuth where I have gone for the first two performances of the Parsifal of my friend Wagner."[30] Their encounter could have occurred when Wagner, looking to recover from some health problems, moved to Southern Italy, in particular to Palermo, and participated in some of the receptions organized by Tina Scalia Whitaker at Villa Malfitano (see section 4.5 "Palermo's *Belle Époque*"). Wagner, moreover, stayed in the Hotel delle Palme, not far from the Palazzo Guccia, with his wife Cosima from November 1881 to February 1882. There, he ended the score of the "Parsifal".

In the meantime, Luigi Cremona, seeking recovery from the many difficult months before, planned to spend a vacation in the village of Zuoz, situated in the Southeastern side of the Swiss Alps. There, surrounded by lakes and dense forests, Cremona proposed to continue his research on the theory of representation of curves and surfaces. He wrote to Guccia on August 15, 1882, inviting him to join the vacation and work on the project. Guccia was in Lucerne, however, working on some of the questions he had discussed with Schoute, Kronecker and Noether, and excused himself from Cremona's invitation. He also intended to travel back to Bayreuth to rest there until mid-September. On his journey to Palermo, he stopped in Rome and discussed with Cremona some problems on the singularity of curves.

Back in Palermo, Guccia was inevitably involved in an unpleasant family situation. It was a legal case resulting from the division of the legacy of his grandmother on his mother's side, Carolina Cipponeri e Guccia, who had died in

---

[29]Letter from Guccia to Cremona, August 29, 1882. Folder Cremona, Archive of the Mazzini Institute of Genoa.

[30]*Ibid.*

1845. The origin of the dispute was the disagreement between Guccia's mother Chiara, her sister Rosalia, and their brother Salvatore, who was legally incapacitated at that time. For many years, the court case had been going on under the instigation of Salvatore's wife, Elisabetta Cannizzaro, who was a sister of the renowned chemist Stanislao Cannizzaro (whom we have already encountered as a scientist of the Risorgimento, Rector of the University of Palermo, and professor of chemistry in the University of Rome). The longstanding legal dispute caused a stressful situation for Guccia, who wrote to Cremona, "She seems to be created for my scourge! This year [...] she intended to obtain from my family something like 600,000 lire. Luckily, in the end, the judges always reject her strange questions and she is obliged to pay the litigation costs. But this does not give me back the time lost running up and down the courts."[31]

At the end of May of 1883, Guccia traveled to Paris. There, he visited Mannheim and his French friends. He also met with George Henri Halphen (1844–1889), who had received the Steiner Prize from the Berlin Academy of Sciences for his work on algebraic curves the year before. Halphen proposed Guccia as a corresponding member of the Société Philomathique de Paris. This was a high-level multidisciplinary scientific and philosophical society, founded in 1788, devoted to promoting tolerance and freedom, and consisting of a limited number of members, who shared their common love for science. Cauchy, Laplace and Poisson had been members of the society. Following Halphen's recommendation, Guccia was elected at the Société's meeting on June 23. After his stay in Paris, Guccia went to Lucerne for vacation.

As Cremona's political and administrate commitments increased, he moved further away from research. This left Guccia alone in Palermo with no one to act as a scientific mentor. Guccia's letter to Cremona of January 1884 with New Year's greetings and a detailed report on his last scientific advances was left unanswered.[32]

Between 1881 to 1883, however, Guccia received many other important influences for his future mathematical research. He experienced great excitement during the intense summer of 1880 in Paris, Reims and London; the following summer he was again in Paris for the auction of the Chasles library; in the summer of 1882 he met Kronecker, Noether and Schoute with whom he discussed mathematical problems; and in the summer of 1883 he met Halphen in Paris. He was twenty-five years old when all of this intense activity started, and by the time he was twenty-eight, he had a very clear international perspective on contemporary mathematical research.

Guccia continued research by himself,[33] working on Cremonian transformations between two planes π_1 and π_2, studying their effect on the transformation of the

---

[31] *Ibid.*, November 12, 1882.

[32] *Ibid.*, January 2, 1884.

[33] The best sources for the evaluation of Guccia scientific work are the papers *L'attività scientifica di Giovan Battista Guccia,* by A. Barlotti, F. Bartolozzi, and G. Zappa, and *Giovan Battista Guccia and the Circolo Matematico di Palermo,* by C. Ciliberto and E. Sallent Del Colombo.

system of lines and rational curves. Under such a transformation, the linear system of lines of π_1 is transformed into a homoloidal net of curves in π_2, and it is possible to associate a finite sequence $(n, \alpha_1, \alpha_2, \ldots, \alpha_{n-1})$, where n is the order of the curves of the homoloidal net and $\alpha_1, \alpha_2, \ldots, \alpha_{n-1}$ are the numbers of base points of the net having multiplicity $1, 2, \ldots, n - 1$. These numbers $\alpha_1, \alpha_2, \ldots, \alpha_{n-1}$ satisfy certain particular conditions. Guccia studied these sequences and, in some cases, wrote down the equations of the corresponding homoloidal net. These results appeared in four papers published in the Rendiconti del Circolo Matematico di Palermo entitled *Formole analitiche per la trasformazione Cremoniana* (items 2, 3, 4 and 9, Appendix 1).[34]

Continuing with the study of the sequences associated with Cremona transformations of a certain order, Guccia extended some results of Ernest De Jonquières from 1885. Guccia's results constituted the paper *Sur les transformations géométriques planes birationnelles,* which was published in the Comptes Rendus de l'Académie des Sciences—and presented precisely by De Jonquières—(item 7, Appendix 1).[35] A similar situation occurred with some results by De Jonquières from 1864 on what he called "hisological curves" arising from a Cremona transformation. De Jonquières and Cremona had obtained some partial results. Guccia conducted a complete study in the paper *Teoremi sulle transformazioni Cremoniane nel piano* that was published in the Rendiconti del Circolo Matematico di Palermo (item 12, Appendix 1).[36]

4.3   The Founding of Circolo Matematico di Palermo

Last's year competition for the chair in mathematics in the Liceo Umberto I opened my eyes to the miserable state of abandonment of the studies of mathematics in Palermo. Whenever you will be willing to direct your attention into this state of affairs, I could provide you with documents and refer to facts that you will find incredible! Young people and professors publish and publish continuously, without having any scientific direction, only motivated by the rush to acquire titles and notoriety. And in Italy there is no one taking care to open their eyes and correct them!

Neither do those Geometers who are erroneously cited. What mainly characterizes the nature of these productions is gullibility, which, in my opinion, in science is more reprehensible than the errors themselves. Meanwhile, there is no doubt that the University supports and hails this unsound scientific activity. None (it must be boldly said) of the fine professors of the continent that have succeeded in a short period of time, such as Dino Padelletti, Alberto Tonelli, or Alfredo Capelli, have managed to change this state of affairs; some, for lack of energy and ability to fight with the elders of the Faculty; others, for natural selfishness not willing to go beyond the limits of their strict duty. The fact is that all of them have left things as they were. [...]

How can it be explained that a book where, among a thousand nonsenses, it is said "let the space of lines of dimension 5" was the subject of a favorable report, signed by Alberto

---

[34] A. Barlotti, F. Bartolozzi, G. Zappa, *op. cit.,* p. xiii.

[35] *Ibid.,* pp. xiii–xiv.

[36] *Ibid.,* p. xiv.

Tonelli and Dino Padelletti, on the occasion of a request for license for university teaching? The sentence with which the work was praised was "the author shows to have an exact knowledge of modern methods."

Moreover, how can it be explained that the engineer F. Paolo Paternò, who was declared ineligible for the competition in the Liceo Umberto I, was afterwards immediately appointed lecturer of Projective Geometry at the University of Palermo?

These considerations are combined in my mind with the impression I received from the Report of the University Commission, that you gave me the opportunity to have in my hands, in which these people have the audacity to request two complementary teachings, in order to be authorized to confer the degree in Mathematics.[37]

This discouraging report of the situation in Palermo was in a letter that Guccia wrote to Cremona on April 13, 1884. The correspondence with his mentor allows us to follow step by step the birth and development of the Circolo from the viewpoint of its creator.

At that time in April 1884, Guccia had no position of any type at the university in Palermo. Moreover, his contacts with the institution were quite scarce, as there was neither a specialized mathematical library, nor a regular research mathematical seminar at the university. This contrasted greatly with the atmosphere he had enjoyed in Rome while working on his thesis, or even more, with the scientific atmosphere he had envisioned in his visits to Paris. What he saw around him in Palermo, as he complained to Cremona, was that people would just "publish and publish continuously, without having any scientific direction," looking only to "acquire titles and notoriety." He could find no trace of interest or devotion to science in his immediate environment.

It was in the above-mentioned letter where Guccia announced to Cremona the beginning of his far-reaching and visionary new plan:

It was then that I had the idea of a Mathematical Circle bringing together, in a comfortable place, equipped with a good reading and studying room (containing at least sixteen periodical publications in mathematics), all people, old and young, that for some reason are interested in mathematics. My warm appeal was heard, and everyone agreed with my plan, that contained two essential points, on which I had no intention of compromising:

1. Absolute exclusion of all matters related to applied mathematics.
2. A priori exclusion of the slightest idea of publication by the Circle, the main purpose of which was (in my mind) to avoid, or better, to avert any publication.

(I will not tell you how much I had to fight, and I still struggle, to persuade these fine people that it is not the case to go around Europe with the <u>Bulletin of the Mathematical Circle of Palermo</u>). This is how this statute has come to be, whose hybrid nature has its origin in the local circumstances that I had necessarily to take into account.[38]

The details of the event that Guccia was explaining to Cremona are the following. In February 1884, he conceived the idea of founding a mathematical society in Palermo, with the aim of having a place where the mathematicians of Palermo could

---

[37]Letter from Guccia to Cremona, April 13, 1884. Folder Cremona, Archive of the Mazzini Institute of Genoa.

[38]*Ibid.*

meet and discuss current research problems. He named it the Circolo Matematico di Palermo. The first meeting of the new society took place on March 2, 1884. The first statute of the society was approved, and the Steering Council was elected. The statute is a humble document which consists of ten pages, including one page for the front cover, one page dedicated to the members of the Steering Council, six pages of regulations, and two pages listing the names and positions of the Circolo's twenty-seven founding members. The statute outlined the aim of the society, which was to increase mathematical studies and discussions in Palermo. The statute also stated that new members had to be nominated by two current members of the society, and then approved during one of the society's meeting sessions, which would be held every two weeks from November to June. The society could also subscribe to journals and buy books to be kept on their premises. The statute was declared to be provisional (Fig. 4.2).[39]

It is remarkable that the mathematicians of the University of Palermo followed the ideas of twenty-eight year old Giovanni Battista Guccia. Some of the founding members of the Circolo included Giuseppe Albeggiani, professor of Mathematical Analysis and former Rector of the University; Gaetano Cacciatore, the Dean of Faculty of Science, full professor of Astronomy, and director of the Palermo Observatory (as well as a friend of Guccia's uncle Giulio Frabrizio Tomasi); Francesco Calderara, full professor of Mathematical Physics; Filippo Maggiacomo, full professor of Geometry; and Alfredo Capelli, associate professor of Algebra and the namesake of the Rouché–Capelli theorem (also known as the Kronecker–Capelli theorem, Rouché–Fontené theorem, or Rouché–Frobenius theorem).

The detailed list of the twenty-seven founders and their profession was:

Alagna, R.	engineer
Albeggiani, G.	mathematician, at the university
Albeggiani, M.L.	mathematician, at the university
Arioti, A.	engineer
Bontade, G.	engineer
Cacciatore, G.	astronomer
Calderera, F.	mathematician, at the university
Capelli, A.	mathematician, at the university
Capitò, M.	architect
Damiani, G.	architect
Fileti, E.	mathematician, at the liceo
Gambera, P.	mathematician, at the liceo
Gebbia, M.	mathematician, at the university
Giardina, A.	mathematician, at the liceo
Guccia, G.B.	mathematician
Guidotti, G.	mathematician, at the liceo

---

[39]Statuto ed elenco dei soci. Circolo Matematico di Palermo, 1884, Palermo. Biblioteca Centrale della Regione Siciliana, Palermo, coll. MISC A 21.21.

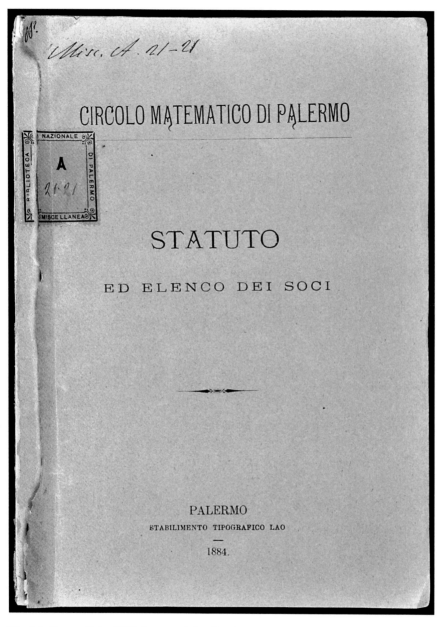

Fig. 4.2 Cover of the 1884 Statute of the Circolo Matematico di Palermo. Courtesy of the Biblioteca Centrale della Regione Siciliana

Maggiacomo, F.	mathematician, at the university
Maisano, G.	mathematician, at the university
Paternó, F.P.	mathematician, at the university
Pepoli, A.	mathematician, at the liceo
Pintacuda, C.	engineer
Pittaluga, G.	mathematician, at the liceo
Politi, G.	engineer
Porcelli, S.	engineer
Righi, A.	physicist
Salemi-Pace, G.	engineer
Scichilone, S.	mathematician, at the liceo

There was a total of seventeen mathematicians in this group, including nine from the University of Palermo, seven from the secondary school (liceo), and Guccia (who was unemployed at the time). In addition, there was one astronomer, one physicist, seven engineers, and two architects. Compared with the founding bodies of other mathematical societies, this group lacked military men, who at that time would have had some knowledge and appreciation for mathematics.

Guccia's letter to Cremona from April 13, 1884, is an invaluable document, as it allows us to understand the founding of the Circolo via the aims, thoughts, doubts, and inside crafting of its creator:

> You cannot imagine how many difficulties I had to overcome, and what effort in simulation I had to assume in order to bring the ship to port. From the very beginning, it was my secret intention not to offend anyone's sensitivity; the purpose that I had determined with the foundation of the Circolo was to improve the situation in Palermo by persuasion and without any violence.
> Therefore, I drafted in my mind the Steering Council just as it was actually elected in the first session on March 20, where due consideration had been given to titles and seniority.[40]

Indeed, a few days later, on March 20, the first Steering Council of the society was elected: Prof. Giuseppe Albeggiani as president; Prof. Francesco Caldarera as vice-president; Prof. Alfredo Capelli and Prof. Michele Luigi Albeggiani as secretaries; the treasurer being Dr. Giovanni Battista Guccia.

It is important to observe the role that Guccia had reserved for himself in his well-thought-out plan for his Society: that of treasurer. In order to guarantee the success of the Circolo as a scientific project, it was of critical importance to take good care of the finances, given that the Circolo was not intended to be just a meeting point for discussing mathematics, but also—and probably most importantly—an effective tool for doing research in mathematics. Guccia explained to Cremona the steps he had taken to financially support the Circolo:

> Now we come to the financial part. If the Circolo had to provide for the rent of the premises and for the costs of its furniture, then it is certain that all of its funds would be absorbed.

---

[40]Letter from Guccia to Cremona, April 13, 1884. Folder Cremona, Archive of the Mazzini Institute of Genoa.

Fig. 4.3 Biblioteca Guccia

> That is how I thought it was necessary to make some sacrifices to come to aid the institution. I offered <u>for free</u> the premises (two good rooms in the ground floor), which I furnished at my expense, and I provided it with everything that was necessary for its use, and I left it at the complete disposal of the members from 11 a.m. to 5 p.m. In this way, the funds of the Circolo, coming from the annual contributions of the members and from the admission fees, will serve exclusively to provide the subscriptions and purchasing of whatever works. It is therefore the creation of a good library one of the main goals of the institution that I have created. Article 14 of the statute is to encourage the members to make donations. But donations from members are not enough! This is the point I allow myself to ask you my warmest plea.[41]

The rooms that Guccia offered to the Circolo for holding the session were on the ground floor of the family palace on Via Ruggiero Settimo 28, and were well situated within the city (Fig. 4.3). The Guccia family palace became the official address of the society, and remained as such for a number of years.

Guccia came from a family accustomed to businesses and financial deals, and thus he knew that the Circolo would need external support in order to survive and achieve its aims. At this point, Luigi Cremona's support came into play. Guccia appealed directly to their common values and viewpoints, their friendship, and the service he had done to Cremona in the past in order to win his mentor's support:

> If you approve this institution and consider it truly useful to a region where (for mathematics) the University does not exist, and the Libraries are lacking everything, I am sure you will not deny me your valuable support, all of your energetic activity. Would it be perhaps

---

[41] *Ibid.*

indiscrete to remind on this occasion that in the summer 1881 I displayed my services for
the Vittorio Emanuele Library?

As this is something for Science, should not one be allowed to appeal to the feelings of
gratitude? Don't you think that this note, if presented with grace, might be able to move the
leaders of the Ministry?

But in this topic, I know that if there is anything to do, you will give me the direction. The
issue is not yet settled, as I have not written to anyone, just because I am looking forward
to your appropriate advice, dictated by your love for Science and by your experience.
So, I have even refrained from addressing the Italian academies in order to obtain their
proceedings and other publications for free.

I have an ongoing (secret) arrangement with the Ministère d'Instruction Publique de France
in order to persuade them and get the works of Lagrange or Laplace.

It is useful to warn you that asking the Municipality or to the Province [of Palermo] for
support is a vain effort or at least dangerous, since these local authorities usually bring
conditions that could degenerate the nature and the purpose of the Circolo. The Circolo
(it is good to know) is in my hands and will follow my views (which I think you will not
disapprove of) up to the moment when it will become poor, and then it will be influenced
by people that will drive it in the wrong direction.

Thus, I rely on you, exclusively on you!! So to start, I think that in your next excursion
to the North [of Europe] you could obtain as a gift to the Circolo the memoirs from the
best Geometers that you will have the opportunity to meet. This would be a very useful
beginning. On your return to Rome, you could tell me what can be done in Italy with the
Ministry, with the Academies, with the Libraries, and privately.

Meanwhile, it would be extremely useful if you could officially send some words to the
Steering Council of the Circolo, to show your approval of the institution, within the limits
of our program. This could follow receiving the statute or receiving a private letter of mine
in which I inform you on the founding of the Circolo. It is well understood that all that I
have written you today is absolutely confidential.[42]

These paragraphs show Guccia's absolute confidence in Cremona, and also show
his certainty that Cremona's support (within and outside Italy) was instrumental for
the success of the project. In fact, Guccia sent Cremona five copies of the statute of
the Circolo Matematico di Palermo along with his letter, writing "in case you deem
it appropriate to distribute them."[43]

The mathematical side of the project got off to a great start. During the first
session on March, the Society both voted for the Steering Council members and had
time for scientific exchange:

Francesco Caldarera presented the communication entitled *On the theory of
harmonic centers,* discussing some point on a memoir of Cremona.

Alfredo Capelli presented a short proof of a known corollary to a theorem of
Green.

The second session of the newly created society was held on April 3, and was
fully devoted to mathematics. There, Caldarera completed his previous communication, which was followed by a discussion with M.L. Albeggiani, Guccia and Capelli.
After that, Guccia posed and discussed a question on Cremonian transformations on

---

[42] *Ibid.*

[43] *Ibid.*

the plane. The rhythm of sessions continued at a good pace of two per month. The third session, on April 17, had communications presented by Maisano, Caldarera, Pepoli, and Capelli. In the fourth session, on May 1, there were communications by Maisano and Gebbia, and a question on geometry proposed by Guccia. The next session was on May 15, with a communication by Guccia, who also posed another question on geometry. The society members also discussed other administrative matters during that session, such as the approval of the budget for the year 1884. Sessions were held on May 29, June 12, and June 26 before a break for summer vacations.

After the intense start to the year with the founding of the Circolo, Guccia spent the summer holiday in Lucerne. His letter to Cremona of July 12 gives a perfect view of their personal relationship, the close connections that Guccia had already established with the group of French mathematicians, and also of Guccia's lifestyle. Guccia was eager to see Cremona and share a vacation with him, and suggested that his mentor join him in the Rigi Mountains near Lucerne. He explained:

> If this is the case, I would recommend you stay at the Rigi-Staffel, as this location is amazing, the food great and the prices very reasonable. I am sure that you and your family will be very well there. This was the place where I wanted to send Mannheim. It is pointless to insist on the many excursions that you could do starting from the Staffel, because perhaps you have already done some of them. What I can add, from my own experience, is that staying at the Rigi for a month is of great benefit to one's health; I do not think that in this respect one can find a better place in Switzerland. Add to this the ease of reaching any other place in Switzerland by train. In case you decide to follow my advice, I would be a faithful companion in the excursions and place at your disposal a small (travel) library where you could find the full series of notes and memoirs by Cremona![44]

Guccia then spent October 1884 in Paris. By then, this was his normal mathematical environment; he was involved in the everyday life (and quarrels) of the Paris mathematical community. A small event illustrates this.

Guccia helped Ernest De Jonquières in the publication of a paper of his. De Jonquières (1820–1901) was a military man—to be precise, an admiral—who was also a geometer that had followed scientifically and collaborated with Michel Chasles. In 1859, he presented a paper on birational transformations to the Institute de France, and this paper was rejected by Chasles. This rejection created a strong controversy, and De Jonquières refused to submit the paper to any other journal, although he did publish a summary of the paper in the Comptes Rendues de l'Académie des Sciences de Paris. In the paper, De Jonquières defined a family of birational transformations that were neither linear nor quadratic. Cremona had not noticed the result at first, as it was published without proofs. However, Cremona, in the second part of his paper *Sulle trasformazioni geometriche delle figure piane*, acknowledged De Jonquières's result and later advertised it. Guccia had met De Jonquières in Paris, and they had long discussions regarding their common interests in geometry. At this point in 1884, De Jonquières was again interested in publishing his full paper, but due to his resentment about the previous rejection, he insisted that

---

[44] *Ibid.*, July 12, 1884.

the publication had to be done outside France. He explained the situation to Guccia, who in turn mentioned the issue to Cremona ("we could give [De Jonquières] a bit of justice"), and suggested publishing the paper in the Giornale di Matematiche. Guccia personally took the manuscript from Paris to Rome, and one year later the paper was published in the journal.[45]

Guccia's splendid and communicative character, along with his clear views on his goals, were very much appreciated by everyone he worked with. Upon meeting Gaston Darboux at the Institute de France in 1884, Darboux congratulated him for the founding of the Circolo Matematico di Palermo. The word about the new mathematical society had already spread around.

On October 22, 1884, while still in Paris, Guccia shared with Cremona his joys and worries concerning the Circolo:

> Regarding the Circolo, I think I have done a good job. Everyone, without exception, has been interested in the Library of the mathematical society of Palermo and donations are coming from everywhere. Let us see what you can now do in Rome!! You know very well that my greatest hopes rely on you, and your influence in the Ministry [of Public Education]. On several occasions you demonstrated that, when you do want, you are able to obtain money for Science.
>
> In Palermo, the situation is serious and should be considered. Never has money been better spent than in support of an institution that has three main purposes of great importance:
>
> 1. To prevent the publication of books and pamphlets that dishonor the country!
> 2. To provide the scholars of the new generation a library that allows them to be aware of Science;
> 3. To establish, through discussion and resolution of questions, a competition, necessary for young people, to create a real love of Science.
>
> Purpose 1, which was the most urgent to attain, produced, from the beginning, the best expectations!
>
> I will try to be more explicit in Rome, when I have the pleasure of meeting you, but for now please believe in my words: It is urgent to get help both in money and in books. It is urgent, since the Government, concerning the University, takes every possible opportunity to commit stupidities. I have known that the Ministry has requested the Faculty of Mathematics of Palermo to set up a list of names to establish, this same year, a degree in Mathematics. It is clear, without possibility of error, that the individuals that will be proposed for teaching mathematics will be the ones that had been battered (due to the desperate fight taken by some members of the Committee) in the competition for the Liceo Umberto I.[46]

The letter expresses a similar discouraging picture as he had reflected before, but this time with a difference: the Circolo existed. Now there was hope for the future of mathematics in Palermo, so Guccia ends his letter with a plea to Cremona: "It is urgent, very urgent, to support the Circolo."

On November 5, Guccia was still in Paris, at the Hotel Continental, planning to attend a dinner organized by the Italian ambassador Luigi Federico Manabrea. Manabrea was a remarkable man of the Italian Risorgimento: engineer, scholar,

---

[45] Ibid., October 22, 1884.

[46] Ibid.

scientist, politician... and early follower of the ideas of Charles Babbage.[47] This was, more or less, the end of his summer traveling.

The scientific sessions of the Circolo resumed on November 20, 1884. By March 1885, one year after the founding, the activity of the Circolo had been intense. There had been seventeen sessions, from May 1884 to March 1885 (roughly two sessions per month), except for the long summer break, which went from July until October. The total number of communications, of different types, was forty-four (some split into several sessions). Guccia was the most productive member, giving twelve communications, and was followed by Maisano, who gave nine. After them, Capelli, Calderera and M.L. Albeggiani each gave five, Gebbia gave four, and Gambera and Pepoli gave two communications each. These figures show that the effort in presenting communications was fairly spread among the members of the society. The content of the communications was diverse: some presented new results on a specific topic (often in Geometry), some regarded extensions and generalizations of recent results, and other presented new proofs of known facts. Members of the Circolo also proposed mathematical questions, as well as solutions to previously posed questions. On other occasions, members reflected on general ideas and principles. The scientific sessions also included bibliographic reviews, where society members made comments, and announcements about articles and books that had recently been published. In these discussions an ample use of recent references was made. This was a first positive outcome of the creation of the library of the Circolo. (Figure 4.3 shows the private library of Guccia).

The number of members in the Circolo increased by seven in 1885 (as eight joined and one left), and this evidenced a good start for the society.

The activity of the Circolo pleased Guccia, who wrote to Cremona: "The Circolo thrives and the mathematical awakening achieved in Palermo is remarkable. As far as I can, I always give all of my strength."[48] This was on March 1885, one year after the founding of the Circolo. He reported to Cremona on the constant rhythm of donations that the Circolo was receiving. The publications missing in the library of the society were precisely those by Cremona. Guccia requested in particular some paper of Cremona on transformations in the (three-dimensional) space to be used by his students.

The students that Guccia refers to in that letter were from a new initiative of his. The University of Palermo had a course in Mathematical Physics, taught by Capelli, and a course on the Theory of Forms, taught by Maisano. As a complement to these courses, he started teaching an unofficial course in Geometry to the same students on the premises of the Circolo.

After his return in mid-November 1884 to Palermo from the summer vacation, Guccia was able to concentrate exclusively on mathematics for a series of months. He later explained to Cremona "Since my arrival in Palermo, I have dedicated myself [to science] completely (I could do it all twelve months of the year!),

---

[47] *Ibid.,* November 5, 1884.
[48] *Ibid.,* March 19, 1885.

neglecting everything else, even the correspondence and the annoyances of business and administration."[49] His letters to Cremona in the following period were mostly devoted to mathematical discussions. He was very much absorbed by research, and was full of ideas and suggestions. Guccia presented theorems and proofs in fine detail to his mentor, and discussed possible new directions of development. Two exchanges illustrate these dialogues.

On March 19, 1885, Guccia sent a letter to Cremona with the following mathematical discussion:

> At present, there are a lot of things to do in Geometry and it is unlikely that Science should deprive herself of your powerful intellect. It is not about developing and applying methods already known (to this service there is a large number of geometers working with prudence and care) rather than creating new ones to urgently repair the gaps that, if ignored, could possibly one day could discredit everything that has been done so far. The geometry of the plane is to be redone: the general curve of order n, being a complete ternary form, is in the domain of the modern algebra that, indisputably, reaches in this field the same results as geometry, but with greater swiftness and often with more rigor. It seems to me that we must redo all the theory of the plane, aiming to curves of order n having some singularity.
>
> It is therefore the geometric definition of a generic singularity of a planar curve that becomes the problem of the utmost importance, to which the efforts of creative minds like yours should tend! The results obtained by Halphen have as a starting point the methods of Cramer and Puiseux. Continuing on that path, it is Algebra that, incidentally, would earn, but not Geometry. [50]

We can see that Guccia was fully aware of the strong potential that the introduction of algebraic methods could bring to geometrical research. However, it seems that he was not willing himself to switch to the new methods, and rather preferred to reinforce the classical geometrical approach.

The never-ending theme of the shortcomings of the University of Palermo was always painful for Guccia. In a letter dated July 6, 1886, he explained to Cremona the latest developments of the situation in the university. The beginning of the letter reveals his outrage: "The University goes badly! [...] It is an insult to Science!"[51] The issue was the last two courses of the university's new degree program in mathematics. He complained about the Faculty, about the Minister (of Public Education), and about his colleagues (even members of the Circolo) who had appointed the people who would teach the new courses, such as Higher Analysis, Advanced Geometry, and Celestial Mechanics. He believed that the mathematical content of those courses was beyond the knowledge of the people proposed to teach them. He exclaimed:

> It is an indecent show. A respectable faculty would have asked for a competitive application process [for filling the positions]. Instead, it [the faculty] wants to distribute the 1,200 lire within the family. Appetite comes with eating, so the professors in charge for the course

---

[49] *Ibid.*

[50] *Ibid.*

[51] *Ibid.*, July 6, 1886.

1885–86 are likely to be (some have already requested it) extraordinary professors for the course 1886–87. All this is done, in order to perpetuate, to make stable, the indecency.[52]

He went on to analyze the duties and incompetence of the members of the Faculty who were appointed to teach these courses. His standards were high and he was very committed to the students: "All four of them [the teachers] are motivated by the deepest cynicism, they are convinced that they have fulfilled their duty and they are indifferent if the young people coming out of this school will be capable of anything!" He complained that they were not even interested in improving their knowledge by consulting the new publications available at the library of the Circolo. The worst for him was that he saw the situation from outside, and was unable to influence the decisions of the university. He was deeply discouraged by the situation, and would have liked Cremona's intervention.

However, Cremona's help was not possible at that time. The reason was that Cremona had been deeply engaged in the university reform project since 1885. This task absorbed so much of his time that he left many of his other duties unattended, in particular, his scientific commitments. One of his neglected scientific duties involved Victor Mannheim. Mannheim was a candidate for the seat in the Académie des Sciences left vacant after the death of the mathematician Joseph Alfred Serret. However, Mannheim was facing strong opposition from Hermite. For Mannheim's application for the position Cremona had to report on a work of Mannheim published in the Rendiconti della Reale Accademia dei Lincei. Cremona seemed to have finished the report, but had not yet sent it for publication. Therefore, Mannheim could not refer to Cremona's report in the memoir he had to present to the Académie in support of his nomination. Mannheim wrote to Guccia asking for his help, and Guccia wrote to Cremona begging him to take some action in two letters dated March 19 and April 10.[53] Guccia was worried not only about the absence of the report, but about the possibility that a defeat of Mannheim's candidacy could be attributed to Cremona, or interpreted as Cremona's criticism of Mannheim's work. We do not know what happened to the report, but Mannheim lost the election in favor of Edmund Laguerre.

The activity of the Circolo from April 1885 to June 1886 involved much consolidation. Fifteen sessions were held and twenty-five communications were presented. These figures are lesser than those in the previous period, however, the communications presented were substantially larger than the previous ones. A novelty was the participation in the sessions of the first foreign mathematician: Thomas Hirst attended two sessions of the Circolo and presented a communication entitled *Sur la congruence Roccella, du troisième ordre et de la troisième classe.*[54] The number of sessions in this period decreased because no sessions were held during the second half of 1885. We have no precise information about the reason for

---

[52] *Ibid.*

[53] *Ibid.*, March 19 and April 10, 1885.

[54] Rend. Circ. Matem. Palermo, 1 (1884–1887), pp. 64–66.

this, but it was most likely related to the cholera epidemic that struck Palermo at the end of the summer of 1885.

The society's internal affairs were also going well. There were eleven new members, the Steering Council was reelected for the second time (the first reelection was in January 1885), and a librarian was elected among the members of the Circolo.

Hiring a librarian had become a necessity, as the library of the Circolo was a successful enterprise. The list of articles received during the first year was sixteen pages long. Two years later, it was twenty-five printed pages long, with articles from ninety-three authors. Among them were Beltrami from Pavia, Catalan from Liège; Eneström from Stockholm, Fiedler from Zurich, Forsyth from Cambridge, Fouret, Hermite, Lemoine and Poincaré from Paris, Goursat from Toulouse, Hölder from Göttingen, Homén from Helsingfors, Kronecker and Weierstrass from Berlin, Luguine and Starkoff from Odessa, Malmsten from Upsala, Mellin from Stockholm, Mayer from Tübingen, Schubert from Hamburg, Schur from Leipzig, and Stolz from Innsbruck.[55]

Between 1886 and 1887, there were two events which troubled Guccia and grabbed his attention (and efforts) away from the Circolo for a while. One was the suicide of the mathematician Ettore Caporali, and the other was the dispute over the water supply in Palermo.

On July 2, 1886, at the age of thirty-one, Ettore Caporali committed suicide. He was a full professor of Geometry, who had graduated from the University of Rome in 1875 under the supervision of Cremona. After some years teaching in secondary schools, he entered the University of Naples, where he became a full professor in 1884. In the following years he had reached the conclusion that his intellectual abilities were declining. Guccia had been with Caporali a few months before his tragic suicide, discussing recent mathematical problems. These conversations, as well as the stories of Guccia's scientific travels, had reanimated Caporali. But he had confessed to Guccia:

> That intelligence had left him; that he was no longer able to create anything; that he was not able to find as before the mental energy needed to undertake any research; that he saw himself reached and surpassed by others; that he browsed around different topics but could not find the strength to delve into anyone of them; that he saw the sacred fire of the science extinguishing in him; he was discouraged.[56]

Guccia had also seen Caporali in Rome while attending some meetings related to a tentative project to create an Italian mathematical society. There, Caporali showed some positive excitement. In any case, Guccia was troubled by trying to understand this sad event. He eventually drew some conclusions concerning the causes of Caporali's suicide, and lamented that "discouragement invaded [Caporali] in recent years, via a slow but persistent brain disease that diminished his mental faculties."[57]

---

[55] Rend. Circ. Matem. Palermo, 1 (1884–1887), pp. 94–118.

[56] Letter from Guccia to Cremona, July 6, 1886. Folder Cremona, Archive of the Mazzini Institute of Genoa.

[57] *Ibid.*

Guccia planned a commemoration of Caporali at the following sitting of the Circolo. For this, he asked Cremona for some personal information on Caporali.

The other matter troubling Guccia at that time was the water supply issue, to which we have already referred. At the end of the summer of 1885, the cholera epidemic resurfaced in Sicily. This time, Guccia's uncle Giulio Fabrizio Tomasi, Prince of Lampedusa, died from the disease on September 17 in Florence, where he had fled with his family to avoid the contagion. The announcement of his death is a true measure of the social importance of the Giulio Fabrizio Tomasi:

> The Princess of Lampedusa, the Duke and the Duchess of Palma, Giovanni Tomasi Baron of Montechiaro and consort, Cavalier Filomeno Tomasi Marquis of Torreta, the Count and the Countess of S. Carlo, Cavalier Francesco Crescimanno Trigona from Barons of Capodarso and his consort Chiara Tomasi, Cavalier Giovanni and his consort Antonietta Tomasi, Misses Carolina, Concettina and Caterina Tomasi, the Prince and the Princess of Niscemi, the Marquis and the Marquise of Favare, the Prince and the Princess of Sirignano, Cavalier Claudio Woghinger and his consort Carolina Caravita, Cavalier Diego Calcagno and his consort Caterina Caravita, Lady Maria Caravita widow Avati, Cavalier Giuseppe Maria Guccia, permeated by the deepest sorrow share with you the irreparable loss of their beloved husband, father, father in law, and uncle Giulio Fabrizio Maria Tomasi Caro Traina, Prince of Lampedusa and Duke of Palma.[58]

It is interesting to note how the Tomasi and the Guccia families were intertwined: there are three Guccia's in this list of eminent people, namely, Maria Stella Guccia e Vetrano, the princess of Lampedusa, Carolina Guccia e Guccia, the wife of Giovanni Tomasi, Baron of Montechiaro, and Giuseppe Maria Guccia, which were, respectively, aunt, sister and father of the mathematician.

As a consequence of this epidemic, the Mayor of Palermo issued the following order in February of 1886: "The sources of drinking water must be covered and well-guarded. By August 1887, the pipelines must be of cast iron from the source, and the pipelines must be reinforced, in order to avoid all the disadvantages of the elevation of the water in the head house."[59] This order of the municipality is of interest to our story since one of the sources of drinking water of the city of Palermo belonged to the Guccia family, and the administrator at that time was precisely Giovanni Battista. Some of the letters he wrote to Cremona even carry the letterhead "Amministrazione delle Acque di Casa Guccia", Via Ruggiero Settimo 28.

Guccia could not avoid getting involved in this issue since it would impose large expenses to the family business if the Mayor succeeded with his intention. So, two months later, in a letter to the City he explained:

> The Municipal Health Committee established that the drinking waters of Palermo are polluted due to the infiltration of microorganisms through the clay pipelines. This resulted in the Mayor's order to replace the clay pipelines by cast iron ones. The implementation

---

[58]G. Lanza Tomasi, *Giuseppe Tomasi di Lampedusa. Una biografia per immagini,* Sellerio, 1998, Palermo, image no. 37.

[59]G.B. Guccia, *Sulla conduttura delle acque potabili: lettera alla giunta speciale di sanità,* Ed. Di Cristina, 1886, Palermo. Biblioteca Scienze Giuridiche, coll. BL M** 16408 U, University of Palermo.

of this order would impoverish hundreds of families of plumbers, and would require the owners of the water companies a total expenditure from ten to fifteen million lira. However, first, the opinion of the Committee is based on the difference in the number of microbes in the water existing between two points of the pipeline in an area of the town notoriously unhealthy; and, second, this method gives clearly false results because it does not take into account the almost instant growth of the microbes and the fact that the clay pipelines are covered by a layer of pozzolan, a lime compound, and sand, with a thickness of almost twenty centimeters.[60]

He included scientific remarks as well, and cited a paper from the Italian Chemical Journal detailing the results of an experiment on the growth of microorganisms in water. A few days later, after pulling some strings in the Paris scientific community, Guccia insisted on his arguments in a second letter where he quoted the opinion of Louis Pasteur (obtained via the intermediation of admiral De Jonquières): "I have never studied a similar question, but I do not think that water contamination propagates through the pipeline."[61]

It seems that his argumentative efforts, as well as the intervention of a specially designated health commission, halted the Mayor's order, and saved Guccia the cost to his business. In any case, Guccia's effort had been intense: he was exhausted, and, moreover, fed up. He spoke about his "most unfortunate capacity as water-supply owner"[62], and he explained to Cremona on August 1886, writing "my time is absorbed in such a way cannot easily be imagined if one is not with me in Palermo."[63] He went on explaining:

> If there were no one else to blame than the dynasty of Borbons of Sicily, I would not hesitate to condemn King Ferdinand I of the Two Sicilies for having given (partially sold) the water supply contract to one of my ancestors, the Marquis Guccia! Similar administrative machinery cannot be described in words: you have to see it! It suffices to say that I would prefer a thousand times to administer by myself, instead of the waters supply, the municipality, the province, and all the charities created and to be created! How many times I have meditated on the project of abandoning everything and rushing into a dark corner in Italy where I would teach Mathematics and devote myself exclusively to science.
> I still hope that I could honorably hand over my property (not a gift to the City, of course!). All my efforts tend to this aim. [...] I do not want to speculate more on this; I don't ask for a penny more than what I receive yearly, but I want to leave at any cost![64]

This was quite a strong statement given that his family owed its prosperity and wealth to the royal concession that he was referring to. His severity must be understood in the context of Guccia's intense wish to concentrate on research and the constant interference that the matters of water supply business caused him.

---

[60]*Ibid.*

[61]G.B. Guccia, *Sulla conduttura delle acque potabili: lettera II alla giunta speciale di sanità*, Ed. Di Cristina, 1886, Palermo. Biblioteca Scienze Giuridiche, coll. BL M** 16408 T, University of Palermo.

[62]Letter from Guccia to Cremona, August 8, 1886. Folder Cremona, Archive of the Mazzini Institute of Genoa.

[63]*Ibid.*

[64]*Ibid.*

An indirect consequence of these quarrels with the municipality was that Guccia was "spontaneously nominated" to run in the political elections for a seat in the Provincial Council. Guccia wrote to Cremona once again in August 1886, explaining these events openly and clearly:

I was induced to accept the nomination for the Provincial Council, which was spontaneously offered to me (although I have never dreamed of entering public administration), and which was contrary to my principles (I said and I will always say that a man of science must stay away from administrative matters), after seeing and touching with my hands, in recent times, the low level that the morality of administrators of public affairs has reached. I realized that I was obliged as an honest citizen to fight against intrusive profiteering! What greatly surprised me in this election was that while for many years I had done everything possible to remain ignored and hidden in my city, always keeping myself, on purpose, away from electoral struggles and politics, avoiding mentioning myself in whatever occasion and with any excuse, the city received my name with lively sympathy.

So it was like this that without having to put fingers in cold water, without spending either a word or a penny, keeping myself absolutely away from the voting campaign, the district of Molo elected me as its provincial representative with 750 votes, a figure not reached before in the district in any voting for provincial councilors. But do not worry; don't think that I want to make a political career.[65]

In the end, Guccia was offered a seat in the provincial government, but he refused. His aim for running for the position was clear in his mind, and he stated: "I will simply restrict myself to contribute with energy, as a simple member of the Council, on issues concerning public instruction."[66] One example of this was related to the recently created University Consortium, supported by the City and Province of Palermo, which aimed to subsidize laboratories and new teaching positions in the universities. In a similar situation, the University of Catania had requested new positions, in contrast to the University of Palermo which had chosen to distribute new teaching tasks among the existing personnel. This kept the University of Palermo from growing, and also deprived the Circolo of the arrival of new intellectual forces.[67]

Guccia's political experience was short-lived. While he was in office, he was still involved in the legal dispute with the Municipality of Palermo over the water supply system, and he had started to teach at the university. In 1889, the time came for him to decide whether he wanted to run for reelection. However, Guccia was already engaged in the competition for a permanent position at the university, and his political career ended.

The third period of activity of the Circolo went from November 1886 to July 1887. The figures reflecting the activities of the Circolo were again very good: twenty-two sessions were held and forty-four communications were delivered. Some of the communications were presented by mathematicians outside of Palermo, including Pasquale del Pezzo from Naples, Corrado Segre from Torino, and Pieter

---

[65] Ibid.

[66] Ibid.

[67] Ibid.

Hendrik Schoute from Groningen (who sent his communication in written form). The number of members of the society increased to sixty-three. Some of the new members were from Italy, including Betti, Brioschi, Cesaro, Cremona, del Pezzo, Peano, Segre, Tonelli, and Volterra, while others were from abroad, such as Catalan from Liege, Fouret and Humbert from Paris, Hirst from London, and Vaněček from Prague.

4.4   The Creation of the Rendiconti

The decision to create a research journal under the auspices of the Circolo was not a straightforward one for Guccia. He had been extremely clear in avoiding any type of article-production machinery when the society was founded. In four years, however, many things had changed for the better. Now, he knew much more about the insides of mathematical research and the world around scientific publishing. Two key advancements in Guccia's mathematical career also occurred during these years: he achieved important mathematical results in his long-standing research, and he began teaching Geometry at the university.

Despite the obstacles Guccia faced, including the legal troubles concerning the water supply, his strong disagreement with the university's mathematics program, and the devastating suicide of Ettore Caporali, he had nevertheless found time for his research.

Guccia also suffered from the bitter side of competition in his mathematical career. Starting from a problem that Kronecker had proposed to him, he proved a related question that concerned the characterization of certain algebraic surfaces having particular plane sections. He wrote to Cremona, "I had no time to see if others had, in any case, done the same proof before me."[68] In December 1885, Guccia presented a long communication to the Société Mathématique de France, where he gave a full proof of a result that Picard had published with an incomplete proof in the Bulletin de la Société Philomathique. Poincaré requested that Guccia clarify certain points of his manuscript, and Guccia heeded his advice. Poincaré, being secretary of the Society, urged him to send the revised version of his communication. Unfortunately, the quarrel with the Mayor of Palermo over the water supply system had just started, and Guccia had to postpone the preparation of the manuscript. In the meantime, Picard reacted quickly and published a correct proof in the Journal de Crelle. This was a sad experience for Guccia. He was playing in the big league, but without his full dedication.

As has been previously explained, Cremona first considered transformations obtained as the product of two quadratic transformations and then the Cremona transformations. The question then was whether each Cremona transformation in the plane could be decomposed into the product of quadratic transformations; or,

---

[68]*Ibid.*, May 12, 1883.

equivalently, whether each homoloidal net, of order n greater than 2, could be transformed into a homoloidal net of order less than n by means of a quadratic transformation.[69] In the case of transformations of order n less or equal than 8, this had been established in 1869 by William Kingdon Clifford, and in the general case, independently, by Max Noether and Jakob Rosanes in 1872. The method of proof was based on the observation, for a homoloidal net of order n greater than 2, that it holds that $n < m_1 + m_2 + m_3$, where m_1, m_2, m_3 are the maximum multiplicity of the curves of the net on its base points. From this relation, Noether deduced that the degree of the net was lowered by applying a quadratic transformation based in the three points of maximum multiplicity.[70]

Noether's method of proof was afterwards used by many mathematicians, such as Bertini, Castelnuovo, del Pezzo, Enriques, Jung, Martinetti and Segre.[71] Guccia, partially modifying Noether's method, proved in the paper *Generalizzazione di un teorema de Noether* (item 14 in Appendix 1) that the more general case of linear systems of curves of genus zero (satisfying the condition that two generic curves meet outside the base points in d points, for $d \geq 1$) could be transformed, by means of quadratic transformations, into one of the following systems:

1. A linear system of order $(d + 2)/2$, having a base point of multiplicity $d/2$ and a simple base point;
2. A linear system of order $(d + s + 1)/2$, $0 < s < d - 1$, having a base point of multiplicity $(d + s - 1)/2$ with s tangent lines common to each curve of the system;
3. A linear system of conics without base points;
4. The net of straight lines of the plane.

These systems were of "minimum order" in the sense that they could not be transformed by quadratic transformations in a system of lower order.[72] In a later paper, *Sulla riduzione dei sistemi lineari di curve ellittiche e sopra un teorema generale delle curve algebriche di genere* (item 16 in Appendix 1) Guccia extended the result to linear systems of curves of genus one, proving that they can be transformed by means of quadratic transformations in one of the following systems:

1. A linear system of cubic curves having $0 \leq \nu \leq 7$ simple base points, not necessarily pairwise distinct;
2. A linear system of quartic curves having two base points of multiplicity 2, not necessarily distinct;
3. A pencil of curves of order $3m$ having 9 base points of multiplicity m.

---

[69] A. Barlotti, F. Bartolozzi, G. Zappa, *op. cit.,* p. xiv.

[70] C. Ciliberto, E. Sallent Del Colombo, 2015, *op. cit.,* p. 187.

[71] *Ibid.*

[72] A. Barlotti, F. Bartolozzi, G. Zappa, *op. cit.,* p. xv.

Guccia ended this paper showing that these systems were also of "minimum order".[73] According to A. Barlotti, F. Bartolozzi and G. Zappa, these two papers on linear systems of curves contain the most important scientific contributions of Guccia.[74]

From these results, Guccia was able to prove that the only rational surfaces in the projective space of dimension three having rational planar sections are the rational ruled surfaces and Steiner's Roman surface (item 15 in Appendix 1).[75] This problem had been solved in 1878 by Picard using the method of resolution of singularity due to Brill and Noether, but Picard's proof contained some unjustified steps.[76]

Guccia presented the paper *Generalizzazione di un teorema de Noether* to the session of the Circolo on June 13, 1886. Not much later, possibly during the summer while walking in the Swiss Alps, he realized that, as a consequence of the announced results, the following theorem on singular points of planar algebraic curves followed:

> The number C of conditions imposed on a given singularity at a point P of an algebraic curve is equal to the number I of all intersections occurring at P of the given curve with another curve that has in P the same singularity reduced in the Plückerian equivalent E ($= \delta + x$) of the given singularity. That is,

$$C = I - E.$$

Excited about this unexpected result, he promptly wrote to Cremona:

> Dearest Professor,
> I cannot resist the temptation of writing to you, the first person, about a theorem on singular points of planar algebraic curves. Certainly you know how many geometers have tried to shed light on this very intricate problem. Almost all of them always came out tired, nauseated and often disillusioned. Within the modest limits of my strength I returned to this topic at least ten times; if in the last five years I often worked in other topics, it was only to rest for a while from the singular points. But I can say that I never abandoned them. You can then understand how glad I am to see largely rewarded the effort and the time spent in these difficult problems, with a result of great importance for the future development of the theory.[77]

It was September 6, 1886. Guccia ended the letter begging Cremona not to inform anyone of the theorem, and added, "A recent misunderstanding, the fault of a common friend of ours, has forced me to be very cautious."[78] Worried about the possible problems regarding priority, Guccia immediately wrote to Halphen asking him to communicate the result to the Académie. On September 18, Halphen replied

---

[73] *Ibid.*

[74] *Ibid.*, p. xiv.

[75] *Ibid.*, p. xvi.

[76] C. Ciliberto, E. Sallent Del Colombo, 2015, *op. cit.*, p. 188.

[77] Letter from Guccia to Cremona, September 6, 1886. Folder Cremona, Archive of the Mazzini Institute of Genoa.

[78] *Ibid.*

that he would be honored to personally submit Guccia's paper to the next session of the Académie.[79] Halphen read the paper at once and commented back to Guccia, "Your result is beautiful and very important."[80] The paper was entitled *Sur une question concernant les points singuliers des courbes,* and was published as a short announcement in the Comptes Rendues de l'Académie des Sciences de Paris 1886 (item 13, Appendix 1). In his next letter to Cremona on October 4, Guccia informed him that the previous result was a consequence of a more general result on systems of curves. He was still worried about the issue of priority, so he asked Cremona to deposit an envelope in the Accademia dei Lincei containing the new results in order to preserve an official record of the receipt date.[81] Since the paper *Generalizzazione di un teorema di Noether* was still being prepared for publication, Guccia added two final sections where the theorem announced in the Comptes Rendues and the more general version on systems of curves were proven. A note was included informing the readers that those sections had been added on October 1886 for the sake of correctness and the issue of priority (item 14, Appendix 1).[82]

The theorem proved by Guccia was the following:

Given a linear system of curves in the plane, of dimension $k > 0$, and assuming that it contains a pencil of curves of genus zero (i.e. all curves of the form $af + bg = 0$, with a and b real numbers and f and g fixed curves of genus zero) or a net of curves of genus one, then we have:

$$k + p - d - 1 = 0,$$

where p is the genus of a generic curve of the system, and d is the number of intersections of two generic curves of the system, away from the base points of the system.

Some time later, Guccia was in Naples with Pasquale del Pezzo, who was also an algebraic geometer. Del Pezzo would later become Rector of the University of Naples, the Mayor of Naples, and senator. He also married Mittag-Leffler's sister Anne Charlotte. Del Pezzo informed Guccia that Corrado Segre had improved his result on systems of curves to more general systems by removing the assumption that the given system of curves contains a pencil of curves of genus zero or a net of curves of genus one. Guccia was very much surprised, and expressed his doubts about the correctness of the Segre's proof. However, when Segre sent him a letter explaining the details of his result, Guccia realized that the proof was correct, and decided to publish Segre's letter in the annual résumé of the activity of the

---

[79]Letter from Halphen to Guccia, September 18, 1886. C. Cerroni, *op. cit.,* p. 119.

[80]*Ibid.,* p. 120.

[81]Letter from Guccia to Cremona, October 4, 1886. Folder Cremona, Archive of the Mazzini Institute of Genoa.

[82]G.B. Guccia, *Generalizzazione di un teorema di Noether*, Rend. Circ. Mat. Palermo, 1 (1884–87) pp. 139–156; note (*) on p. 154, paragraph 14 on p. 154, paragraph 15 on p. 156, and Theorem X on p. 156.

sessions of the Circolo for the year 1887.[83] Segre's result is the so-called theorem of completeness of characteristic series: a complete system of curves of genus p that intersect outside of the base points in $d > 2p - 2$ points has dimension $k = d - p + 1$.[84] It was the starting point for the fruitful research on linear systems of curves by Guido Castelnuovo. It is ironic that, after so many precautions to ensure priority, Guccia's role has been forgotten and the theorem of completeness is only attributed to Segre.

Guccia applied the above results to obtain further results on singular points of algebraic surfaces with rational or elliptic curve sections, which were published in two papers, *Théorème sur les points singuliers des surfaces algébriques* and *Un teorema sulle curve singolari delle superficie algebriche* (items 20 and 21, respectively, Appendix 1).[85] As expressed by Ciliberto and Sallent Del Colombo, these results "played a fundamental role in the classification of algebraic surfaces later carried out by Castelnuovo and Enriques. In particular, they formed the basis for Castelnuovo's famous 'criterion of rationality' for algebraic surfaces (1894), which paved the way for classification."[86]

It is worth mentioning that in 1901 Segre noted that the proof of the original result of Noether contained an error; he found a homoloidal net whose degree could not be lowered with any single quadratic transformation. In the same year, Castelnuovo gave the correct characterization of Cremona transformations: they are obtained by composing linear and quadratic transformations. In 1902, a student of Castelnuovo filled in the errors caused in some Guccia's articles by the application of Noether's method, thus demonstrating the validity of the results contained in Guccia's articles.[87]

The academic year 1886–87 at the university had already been going for a few months when in February 1887 the Dean of the Faculty of Mathematics, Giuseppe Albeggiani, visited Guccia at his house. He explained the exceptional situation occurring at the School of Mathematics, where there was no teaching in Higher Geometry, and described the threats that the school was facing because of that. The aim of the visit and the conversation was to ask Guccia to teach this course in Higher Geometry. Guccia declined, pointing out the poor preparation of the students in Geometry and the fact that the school year had already started several months before. Shortly after that visit, however, Guccia received a letter from the Rector of the university begging him with great insistence to consider the consequences for the School of Mathematics if he refused to teach the course. Guccia later accepted the position, and explained his feelings about it to Cremona, writing: "In short, I felt

---

[83]C. Segre, *Sui sistemi lineari di curve piane algebriche di genere p*, Rend. Circolo Mat. Palermo, 1 (1887), pp. 217–221.

[84]C. Ciliberto, E. Sallent Del Colombo, 2015, *op. cit.,* p. 188.

[85]A. Barlotti, F. Bartolozzi, G. Zappa, *op. cit.,* pp. xiv–xvii.

[86]C. Ciliberto, E. Sallent Del Colombo, 2015, *op. cit.,* p. 188.

[87]*Ibid.*

so much surrounded that I ended yielding."[88] In any case, he made it clear that he would not allow students to take the exam at the end of the course, as he did not believe that they could learn sufficiently in the topic after so few months of study. In any case, Guccia was now an official member of the university, although at the lowest level of professor in charge of a course.[89]

In these favorable circumstances, Guccia devoted all his enthusiasm to the Circolo. He was aware of the achievements of his project, but he was also aware of its weaknesses. Firstly, it was necessary to enhance the statute of the society, as the initial one, approved at the time of the founding, was insufficient for the further development that he envisioned for the society.

Secondly, the issue of a periodical scientific publication had become important. The Circolo had been printing a small booklet with the minutes of its sessions and with other useful information. The first booklet corresponded to the period from March 1884 to March 1885, and consisted of forty-four pages. The second one corresponded to the period from April 1885 to April 1886; it consisted of forty-eight pages. The third booklet went from May 1886 to June 1886, and consisted of sixty-four pages. The last one went from November 1886 to July 1887, and it had 250 pages. It was not a formal journal, as it lacked an editorial committee and publication rules. However, while the first booklet's intention was to provide an account of the mathematical content of the society's sessions, the second booklet began to include proper scientific articles. Guccia tried to exchange this in-house publication with the journals of other scientific institutions. He had some success, but the reactions were not always positive: the Accademia dei Lincei in Rome refused the exchange request (while they regularly sent the Atti and the Rendiconti of the Accademia to the Technical Institute of Palermo). Reacting to the rejection of his Italian peers, Guccia exclaimed, "The Circolo must be content to receive only the publications of foreign Academies!"[90] The need for a proper mathematical journal for the Circolo was evidently building up in his mind, and in August 1886, he wrote to Cremona, "I will be obliged to ask for help from my foreign friends, making the Rendiconti an international journal!"[91]

These were the circumstances under which the General Assembly of the Circolo met in February 26, 1888 to renew its statute. Guccia had proposed this meeting during the session held on January 8 of that year. During that session, he justified the renewal of the statute by citing the rapid development of the society and its increasing membership. A five-member commission had been appointed, chaired by Guccia, to present a new statute. As stated in the proceedings of the session, the statute was discussed and unanimously approved in that same session. A careful

---

[88]Letter from Guccia to Cremona, March 5, 1887. Folder Cremona, Archive of the Mazzini Institute of Genoa.

[89]Annuario della R. Università degli studi di Palermo, Anno Accademico 1887–88, p. 44.

[90]Letter from Guccia to Cremona, March 5, 1887. Folder Cremona, Archive of the Mazzini Institute of Genoa.

[91]*Ibid.*

analysis of its content is interesting, as it reveals many of the ideas, projects, and worries that Guccia had in mind.

The statute consisted of forty-five articles, which were organized into several sections:

Aim of the society—Seat
Members—Admission—Contribution
The Presidency
The Steering Council
The sessions
The Rendiconti
The library
Revenue and expenses

The statute declared that the aim of the Circolo Matematico di Palermo was to develop and disseminate mathematical sciences throughout all of Italy. This aim expanded the scope of the society beyond the city of Palermo and the region of Sicily. The Circolo would hold scientific sessions in its seat on the second and fourth Sunday of every month, except for the vacation months of September and October. The topics of discussion in the sessions could only be scientific or related to the aims of the society. With such measures, Guccia was preventing the Circolo from getting involved in issues unconnected to science, such as local or political matters.

The aim of having an international society was clear when membership of the society was discussed. The statute defined two categories of membership: Palermo resident and non-resident. The number of members was unlimited, and foreigners were allowed to belong to the society. Admission of a new member remained as before: it required the proposal of two members and the approval by the majority of members present in the following session.

The internal organization of the society was twofold. The Presidency office was in charge of administration, and consisted of the president, the vice president, two secretaries, two vice secretaries, the treasurer, and two librarians. They were to be elected from among the resident members, and would serve for a two-year period. The scientific oversight of the society was in the hands of the Steering Council, consisting of twenty members, five of which were resident members and fifteen of which were non-resident members. The Steering Council was elected for a three-year period. In order to allow non-resident members to participate in this election, the voting was to be done by mail. The composition of the Steering Council shows the drive towards internalization and highly qualified scientific leadership.

The society would publish a periodic journal entitled Rendiconti del Circolo Matematico di Palermo (Figs. 4.4 and 4.5). The Steering Council would act as Editorial Board of the Rendiconti, and one its members would be the Director of the journal. The Rendiconti would include reports on the discussions held in the sessions, the original notes and the memoirs communicated by members and accepted by the Editorial Board, and bibliographic reviews of mathematical publications (both national and international, as well as periodic and non-periodic) received in exchange by the society. It would only consider unpublished manuscripts, written

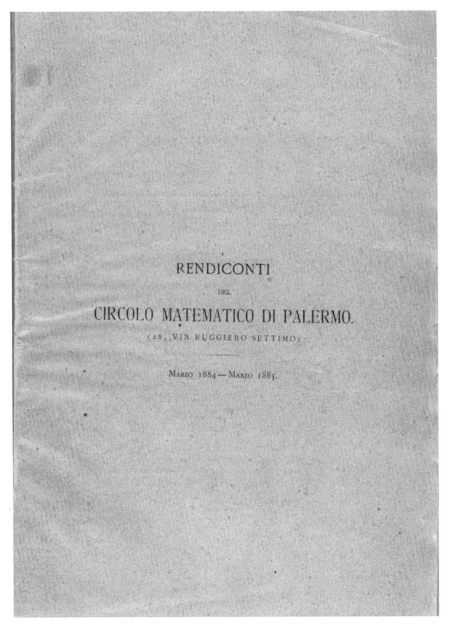

RENDICONTI

DEL

CIRCOLO MATEMATICO DI PALERMO.

(28, VIA RUGGIERO SETTIMO)

MARZO 1884 — MARZO 1885.

Fig. 4.4 Cover of the first volume of the Rendiconti del Circolo Matematico di Palermo. Courtesy of the Circolo Matematico di Palermo

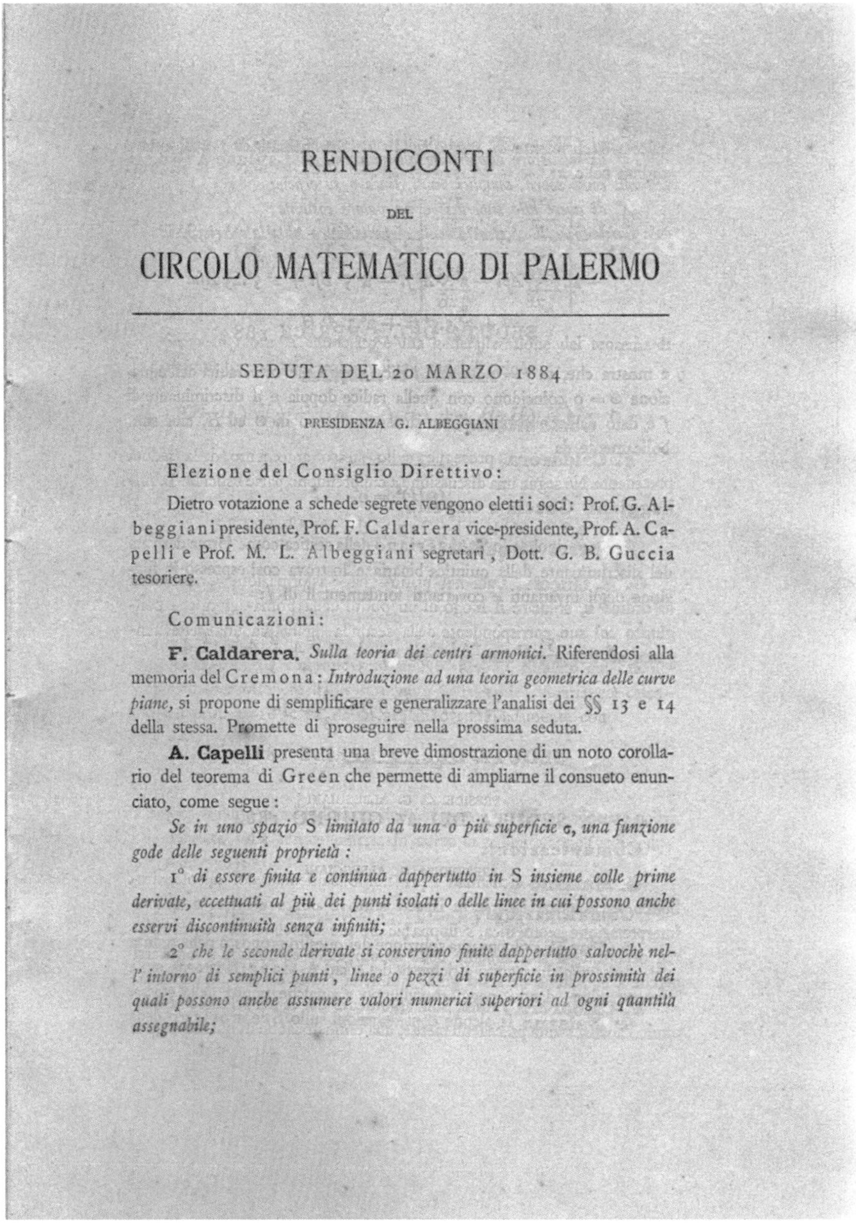

RENDICONTI

DEL

CIRCOLO MATEMATICO DI PALERMO

SEDUTA DEL 20 MARZO 1884

PRESIDENZA G. ALBEGGIANI

Elezione del Consiglio Direttivo:

Dietro votazione a schede segrete vengono eletti i soci: Prof. G. Albeggiani presidente, Prof. F. Caldarera vice-presidente, Prof. A. Capelli e Prof. M. L. Albeggiani segretari, Dott. G. B. Guccia tesoriere.

Comunicazioni:

F. Caldarera. *Sulla teoria dei centri armonici.* Riferendosi alla memoria del Cremona: *Introduzione ad una teoria geometrica delle curve piane,* si propone di semplificare e generalizzare l'analisi dei §§ 13 e 14 della stessa. Promette di proseguire nella prossima seduta.

A. Capelli presenta una breve dimostrazione di un noto corollario del teorema di Green che permette di ampliarne il consueto enunciato, come segue:

Se in uno spazio S limitato da una o più superficie σ, una funzione gode delle seguenti proprietà:

1° di essere finita e continua dappertutto in S insieme colle prime derivate, eccettuati al più dei punti isolati o delle linee in cui possono anche esservi discontinuità senza infiniti;

2° che le seconde derivate si conservino finite dappertutto salvochè nell'intorno di semplici punti, linee o pezzi di superficie in prossimità dei quali possono anche assumere valori numerici superiori ad ogni quantità assegnabile;

Fig. 4.5 First page of the first volume of the Rendiconti del Circolo Matematico di Palermo. Courtesy of the Circolo Matematico di Palermo

in Italian, Latin, Spanish, French, German, or English. Members of the society were entitled to a free copy of the journal. The Rendiconti's policies reflect Guccia's clear intention to prevent the society's journal from being used by those looking to "acquire titles and notoriety." The statute also regulated the use of the library, with the possibility of borrowing books. This issue would become controversial in the coming years.

Membership fees were set at an annual fee rate of fifteen lire. For resident members, there was also an admission fee of ten lire. The seat of the society was set in Palermo by Article 2 of the statute, and this seat was declared to be "immovable." The inclusion of this provision is interesting, and must be interpreted within its historical context. There was no nationwide mathematical society in Italy at the time; the first nationwide society wasn't created until thirty-four years later in 1922. Fixing the seat of the society in Palermo could have been directed at preventing attempts to move the seat to another Italian city, most probably Rome, if the Circolo was transformed into the Italian mathematical society.

When the second statute was developed in March of 1888 to regulate the activity of the Circolo, the society already had one hundred and two members. There were already more non-resident members than resident members: there were fifty-six non-resident members compared to forty-six resident members. Among the non-resident members, fifty were Italians, and six were foreign. The foreigners included E. Catalan, who became a member in December 1886; T.A. Hirst, J.S. Vaněček, and M.N. Vaněček, in January 1887; and G. Humbert and G. Fouret in December 1887. Among the Italian non-resident members there was G. Battaglini and V. Cerruti (December 1886); C. Segre (July 1887); F. Brioschi and L. Cremona (November 1887); L. Belzolari, E. Betti, G. Peano and V. Volterra (December 1887); C. Arzelà, E. Beltrami, E. Bertini, F. Casorati, S. Pincherle, F.P. Ruffini and G. Veronese (March 1888).

The first Steering Council (acting as Editorial Board of the Rendiconti) was elected in March 1888 and consisted of the resident members G. Albeggiani, M.L. Albeggiani, Caldarera, Gebbia, and Guccia, and the non-resident members Battaglini, Beltrami, Bertini, Betti, Brioschi, Casorati, Cerruti, Cremona, del Pezzo, de Paolis, D'Ovidio, Jung, Pincherle, Segre, and Volterra. All of the Steering Council members were Italians. Guccia was appointed as Director of the journal. Volume I of the Rendiconti compiled the previous four booklets containing the content of the sessions in the period from March 1884 to July 1887, and was 406 pages long.

In a letter to Cremona from May 6, 1888, Guccia dwelled on the recent election of the Editorial Board of Rendiconti, expressing his hope that the elected members would be aware of the importance of their role, and cooperate with him in the delicate task of reviewing the papers submitted for publication. He explained:

> I am convinced that there are five or six members of the Board who can do a great job; some others I do not trust too much, because we are all by nature weak, and it is difficult to resist insistent requests. [. . .]
> The number of manuscripts submitted so far (after the definitive constitution of the society) exceeds that which can be published in Volume II, that is, until December 1888. So we have

the difficult task of making a good choice! Up to this day, after review, I have rejected ten manuscripts.[92]

We should pause here in our story to consider the relationship between Giovanni Battista Guccia and his mentor Luigi Cremona. The two were very close since the beginning of their relationship, and exchanged letters on a regular basis (taking advantage of the speed of the mail in those days—unexpected nowadays). These letters contained a mixture of mathematics, comments on their relations with other mathematicians, discussions of administrative and political issues, and also a number of personal matters. Cremona regularly commented to Guccia about many of his personal issues, even family affairs. For example, Cremona informed Guccia on February 1887 about the intention of his son Vittorio to emigrate to the United States of America. Some three weeks later, Guccia replied and expressed his opinion on the subject, hoping that Vittorio would stay in Europe.[93] Another example is from May 1888, when Guccia wrote to Cremona informing him that their common friend Hirst had not yet received an invitation to Cremona's wedding (his second marriage).[94] Cremona and his wife visited Guccia in Palermo on June 1889. Guccia would also tell Cremona about university affairs in Palermo, the running of the Circolo, the visit of Mittag-Leffler and his family for several days in Palermo, and about his family businesses. Certain more personal matters, such as the death on November 1886 of his aunt Maria Stella Guccia e Vetrano, were left aside (or those letters were lost). This shows that from Guccia's perspective the relationship with his mentor was a discreet and respectful one, different from a friendship.

4.5   Palermo's *Belle Époque*

The European economy experienced a period of rapid development between 1880 and 1914. This development took advantage of the long period of peace in Europe after 1870. As a consequence, industrialization deepened, the railway network continued its expansion, commerce flourished, and the financial markets were booming. All these improvements were taking place at the same time as imperial expansion was becoming the backbone of European politics and economy.

For Sicily, it was also a period of strong economic development and social progress. Sicily became an active exporter of sulfur, tuna fish, and Marsala sweet wine. This activity in turn spurred the development of foundries and shipyards, which supported the business of shipping companies. The industrialists and businessmen Ignazio Florio[95] and Joshep Whitaker played a prominent role in the country's economic development at that time.

---

[92] *Ibid.*, May 6, 1888.

[93] *Ibid.*, March 5, 1887.

[94] *Ibid.*, May 6, 1888.

[95] O. Cancila, *I Florio. Storia di una dinastia imprenditoriale*, Bompiani, 2008, Milan.

A public recognition of the economic and industrial development of Sicily came in 1891 with the celebration of the IV National Exhibition in Palermo. Beginning in 1861, these national exhibitions were organized by the Kingdom of Italy to demonstrate advances in industry and commerce and strengthen the country's national spirit. They were held every ten years: the first was held in Florence in 1861, and the following ones in Milan in 1871 and 1881. The exhibitions were a great success in terms of attendance of national and international exhibitors and citizens, and in terms of the economic revenue they generated. In the Palermo National Exhibition there was a special exhibit devoted to electricity and its novelties. Guccia certainly visited the exhibition. We have already mentioned that a few years later, Guccia installed electric heating and an elevator in his palace.

This wealth was reflected in Palermo's daily life. The way of life for high classes consisted of free spending, elegance, refinement, and a widespread joy of living. They enjoyed hosting brilliant receptions, elegant dances, and nouvelle happenings such as "tableaux vivants", "cotillions" and "flowers courses". The elites exhibited themselves riding elegant carriages in the Foro Italico and on Via Libertà, and attending the horse races in the park "La Favorita". This atmosphere was also reflected in Palermo's architecture. The dominant architectural style of the period of the time was known as the Liberty style, and featured the architect Ernesto Basile, painter Ettore De Maria Bergler, and the designer Vittorio Ducrot as its main protagonists. The Liberty style buildings were defined by their curved facades, vine motifs, and extravagantly painted and glazed windows. It was a unique and unrepeatable time for the city that was later remembered as the *Belle Époque* of Palermo.

Two women from the high bourgeoisie played a prominent role in representing locally and internationally the sophisticated atmosphere of Palermo. They sought to give Palermo a high European style. One of these women was Franca Jacona di San Giuliano. She was married to Ignazio Florio, and was known as Franca Florio. She was a charming and generous lady whose dresses were made by Worth, a well-known Parisian tailor, and whose jewelry was crafted by Cartier and Lalique. A painting by Giovanni Boldini (which was recently sold in public auction for more than a million euros) shows Franca's superb "Florio Pearl Necklace" consisting of three hundred and forty-five pearls of enviable caliber (Fig. 4.6). The Florios organized balls and receptions in their home Villino Florio all'Olivuzza and in the family property of Villa Igiea. These were fantastic and luxurious gatherings that would commonly end with gifts for the guests: bracelets and trinkets for the ladies and necklaces and "petits crayons" for the men, all in silver.[96]

The other woman was Tina Scalia, daughter of colonel Alfonso Scalia, who was a member of Garibaldi's Expedition of the Thousand. As a promising soprano in her youth, Tina had sung for Richard Wagner. She was married to the ornithologist and archaeologist Joseph Whitaker, heir to the immense fortune of his granduncle Benjamin Ingham. The couple lived in Villa Malfitano. The receptions that they

---

[96] *Ibid.*, p. 312.

Fig. 4.6 Franca Florio, portrait by Giovanni Boldini

organized hosted many characters from the European aristocracy, including the King of England Edward VII and Queen Mary. The most appreciated artist who visited Villa Malfitano was Richard Wagner. Tina Scalia was appreciated by her contemporaries for the extraordinary versatility of her spirit and for her wide range of interests in everything from singing to history. She would perform for her guests during receptions, and she published the history book, *Sicily & England. Political and Social Reminiscences, 1848–1870.*

Members of the European aristocracy abroad also started spending winters in Palermo. Czar Nicholas I was among the first to do so in 1845.[97] They benefited from the mild Mediterranean climate, and enjoyed the friendly reception of Palermo's upper classes. The wealthy families of Palermo also travelled around Europe during the summer, and visited their new friends. It was a life of luxury and exhibition. The Florios used their own railway car furnished as an apartment during their travels through Europe, visiting everywhere from St. Moritz and St. Petersburg, and from London to Budapest. The intertwining of Palermo's upper class with the international European aristocracy was reflected by the attendance of the Florios, the Whitakers, and other wealthy families of Palermo at the funeral of Queen Victoria in London in 1901.

Another important family of Palermo during the "fin de siècle" was that of the Princess of Niscemi and her husband Corrado Valguarnera Tomasi (the woman and man behind the characters of Angelica and Tancredi in the novel *The Leopard*). Corrado Valguarnera was a nephew of Giulio Fabrizio Tomasi. He was one of the leaders of the conspiracy preceding the 1860 revolution in Sicily that was followed by Giuseppe Garibaldi's fight against foreign domination.[98] Every year, from mid-August to the end of October, the Princess of Niscemi and her family travelled around Europe, and ended their travels with a stay in Paris.

Two highlights of Palermo's exposure to aristocrats, businessmen, statesmen, and artists of the world were the openings of the Teatro Massimo in 1897 and of the Grand Hotel Villa Igiea in 1900. The Teatro Massimo became "the meeting place for the high society of Palermo, which occupied the second row of the seats; [...] the high bourgeoisie of lawyers, industrialists and dealers were satisfied with a seat in the third row; for the middle class of military, state and municipal officials was reserved the first row; in the fifth row was placed the lower and the impoverished nobility. It demonstrated the depth of class divisions in the city."[99] In this regard, it is appropriate to quote the journalist and novelist Edmondo de Amicis. In 1906 he wrote

> Palermo is a large city [...] The new elegant neighborhoods, the new ample tree-covered squares, the new magnificent public promenades are places of true delight comparable to those in Paris and London [...] However, Palermo is also a scene of dramatic contrasts [...] Only a few steps away from the marine boulevard of Foro Italico [...] where hundreds of aristocratic carriages move on, you encounter [...] all the miseries and calamities of the adventurous and brutal life of the past centuries. Walking out of the huge labyrinth of dark and dirty alleys [...] where swarms of extremely poor population teems in thousands of fetid hideouts, similar to those used by the Arabs in the ninth century, you find yourself in

---

[97] *Ibid.,* pp. 203–205.

[98] Speech by the President of the Senate Giuseppe Scarano, *Senato del regno. Atti parlamentari. Discussioni, 27 gennaio 1903.*

[99] O. Cancila, *Palermo,* pp. 317–320.

front of the Teatro Massimo, the largest and most beautiful theatre in Italy [. . .] constructed at the time when Palermo lacked a hospital for its most basic needs.[100]

The Grand Hotel Villa Igiea is situated at the foot of Mount Pellegrino, near the Acquasanta thermal spring. Ignazio Florio built it according to the plans of the architect Ernesto Basile. It displays a grand style with specially designed furniture, drawn carpets, and a splendid floral decoration by the renowned painter Ettore De Maria Bergler. This was a luxury hotel that soon became a major attraction for international tourism and a symbol of the *Belle Époque* of Palermo.

Giovanni Battista Guccia was closely connected to this atmosphere of wealth. Franca Florio was the niece of his sister, Maria Stella, who married Salvatore Jacona di San Giuliano, uncle of Franca. It is reasonable to assume that the Guccia family was also influenced by the Valguarneras. The Guccias and the Valguarneras shared a close relationship via the Prince of Lampedusa. They also shared an interest in horse racing, as Giovanni Battista Guccia's father Giuseppe Maria Guccia was one of the organizers of the "La Favorita" horse races.[101] The strong anticlericalism of Corrado Valguarnera may have influenced that of Giuseppe Maria Guccia.[102] It is even possible that Corrado Valguarnera introduced Luigi Cremona to his uncle, the Prince of Lampedusa, before the meeting of the Società Italiana per il Progresso delle Scienze. Valguarnera and Cremona were both men of the Italian Risorgimento (and both became senators of the Kingdom of Italy almost simultaneously).

Giovanni Battista Guccia was likely introduced to Richard Wagner during one of the parties in Villa Malfitano Whitaker. Even Guccia's summer holiday programs, which were so important for the international development of the Circolo, were quite similar to that of the Princess of Niscemi.

Guccia's relationships with the Florios, the Whitakers, the princes of Niscemi and other members of Palermo's high society reflect his personal involvement in the aristocratic activity of the *Belle Époque* of Palermo. We cannot be sure of the extent of Guccia's activity, due to the scarcity of documents related to his personal life. However, we do not believe that Guccia was very much involved in Palermo's aristocratic circles, since Guccia was completely devoted to the Circolo and to the Rendiconti. Yet, despite this speculation, two documents indicate that Guccia did participate in the social life of the city, at least during the period prior to the Rome International Congress. The first document is a letter to Gösta Mittag-Leffler written on Christmas Day 1905. In the letter, Guccia mentions the possibility of Mittag-Leffler and his wife visiting Palermo and, in that event, offered to introduce Mittag-Leffler to the social life of the city, writing, "[I will] accompany you with

---

[100]E. de Amicis, *Ricordi di un viaggio in Sicilia*, Giannotta Editore, 1908, Catania. Newly edited: Il Palindromo, 2014, Palermo.

[101]L.M. Majorca Mortillaro, *op. cit.,* p. 50.

[102]Regarding the anticlericalism of Giuseppe Guccia, see his publications *La quistione religiosa in Italia risoluta a Campo dei Fiori o il vero cristianesimo,* Tipografia Diretta da Sani Andó, 1891, Palermo, and *Ancora una parola sulla convenienza e possibilità di una Riforma Religiosa*, Pensiero Italiano, 16 (1894).

great pleasure, in order to get you acquainted with our beautiful ladies and to show you our beautiful salons."[103] The second document is a letter to Andrea Guarnieri, senator of the Kingdom of Italy. In the letter, Guccia refers to a previous conversation with Guarnieri during a glamorous party organized in the Palazzo Butera. The conversation concerned the creation of an international award—the Guccia Medal—for the International Congress to be held in Rome in two years' time.[104]

We end this brief section by pointing out that on the eve of World War I the economic situation worsened considerably. The European upper classes were no longer in Palermo; the local wealthy families closed their palaces and moved out of the city for prolonged periods of time. Without the opulence of the Florios, the Whitakers, and many other wealthy families, the *Belle Époque* of Palermo soon came to an end. Many architectural works created in the Liberty style during this period remained for some time. During the 1950s, however, the city's uncontrolled urban development (known as "Il sacco di Palermo", in English: The sack of Palermo) severely damaged the memory of that unrepeatable time.

---

[103]Letter from Guccia to Mittag-Leffler, December 28, 1905. Folder Mittag-Leffler, Archive of the Circolo Matematico di Palermo.

[104]Letter from Guccia to Guarnieri, January 6, 1906. Folder Guarnieri, Archive of the Circolo Matematico di Palermo.

Chapter 5
The Projects of Guccia: Second Stage

This chapter is devoted to the most ambitious of Guccia's goals: developing the Circolo into an international association of mathematicians. We start by reviewing the relation of Guccia with two important mathematicians: Vito Volterra and Henri Poincaré. Guccia's close relationships with these two mathematicians were a great influence on him and his projects for the Circolo. In the period from 1889 to 1908, several important events also occurred: he applied to a professorship in the University of Palermo through a somewhat turbulent process, and he had to face a rebellion inside the Circolo that threatened the nature of the society as Guccia had conceived of it. On both issues, he succeeded. The Circolo experienced a burst of success after the Heidelberg International Congress of 1904: its membership increased, and its fame expanded worldwide. These were the contour conditions for Guccia's visionary project: transforming the Circolo into the international association of mathematicians. The issue was discussed and decided at the Rome International Congress of 1908.

5.1 Two Important Encounters: Volterra and Poincaré

I remember, more or less twenty-nine years ago, I was on my first visit to this enchanting island. Having arrived at Palermo, after admiring the superb beauty of Nature and the splendid art works of this marvelous city, a friend of mine, Ignazio Conti, who unfortunately is no longer with us, invited me to visit a recent Palermitan scientific institution. We went to via Ruggiero Settimo and I was introduced to G.B. Guccia.

Few were then, although willing and eager, the members of the Circolo Matematico, the first issue of its publication had not yet started, and the library was modest. But great was the enthusiasm animating Guccia. Full of hope and ardor, he had ever since the vision of the vital and useful things coming from that modest beginning. His sincere and deep faith

© Springer International Publishing AG, part of Springer Nature 2018
B. Bongiorno, G. P. Curbera, *Giovanni Battista Guccia*,
https://doi.org/10.1007/978-3-319-78667-4_5

1904

Fig. 5.1   Vito Volterra. Courtesy of the Accademia Nazionale dei Lincei

revealed the type of energy that creates great institutions, and gives rise to lasting and fruitful works. I was impressed since that first visit. [1]

This was Vito Volterra's recollection of his first encounter with Giovanni Battista Guccia. Volterra's recollection was delivered in 1914 during the celebration commemorating the thirtieth anniversary of the foundation of the Circolo. Edmund Landau's speech with which we started our story was delivered at the same occasion.

Volterra (Fig. 5.1) had travelled to Sicily after he was appointed by the Ministry of Public Education to act as external examiner for secondary school students in the Technical Institute of Catania. To travel to Catania, he first took the train from Pisa to Naples, where he boarded the steamship to Palermo. Once in Sicily, he continued by open carriage to Catania. After the exams had taken place, he returned

---

[1] *Discorso del Senatore Prof. Dr. Vito Volterra*, XXX anniversario della fondazione del Circolo Matematico di Palermo, Adunanza solenne del 14 aprile 1914, Suppl. Rend. Circ. Mat. Palermo, 9 (1914), pp. 17–19.

to Palermo and waited for the steamship back to Naples. It was at this point in his return journey that he contacted Ignazio Conti. It was June 1885, and the Circolo had been founded one year before. Conti was studying mathematics at the University of Pisa (where he met Volterra), but had started his studies in his hometown at the University of Palermo. Conti had been a member of the Circolo since February 1885. In fact, he was the third member to enter the society after the twenty-seven founding members. He was a good example of the support that the Circolo received from the mathematics community in Palermo. He graduated in 1887 from Pisa with honors, and eventually became professor of mathematics in the Technical Institute of Palermo. He had two papers published in the first issue of the Rendiconti.

The family origins of Volterra were quite different from those of Guccia.[2] He was born in 1860 into a Jewish family in Ancona, a city on the Adriatic Sea in the center of Italy. His father died when he was five years old, leaving the family in difficult economic circumstances. Support from his uncle helped the family, and allowed young Vito to attend the Technical Institute in Florence, where he immediately exhibited his mathematical talent. In 1878, contrary to the family's needs and his mother's wishes, he enrolled in the University of Pisa. His good academic record and his high marks in the entrance exams for the Scuola Normale Superiore di Pisa granted him a scholarship from that prestigious institution. There he met Enrico Betti, who would become his mentor. In 1882, he graduated with a degree in Physics. By the year of his graduation, he had already published two deep mathematical research papers on mathematical analysis: *Sui principi del calcolo integrale* and *Alcune osservazioni sulle funzioni punteggiate discontinue*.[3] One year later, in 1883, he was appointed professor of Rational Mechanics in the University of Pisa.

After their first encounter in Palermo, the relationship between Guccia and Volterra developed quickly. Upon his return to Pisa, Volterra sent copies of three of his papers as a donation to the library of the Circolo (they appear listed in the first issue of the Rendiconti). Two years later, in December of 1887, Volterra joined the Circolo as member number 87. A few months later, he published a paper in the first volume of the Rendiconti, *Sulla teoria delle equazioni differenziali lineari*. Right after the formal creation of the Rendiconti in March 1888, Volterra was elected member of the first Editorial Committee. He was one of the members of the committee that Guccia trusted most for the task. Soon after his election, Guccia requested Volterra to report on two papers submitted to the journal. Volterra reviewed the papers thoroughly, and told Guccia that he should not accept the papers. He wrote to Guccia, "I am pleased to know that there is a great deal of

---

[2]J. Goodstein, *The Volterra Chronicles*, American Mathematical Society, 2007, Providence.

[3]A. Guerraggio, G. Paoloni, *Vito Volterra*, Franco Muzzio Editore, 2008, Padua (in Italian). Published in English by Springer, 2012, Heidelberg, Section 1.4.

material to be published in the Rendiconti of the Circolo. This is the result of your hard work in promoting the institution."[4]

Guccia was always attracted by mathematical geniuses, and so he was deeply impressed when he first met Volterra in 1885. Ever since this first encounter, Guccia supported Volterra's mathematical career as much as he could. In particular, Guccia took care to introduce Volterra to his mathematical acquaintances across Europe. When Mittag-Leffler and his family visited Guccia in Palermo in June 1888, for example, Volterra expressed his interest in having Mittag-Leffler stop in Pisa on his return to Sweden. Guccia did not have much difficulty suggesting this visit to Mittag-Leffler, given that the Swedish mathematician had been to Pisa to attend Ulisse Dini's seminar. During this seminar, Volterra, who was still an undergraduate at the time, had presented an example of a function which is everywhere differentiable, and whose derivative is bounded but not Riemann integrable.

Guccia was worried that Volterra's work may not receive the attention it deserved in Europe if it was not advertised properly. He advised him, "We Italians, you will pardon me, have the great defect of not knowing how to value our scientific production: we keep it under lock and key, hidden, in ways and means that foreigners never succeed in recognizing and appreciating."[5] Guccia suggested that Volterra send reprints of his papers to Halphen, Darboux, Picard, Appel, and Humbert, and convinced him to visit Paris, writing introduction letters for him to Darboux, Jordan, Humbert, Poincaré, and Mannheim. Guccia, who was much more socially experienced than Volterra, also gave him advice on how to approach these mathematicians and interpret their reactions. Guccia was also instrumental in the encounter between Volterra and Mittag-Leffler in the Engadine mountains in Switzerland. During their meeting, Mittag-Leffler suggested Volterra travel to Germany. Volterra followed this suggestion, and after three weeks in Germany, he met some of the best German mathematicians, including Karl Weierstrass (and Sofia Kovalevskaya), Felix Klein, Hermann Schwarz, and George Cantor.

Guccia and Volterra had a close and warm personal relationship that involved mutual aid and benefits. Volterra immediately congratulated Guccia when, in 1889, he obtained the position of professor in Higher Geometry at the University of Palermo. Volterra wrote, "I learned this news with great pleasure, though I had no doubt about the outcome of the competition."[6]

Volterra's support of the Circolo's activity was very important for Guccia. Volterra frequently referred young Italian mathematicians to the society. As a result of his efforts, the society gained many new members over the years: Carlo Bigiavi,

---

[4]Letter from Volterra to Guccia, May 5, 1888. Folder Volterra, Archive of the Circolo Matematico di Palermo.

[5]Letter from Guccia to Volterra, December 4, 1887. Vito Volterra Collection, Archive of the Accademia Nazionale dei Lincei.

[6]Letter from Volterra to Guccia, December 1, 1889. Folder Volterra, Archive of the Circolo Matematico di Palermo.

Aristide Fiorento and Giuseppe Lauricella joined the society in 1893; Ermenegildo Daniele and Edgarde Ciani in 1898; Emilio Almansi in 1899; Filadelfio Insolera in 1903; Robert d'Adhémar in 1905; and possibly many others. Volterra also helped Guccia convince Ulisse Dini and Luigi Bianchi to join the Circolo by suggesting that Guccia take a prudent approach to the two mathematicians. Bianchi later became a member at the end of 1893. Dini waited several years until the Circolo and the Rendiconti had spread internationally, and became member number 282 in 1900.

Volterra's main contribution to the Circolo was his hard (and rigorous) work as a member of the Editorial Committee of the Rendiconti. However, Volterra would occasionally recommend the manuscripts of young mathematicians for publishing after conducting a general assessment rather than a thorough evaluation. Volterra would also request speedy publications for these papers. Fast publication of results was a widespread problem in the mathematical community at the time. When appointing new professors, universities required young mathematicians to present reprints of their published papers to their selection committees. Guccia had complied with this before, as Volterra recognized in one of his letters: "I know of your exquisite courtesy with respect to these new guests, that has made the Circolo very valuable to young people, providing an early knowledge of their publications and allowing them to present themselves to the competitions."[7] Regardless, Volterra's recommendations for publication did not always succeed, since Guccia was dedicated to maintaining the journal's standard of quality.

In many ways, Volterra occasionally acted as the Circolo's official representative in Northern Italy. In 1890, for example, he wrote to Guccia explaining that the National Library of Florence only had the first two fascicles of the Rendiconti, and had no money to purchase the remaining volumes. In his letter, Volterra hoped that the Circolo could give the volumes to the Library for free. Guccia accepted Volterra's suggestion, and sent the volumes to Florence.

Volterra and Guccia's relationship remained amicable until 1907 and 1908. Then, two important events took place: the meeting of the Società Italiana per il Progresso delle Scienze in Parma, and the International Congress of Mathematicians in Rome. These two events had a quite negative effect on the understanding between Guccia and Volterra, as we will later consider.

The other important mathematician in the life of Giovanni Battista Guccia was Henri Poincaré (Fig. 5.2). Their first encounter was during the meeting of the Société Mathematique de France, held in Paris in December of 1885. There, Guccia presented a communication that attracted Poincaré's attention. At that time Poincaré was the secretary of the society, and requested Guccia to send the manuscript of the communication to be published in the Bulletin of the society. Guccia commented to Cremona that Poincaré had "a versatile and deep talent, more than any other French mathematician."[8] As we have already commented, Guccia was unable to

---

[7]*Ibid.*, October 20, 1898.

[8]Letter from Guccia to Cremona, August 8, 1886. Folder Cremona, Archive of the Mazzini Institute of Genoa.

Fig. 5.2 Henri Poincaré. Courtesy of the Archives Henri-Poincaré

write the manuscript due to the administrative complications related to the water supply business of his family. We have no precise information on any subsequent encounters between Guccia and Poincaré in Paris, although they could have easily occurred, as Guccia visited Paris quite frequently and was very much involved in its mathematical life.[9]

A project in which Guccia and Poincaré coincided was related to one of the big worries of mathematicians in the late nineteenth century: the unexpected exponential growth of mathematical knowledge that both induced a constant increase in the number of mathematical journals, and revealed the increasing specialization of the different fields of mathematics. This was perceived as a serious threat to research, since such decentralized growth could make it more difficult for working mathematicians to keep up with new results. This problem would not be solved by controlling or mitigating the increase in mathematical activity, but rather by developing appropriate ways of managing it. This is where bibliographical tools entered the scene, with the aim of aiding researchers in this new and more difficult to manage environment.

We have already commented that the main tool in this regard was the Jahrbuch über die Fortschritte der Mathematik, created in 1868. However, it soon suffered

---

[9]The main source to follow the relations between Guccia and Poincaré is their correspondence; most of it is in the Folder Poincaré of Archive of the Circolo Matematico di Palermo and in the Archives Henri-Poincaré of the Université de Lorraine.

from the growing lapse between the publication of articles and their bibliographic registration.

On March 1885, the Société Mathématique de France launched the long-term project for creating a new bibliographical tool: the Répertoire Bibliographique des Sciences Mathématiques. The project differed from other bibliographical projects in that it was based on a previously established logical classification of mathematics: the works were to be classified, not according to the author's names, but by the logical order of their subject (organized into domains, classes, subclasses, divisions, sections, and subsections). The mathematical domains considered were Mathematical Analysis, Geometry, and Applied Mathematics. Classes A through J corresponded to the first domain, Mathematical Analysis; classes K through Q to the second, Geometry; and classes R through X to the third, Applied Mathematics. The following example can illustrate the preciseness—and difficulties—of the classification: the code L14cá corresponded to the class of conics and second-degree surfaces (code L), the subclass of conics (code 1), the division of tangents (code 4), the section dealing with tangents satisfying specific conditions (code c), and the subsection of right angles (code á).[10]

The first Congrès International de Bibliographie des Sciences Mathématiques took place in Paris in 1889. It was decided that an inventory would be made of all memoirs in pure and applied mathematics published between 1800 and 1889, as well as those in the history and philosophy of mathematics from 1600 to 1889. The titles of the works written in languages other than French (Italian, German, English, Spanish or Latin) were to be translated into French. The sixty-six pages of the proceedings of the Congress mostly outline the full classification scheme of mathematics.[11] Over the years, the project developed and grew: it involved some fifty mathematicians from sixteen countries who classified more than three hundred mathematical journals. Between 1885 and 1912, more than 20,000 bibliographical references to mathematical works were listed and systematically classified. Publication started in 1894 and ended in 1912. Poincaré ran the committee in charge of the Répertoire from the very beginning. At the 1889 congress, a permanent commission of seventeen members was established, chaired by Poincaré. Guccia was chosen as the representative for Italy, and was thus responsible for the information coming from his country.

Several documents regarding the bibliography project are kept in the Archive of the Circolo Matematico di Palermo, including the first circular letter, dated May 1885, and entitled "Projet de répertoire bibliographique"; an announcement of the 1889 congress; an internal communication of the permanent commission; and a brochure containing the classification scheme (Fig. 5.3). A third circular letter from

---

[10]L. Rollet, P. Nabonnand, *An Answer to the Growth of Mathematical Knowledge? The répertoire Bibliographique des Sciences Mathématiques,* European Math. Soc. Newsletter 47 (2003), pp. 9–14, p. 11.

[11]*Congrès international de bibliographie des sciences mathématiques tenu à Paris du 16 au 19 juillet 1889. Procès-verbal sommaire,* Exposition universelle, 1889, Paris.

the Italian committee of the Répertoire, addressed to the Italian mathematicians who were collaborating with the répertoire, also shows Guccia's deep involvement in the project. The letterhead carried as the official address of the committee was via Ruggiero Settimo, the Palazzo Guccia, where the Circolo was located. Overall, the Répertoire's project was a large-scale, ambitious endeavor that must be understood within the context of addressing the increasing specialization and internationalization of mathematics that took place at the end of the nineteenth century.

Another example of such attempts to catalogue mathematical publications was the work of Georg Valentin, chief librarian of the Berlin Library and the German representative on the Répertoire's commission. Valentin was a doctoral student of Karl Weierstrass. In 1885, he began compiling all mathematical publications that had appeared since the invention of the printing press. He worked on this project for over forty years, and traveled throughout the world. He was able to obtain financial support due to the backing of the leading German mathematicians of the time. By the time of his death in 1926, he had compiled more than 200,000 references. The idea of publishing his catalog was discussed at several International Congresses of Mathematicians (1897 in Zurich, 1928 in Bologna),[12] and was considered by some academic bodies (the Prussian Academy of Sciences, the Mathematical Association of America), but the catalog was never published. The only existing copy of Valentin's catalogue was sadly destroyed by a bomb on February 1944.[13]

Guccia collaborated on the Répertoire's project out of his loyalty to his French colleagues and friends, and not because of its importance for the development of the Circolo. His commitment to this project could have played an important role in building a close relationship with Poincaré. In the years that followed, Guccia and Poincaré exchanged letters regularly (fifty-three are recorded in the Archives Henri-Poincaré), mostly related to the Rendiconti, either concerning the publication of papers of Poincaré or the refereeing of papers. Poincaré published fifteen papers in the Rendiconti from 1888 until his death in 1912. Fifteen may seem a small figure, but it is only small in comparison to the 185 papers that he published in the Comptes Rendus de l'Académie des Sciences de Paris, and the 25 published in Acta Mathematica. Moreover, Poincaré published more papers in the Rendiconti between

---

[12]G. Eneström, *Über die neuesten mathematisch-bibliographischen Unternehmungen. II. Die allgemeine mathematische Bibliographie des Herrn G. Valentin.* In: F. Rudio (ed.), *Verhandlungen des ersten Internationalen Mathematiker-Kongresses: in Zürich vom 9. bis 11. August 1897,* B.G. Teubner, 1898, Leipzig, pp. 285–286; and R.C. Archibald, *Georg Hermann Valentin (1848–1926).* In: *Atti del Congresso Internazionale dei Matematici: Bologna, 3–10 Settembre 1928 (VI),* Nicola Zanichelli, 1929–1932, Bologna, pp. 465–470.

[13]A. Jackson, *Chinese acrobatics, an old-time brewery, and the "Much needed gap": The life of Mathematical Reviews,* Notices of the American Mathematical Society, 44 (1997), pp. 330–337, p. 336.

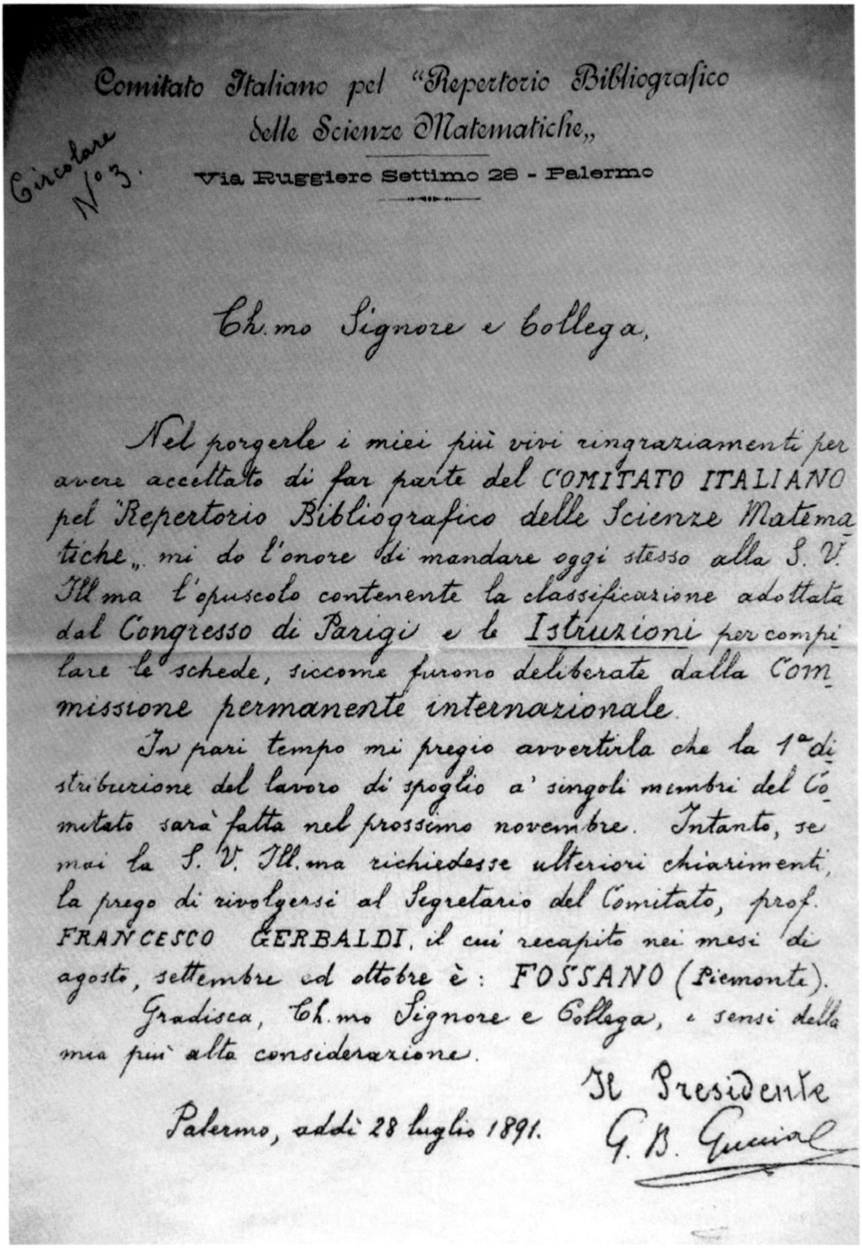

Comitato Italiano pel "Repertorio Bibliografico delle Scienze Matematiche,,
Via Ruggiero Settimo 28 - Palermo

Circolare N° 3.

Ch.mo Signore e Collega,

Nel porgerle i miei più vivi ringraziamenti per avere accettato di far parte del COMITATO ITALIANO pel "Repertorio Bibliografico delle Scienze Matematiche,, mi do l'onore di mandare oggi stesso alla S. V. Ill.ma l'opuscolo contenente la classificazione adottata dal Congresso di Parigi e le Istruzioni per compilare le schede, siccome furono deliberate dalla Commissione permanente internazionale.

In pari tempo mi pregio avvertirla che la 1a distribuzione del lavoro di spoglio a' singoli membri del Comitato sarà fatta nel prossimo novembre. Intanto, se mai la S. V. Ill.ma richiedesse ulteriori chiarimenti, la prego di rivolgersi al Segretario del Comitato, prof. FRANCESCO GERBALDI, il cui recapito nei mesi di agosto, settembre ed ottobre è: FOSSANO (Piemonte).

Gradisca, Ch.mo Signore e Collega, i sensi della mia più alta considerazione.

Il Presidente
G. B. Guccia

Palermo, addì 28 luglio 1891.

Fig. 5.3 Communication from the Italian committee of the Répertoire Bibliographique des Sciences Mathématiques. Courtesy of the Circolo Matematico di Palermo

1899 and 1912 than in other foreign mathematical journals. He published ten papers in the Rendiconti versus four in Acta Mathematica, for example.[14]

Some of Poincaré's papers in the Rendiconti were quite important. One was *Sur les équations de la physique mathématique* published in 1894.[15] Another one was *Cinquiéme complèment à l'Analysis situs*, published in 1904, where his famous conjecture on the three-dimensional sphere was stated.[16] This conjecture was solved by Grigori Perelman almost one hundred years later. On June 5, 1905, at a session of the Paris Academy, Poincaré presented a four-page note entitled *Sur la dynamique de l'électron*; the note was released a few days later (this is the version of the work of Poincaré that supposedly reached Albert Einstein in Bern). Poincaré personally gave Guccia the extended version, a 47 page-long manuscript, in September 1905. The paper appeared in the Rendiconti in 1906.[17] Another important paper that Poincaré published in the Rendiconti was *L'avenir des mathématiques,*[18] which was his plenary lecture at the International Congress of Mathematicians held in Rome in 1908. This paper was also published in the Bulletin des sciences mathématiques, and in the Revue générale des sciences pures et appliquées. Near the end of his life, Poincaré sent a letter to Guccia asking him to publish his last paper, even though the research contained in the paper was not complete.[19] Guccia agreed, and the paper *Sur un théorème de géométrie* was published in 1912 in the Rendiconti.[20] Guccia also published Poincaré's letter in the Supplemento ai Rendiconti.[21]

The letters exchanged between Guccia and Poincaré contain many personal details and comments on the social and political events of the time. Poincaré often followed Sicilian events; for example, in a letter to Guccia from July 30, 1904, he commented on the judgment of acquittal for Raffaele Palizzolo, the suspect of a Mafia's murder, and wrote: "Well, your Palizzolo triumphs"[22]. We know that Poincaré visited Guccia in Palermo during a trip with his cousin Raymond Poincaré (who later became prime minister and president of France). The Rome International Congress of Mathematicians, held on April of 1908, was a favorable occasion for their friendship. Poincaré spent the full congress ill in bed in the Grand Hotel de la

---

[14]P. Nabonnand, L. Rollet, *Éditer la correspondance d'Henri Poincaré.* In: F. Henryot (ed.), *L'historien face au manuscrit: Du parchemin à la bibliothéque numérique*, Presse universitaire de Louvain, 2012, pp. 285–304.

[15]Rend. Circ. Mat. Palermo, 8 (1894), pp. 57–155.

[16]Rend. Circ. Mat. Palermo, 18 (1904), pp. 45–110.

[17]Rend. Circ. Mat. Palermo, 21 (1906), pp. 129–176.

[18]Rend. Circ. Mat. Palermo, 26 (1908), pp. 152–168.

[19]Letter from Poincaré to Guccia, December 9, 1911. Folder Poincaré, Archive of the Circolo Matematico di Palermo.

[20]H. Poincaré, *Sur un théorème de géométrie*, Rend. Circ. Mat. Palermo, 33 (1912), pp. 375–407.

[21]Suppl. Rend. Circ. Mat. Palermo, 8 (1913), pp. 28–29. Guccia's response is dated December, 12, 1911, see S.A. Walter et al. (eds.), *Henri Poincaré Papers*, http://henripoincarepapers.univ-lorraine.fr/chp/.

[22]Letter from Poincaré to Guccia, July 30, 1904. Folder Poincaré, Archive of the Circolo Matematico di Palermo.

Minerve and Guccia took care of all of his accommodation. In a series of letters, Poincaré thanked Guccia "for all the trouble that you have gone through for me; I am grateful for the sincere friendship you have shown in these circumstances." He also informed Guccia of the medical instructions for the return trip, writing: "Mr. Mazzoni has given me a list of doctors in the various cities where I will stay [Florence, Milan, Lausanne]"; and commented on his arrival in Paris "We arrived Friday evening [. . .] the doctor examined me and confirmed the diagnosis of Mr. Mazzoni [. . .] within fifteen days I could recover completely."[23] Poincaré's daughter and son maintained a relationship with Guccia after the death of their father, and they congratulated Guccia on the occasion of the celebration of the founding of the Circolo.

The close friendship between Guccia and Poincaré turned out to be very important to Guccia's long-term project regarding the internationalization of the Circolo. This was a goal that Poincaré also seemed to share. We will review these events in detail when we discuss the 1908 International Congress of Mathematicians held in Rome.

5.2 Development and Crisis

The years between 1889 and 1894 were critical for Guccia in terms of his position at the university as well as his management of the Circolo. In both cases, there was an advance and a crisis, which resulted in a positive development.

The summer of 1889 was, as usual, full of activities and travelling for Guccia. First, Luigi Cremona and his wife, and then Camille Jordan, visited him in Palermo. There Jordan attended a session of the Circolo, where he was elected "by acclamation" as a non-resident member. In that same session, the Circolo accepted an invitation to attend the Bibliography congress in Paris, and designated Camille Jordan as its representative. That session also held a remembrance for the French geometer Georges Halphen, who had recently passed away and who Guccia considered as his "second mentor."[24] Guccia also travelled with Jordan to Trapani that summer, and they visited the spectacular Greek ruins of Selinunte and Segesta.[25] Guccia had planned to go to Paris in September, but changed his plans and first went to Heidelberg. He rerouted his trip in order to continue his previous work on surfaces that was related to results of Max Noether. Guccia had exchanged several letters with Noether and they had agreed to meet in Heidelberg during the annual meeting of the society of German natural scientists and medical doctors (Gesellschaft Deutscher Naturforscher und Ärzte). Guccia brought a presentation

---

[23] *Ibid.*, April and May, 1908.

[24] Letter from Guccia to Cremona, July 5, 1889. Folder Cremona, Archive of the Mazzini Institute of Genoa.

[25] *Ibid.*

letter from Cremona to the meeting, which he gave to Leo Königsberger, the mathematician who presided over the mathematics section of the congress. There, Guccia met many German mathematicians: Georg Cantor, Walther von Dyck, Eugen Netto, Alfred Pringsheim, Theodore Reye, Ernst Schroeder, Arthur Schönflies, Heinrich Wolf, and many others. He was happy to see how well known the Circolo Matematico di Palermo was among the German mathematicians, and was pleased with the congratulations he received for founding the society. Some participants requested the statute of the Circolo Matematico di Palermo from him (which he was obviously carrying with him). He was delighted to receive requests for membership to the Circolo. He was very much interested in procuring papers for publishing in the Rendiconti during the meeting, such as some work on set theory by Cantor.[26]

In Heidelberg, Guccia witnessed the discussions regarding the creation of the Deutsche Mathematiker-Vereinigung (the German Mathematical Society). He was the only foreigner attending the meeting. As a result of those discussions, the Deutsche Mathematiker-Vereinigung was founded one year later in 1890 at the following meeting of the German natural scientists in Bremen.[27]

Aside from the meeting, Guccia aimed to discuss problems related to the resolution of singularities of curves with Max Noether. On his arrival in Heidelberg, Guccia was invited to lunch at Noether's home in Mannheim (which is just twenty kilometers away from Heidelberg). There they spent five hours discussing mathematics. Guccia was happy about that! After the congress, Guccia spent four days in Mannheim, and was invited to lunch and dinner each day at the luxurious palace of the Noether family (which, as he explained to Cremona, "by its comfort and luxury only a Roman prince could have.").[28] Before leaving Germany, Guccia made sure to obtain a book with the addresses of all German mathematicians, as he was always thinking of advertising the Circolo.

The mathematical aim of the visit was related to the paper that in 1874 Alexander von Brill and Max Noether had published in Mathematische Annalen entitled *Über die algebraishen Functionen und ihre Anwendung in der Geometry*, where they proved that any plane curve could be birationally transformed with a Cremona transformation into a curve having only ordinary multiple points, i.e., points of multiplicity m with m distinct tangent lines.[29] Guccia had been interested in the result and had worked previous years on singularities of curves and of algebraic surfaces. For this, he defined the concept of "compound singularities" (in Italian, "singolarità composte") and studied their behavior under different hypotheses. Six papers came out of that work. Two were published in the Comptes Rendus

---

[26] *Ibid.*, October 2, 1889.

[27] H. Gispert, R. Tobies, *A comparative study of the French and German Mathematical Societies.* In: C. Goldstein, J. Gray, J. Ritter (eds.), *L'Europe mathématique: Histoires, Mythes, Identités,* Maison des sciences de l'homme, 1996, Paris, pp. 409–430.

[28] Letter from Guccia to Cremona, October 2, 1889. Folder Cremona, Archive of the Mazzini Institute of Genoa.

[29] J. Gray, 1989, *op. cit.,* p. 371.

in 1888: *Sur l'intersection de deux courbes algébriques en un point singulier*, and *Théorème général concernant les courbes algébriques planes* (the first one presented by Halphen); and four published in the Rendiconti della Reale Accademia dei Lincei in 1889: *Sulla classe e sul numero dei flessi di una curva algebrica dotata di singolarità qualunque, Su una proprietà delle superficie algebriche dotate di singolarità qualunque, Sulla intersezione di tre superficie algebriche in un punto singolare e su una questione relativa alle transformazioni razionali nello spazio*, and *Nuovi teoremi sulle superficie algebriche dotate di singularità qualunque* (items 22 to 27, Appendix 1).[30] During his visit, Guccia wanted to discuss an argument in one of his papers that Noether had objected to.[31]

However, Guccia also devoted part of his time and attention that summer to the upcoming competitions in Italy for professorships in mathematics. In early 1889, two openings for professorships came up in Southern Italy. One was for professor of Higher Geometry and the other one was for Analytic Geometry. The first one was the one that mostly suited Guccia, due to his research specialization. However, the position was in the University of Naples. The University of Palermo was assigned the position in Analytic Geometry. Guccia could not apply for the professorship in Naples, given his responsibilities to his family business and the Circolo in Palermo. Therefore, he only applied to the professorship in Analytic Geometry, despite the fact that this was not his primary research topic. He communicated this decision to Cremona in a letter.[32] Later in the year, however, the High Council for Public Education (the government advisory board for public education) created a professorship in Higher Geometry for the University of Palermo. In a short autobiography that Guccia wrote many years later, he attributed the creation of the Higher Geometry position in Palermo to Brioschi and Cremona, but no other evidence exists to support his assumption.[33] In the end, Guccia decided to apply to the Higher Geometry position at the University of Palermo, and withdrew his application for the Analytic Geometry.

The commission appointed to judge the competition for the Higher Geometry professorship was led by Luigi Cremona, and consisted of Felice Casorati, Eugenio Bertini, Enrico D'Ovidio and Riccardo de Paolis. The competition took place on October 1889, and Guccia won the position. By November 9, 1889, the first step for securing his position at the university was achieved: he was "professore straordinario" of Higher Geometry of the University of Palermo.

---

[30] A. Barlotti, F. Bartolozzi, G. Zappa, *op. cit.*, pp. xvii–xviii.

[31] *Ibid.*, p. xviii. In Italy the study of singularities of surfaces was also carried on by Corrado Segre and by Pasquale del Pezzo; see P. Gario, *Resolution of Singularities of Surfaces by P. del Pezzo. A Mathematical Controversy with C. Segre*, Arch. Hist. Exact Sci., 40 (1989), pp. 247–274.

[32] Letter from Guccia to Cremona, July 5, 1889. Folder Cremona, Archive of the Mazzini Institute of Genoa.

[33] In 1907 Guccia wrote a short curriculum vitae for an album published by the Publisher Adolf Eckstein from Berlin, in a series named *Mondo Intellettuale*. The volumes are not kept, but the draft letter is. Folder Guccia, Archive of the Circolo Matematico di Palermo.

In 1890, Guccia published the textbook *Teoria generale delle curve e delle superficie algebriche*, based on his lectures, where he presented the general theory of algebraic surfaces (item 45, Appendix 1). In the textbook, Guccia devised a new setting for the study of projective properties of algebraic curves. His aim was to base the theory on a new definition of the polar of a point with respect to a given curve that had a more geometrical nature. The study for linear systems of planar curves was done in three papers that appeared in the Rendiconti in 1893: *Ricerche sui sistemi lineari di curve algebriche piane, dotati di singolarità ordinarie* (two notes), and *Una definizione sintetica delle curve polari* (items 32, 33 and 35, Appendix 1).[34] The results were then extended to twisted curves defined on a surface in three papers from 1895: *Sur une question concernant les points singuliers des courbes gauches algébriques*, and *Sur les points doubles d'un faisceau de surfaces algébriques* (published in the Compte Rendus and presented by Jordan); and *Sur une expression du genre des courbes gauches algébriques douées de singularités quelconques*, published in the Bulletin de la Société Mathématique de France (items 36 to 38, Appendix 1).[35] Two other papers of that period were published in the Rendiconti in 1893 and 1894 and were concerned with involutions, *Due proposizioni relative alle involuzioni di specie qualunque, dotate di singolarità ordinarie,* and *Sulle involuzioni di specie qualunque, dotate di singolarità ordinarie* (items 31 and 34, Appendix 1)[36] Eight papers came out of this period of work.

Three years after Guccia's appointment, the university had the legal right to turn his professor straordinario position into a full professorship, or that of "professore ordinario." The university thus decided to convert the position into a full professorship, and thus Guccia had to be judged by another commission to obtain the new position. This second examination was called a "promozione." The "promozione" was not a competitive exam, but an assessment of the candidate's activity in research, teaching and administration since the first evaluation. Every year, the Ministry of Public Education named a national commission for each area to administer these evaluations. If Guccia passed this second examination, he would become "professore ordinario" and would enjoy state tenure.

The administrative process followed its normal course, and the commissions to judge the applications for "professore ordinario" were appointed. In the area of Higher Geometry, the candidates were Pasquale del Pezzo, from the University of Naples, and Giovanni Battista Guccia, from the University of Palermo. The commission appointed for this area consisted of Ferdinando Aschieri, Eugenio Bertini, Enrico D'Ovidio, Corrado Segre and Giuseppe Veronese. In the area of Analytic Geometry, the candidate was Francesco Gerbaldi, from the University of Palermo. For some unknown reason, the commission appointed for Analytic Geometry turned out to be the same as the one for Higher Geometry. Consequently, the same commission was in charge of the "promozione" of the three candidates

---

[34] A. Barlotti, F. Bartolozzi, G. Zappa, *op. cit.,* pp. xviii–xx.

[35] *Ibid.,* pp. xviii–xx.

[36] *Ibid.,* pp. xviii–xx.

del Pezzo, Gerbaldi and Guccia. The commission met on October 6, 1893, and, unexpectedly, rejected the applications of the three candidates. This outcome was possible but it was certainly unusual, and extremely rare for three candidates to be rejected at the same time. A sequence of administrative events then followed, which involved several successive interventions of the Ministry of Public Education and the High Council of Public Education. A formal error in the procedure had occurred: it was legally required that each member of the commission produce a written report, and an additional final report by the full commission; instead, the commission only produced the last joint report. This motivated the decision to be annulled, the commission dissolved, and a new commission appointed.

What had occurred? The facts written in the documents specify that the initial commission made a severe criticism of the work and memoirs of the candidates. Curiously enough, two members of the commission for the "promozione," Eugenio Bertini and Enrico D'Ovidio, had been members of the previous commission that had appointed Guccia three years before. It was Corrado Segre (Fig. 5.4) who wrote the joint report presented by the commission in charge of evaluating the "promozione", and it seems that he was the member of the commission who led the others to the negative judgment.

A detailed analysis shows that several underlying conflicts coincided on the occasion of the "promozione" for the three candidates. One was the academic conflict between the older generation of mathematicians, represented by Brioschi and Cremona, which had been in control of the academic scene for a long time, and a younger generation. Some mathematicians from this younger generation

Fig. 5.4 Corrado Segre. Courtesy of the Department of Mathematics "Giuseppe Peano", University of Turin

sought a new academic order which they could lead. In order to establish their leadership, however, they relied on positions available at the universities. Segre was a leading member of this group. The three candidates were seen as belonging to the apparatus of the old establishment. Another conflict was the quest for rigor in mathematical research. At the time, that process was taking place in other areas of mathematics, particularly in Analysis. Corrado Segre was a champion of this new trend in Geometry. He fought ardently against vagueness and ambiguity in mathematical reasoning, a fault that he saw in the work of the candidates. In regard to rigor, the position of Segre was extreme. He memorably stated: "I fervently advise rigor, rigor, rigor."[37] Segre's colleagues believed that his obsession with rigor had an overall negative effect on his career, and limited his mathematical production. Lastly, a battle over the control of the Steering Council of the Circolo Matematico di Palermo (and therefore over the control of the Rendiconti) posed another conflict that influenced the decision.

Letters by all people participating in the "promozione" show that all three of these conflicts coincided to influence the judgements of the commission. In his letters to Guido Castelnuovo during the meetings of the commission—on October 1893–, Segre spoke about the "defeat of the protégé of Cremona."[38] Here Segre was of course referring to Guccia. As a consequence of the rejection of the three candidates, there was a great deal of turmoil among mathematicians, but also beyond the mathematical community. Tensions resulting from the Unification surfaced during this controversy, as the judges of the commissions were all Northerners while the candidates came from the South of Italy (although Gerbaldi was born and educated in the North[39]). The feeling of unbalance between the North and the South of Italy was still a very sensitive public issue. As was explained above, the Minister of Public Education decided to overrule the commission of the "promozione" and appointed a new commission, of which Cremona was a member. The result this time was positive for the three candidates. The issue was still not over because the High Council of Public Education intervened once more. After one year of intense and cumbersome administrative activity, the Minister of Public Education made the final decision: it assumed the favorable judgment made by the second commission, and the three candidates were promoted to "professore ordinario". Nevertheless, this controversy had damaged many personal relationships along the way.[40]

A second important conflict for Guccia during this period was related to the Circolo and its journal. By the time of the affair of the "promozione," the Rendiconti was starting to become a prestigious and influential journal. In the next chapter, we

---

[37] Letter from Segre to Castelnuovo, May 27, 1893. Guido Castelnuovo Collection, Archive of the Accademia Nazionale dei Lincei.

[38] *Ibid.*, November 5, 1893.

[39] L. Martini, *Algebraic research schools in Italy at the turn of the twentieth century: the cases of Rome, Palermo, and Pisa*, Hist. Math., 31 (2004), pp. 296–309, p. 302.

[40] C. Ciliberto, E. Sallent Del Colombo, *Pasquale del Pezzo, Duke of Caianiello, Neapolitan mathematician*, Arch. Hist. Exact Sci., 67 (2003), pp. 171–214.

will review the development of the journal, the reasons for its success, the role of the journal in the success of the mathematical society, and Guccia's role in running the journal.

The journal was run by the Steering Council of the Circolo which, as explained before, acted as the editorial board. The first council was elected for the period 1888–1890 and consisted of:

Five resident members: G. Albeggiani, M.L. Albeggiani, Calderera, Gebbia and Guccia.

Fifteen non-resident members, coming from Pisa (Betti, de Paolis and Volterra); from Pavia (Beltrami, Bertini and Casorati); from Rome (Cerruti and Cremona); from Turin (D'Ovidio and Segre); from Milan (Brioschi and Jung); from Naples (Battaglini and del Pezzo); and from Bologna (Pincherle).

Guccia was appointed delegate of the council for the direction of the Rendiconti, and acted as the Rendiconti's managing editor. Three years later, the Council was re-elected; the same members were to serve for another three year period, with the exception of Felice Casorati, who died in September of 1890 and was replaced by Henri Poincaré. Guccia continued as managing editor.

The Council members for the 1894–1896 term were to be elected in January 1894. As early as December 1893, Segre had expressed great interest in the elections in his letters to Castelnuovo. It seems that his aim was to form a group that would be able to win a sizable part of the council. With this intention in mind, Segre was carefully following the movements of Guccia, who at the time was contacting mathematicians in order to form an (official) proposal that would gather sufficient backing from the members of the society. Segre's letters to Castelnuovo reveal a hostile attitude towards Guccia, and comment sarcastically on the daily events and internal activity of the Circolo.[41] This was a change from the tone of the communications between Segre and Guccia in earlier times. On November 28, 1887, Segre wrote to Guccia

> I regret that you did not stop here [in Turin] for a few days. I would gladly have heard from you the news from France. I hope you will tell me by letter. Have you heard about any important work in Geometry? Did you see Hirst and tell him that, if he comes to Italy, he could stop in Turin?[42]

Things were different now. Segre and his colleagues were now interested in controlling this prestigious research journal. The Annali di Matematica Pura ed Applicata, although much older than the Rendiconti, had not acquired the same level of quality and prestige.

The elections for the Council took place on January 21, 1894, and, despite the efforts of Segre and his followers, Guccia's proposal received the most votes. In contrast with the previous council, eight out of the fifteen non-resident members

---

[41] Letters from Segre to Castelnuovo, October 27, November 5, 1893, and January 6, January 20, January 24, 1894. Guido Castelnuovo Collection, Archive of the Accademia Nazionale dei Lincei.

[42] Letter from Segre to Guccia, November 28, 1887. Folder Segre, Archive of the Circolo Matematico di Palermo.

were newly elected. Some replacements were necessary, as three members of the previous Council had passed away (G. Albeggiani, Betti and De Paolis), and one did so three months later (Battaglini). There were two new resident members and six non-residents. One of the latter was from abroad, the Swedish mathematician Gösta Mittag-Leffler, founder and editor of the journal Acta Mathematica. The other five new non-resident members were Italians: Bianchi, Capelli, Loria, Maisano and Peano. Thus, four members of the previous Council were not renewed. Three of them belonged to the commission of the "promozione": Bertini, D'Ovidio and Segre. This was an expected consequence of the conflict around the "promozione", which was still ongoing at that moment (the commission of the "promozione" met on October 1893 and the final decision by the Minister was taken on November 1894). The elections were a challenge to Guccia's role in the Circolo, but the challenge then turned into a failure. Two weeks before the elections took place, Segre was already joking to Castelnuovo about the "condolences for my expulsion from the Steering Council."[43] The fourth member that was replaced was Francesco Caldarera, one of the founding members of the Circolo and president of the society at that time.

It is worth mentioning that Segre had a negative attitude towards the project of the Répertoire; he disliked the classifications that had being adopted, and had decided not to collaborate in the project.

Some years later, another conflict erupted within the Circolo regarding the library. This conflict had fundamental implications for mathematical research, and involved members related to those in the Rendiconti conflict. The Circolo's first statute from 1884 included several articles regarding books and periodicals: Article 12 stated that the society could use its funds to subscribe to journals, Article 13 allowed the society to purchase mathematical works, and Article 14 regulated donations of books and journals. The main article related to the library, however, was the terms of use established for the library materials. These terms were presented in Article 15, which established that library materials could be used by the members of the society on the premises of the Circolo, and established that under special regulations, to be approved by the society, the library materials could also be used outside the premises of the society.

When the second statute was approved four years later in 1888, the library's collection of books, articles and journals, obtained through donations and exchanges, was substantial. Taking this into account, the second statute included a full section devoted to the library. The statute declared that books, memoirs and periodicals that were bought, received through donation or obtained through exchange were the property of the society (Article 37), and it also established that membership dues could not be used to make purchases for the library (Article 39). The key issue was considered in Article 38, which established (once again) that special provisions had to be approved by the society in order for members to check books out from

---

[43]Letter from Segre to Castelnuovo, January 6, 1894. Guido Castelnuovo Collection, Archive of the Accademia Nazionale dei Lincei.

the library premises. Some months later, in October 1888, the society discussed and approved the regulations for the library. We have not been able to find these regulations in the Archive of the Circolo.

Between 1888 and 1899, the library's collection grew substantially. The library had acquired many books through donations, and through the regular exchanges between the Rendiconti and other journals. As a result, some new "Special Regulations for the Library" were discussed and approved in an extraordinary session held on January 15. The regulations were extensive and thorough: there were eighteen articles devoted to the subject. The main issue was the possibility of members of the Circolo borrowing books (at most two for a maximum period of two weeks), reprints of articles (at most five for a maximum period of one month), and full volumes of journals (at most one for a maximum period of one week). It was not possible to borrow recent articles and journal issues, rare and didactic books, dictionaries, and any other publication needed for the running of the Rendiconti. A deposit of twenty-five lire was established as a guarantee, and strict conditions were set on its recovery to prevent misuse of the borrowing system. A director was also appointed to oversee the library. This post was different to that of the two Librarians, who were members of the Presidency. The director was paid a salary, and was in charge of the daily running of the library (which was to be open for members all days of the week from 14 to 17 hours except for Thursdays and Sundays).

The salary of the director of the library soon became a burden on the finances of the Circolo, and when the post became vacant in September 1900, a new director was not appointed. In fact, the salary of the director had only been paid thanks to a special donation to the Circolo made directly by Guccia. Six months later, the Circolo decided not to appoint another director, and instead had a member of the society take over those duties. This remained the case until the session of January 24, 1904, when Guccia proposed that the existing regulations for the library be revoked and replace with new ones that would prohibit borrowing library materials. A lengthy (and probably heated) discussion took place during the session, and was followed by a vote: the proposal was rejected with ten votes in favor of Guccia's suggestion, and fourteen against. Elections for the presidency of the Circolo took place immediately after this vote. Before the election, Guccia declared, after thirteen terms holding different positions within the presidency, that he did not want to continue his participation. The result of the voting was clear: the positions of president, two vice-presidents, a secretary, two vice-secretaries, and two librarians went to members who had opposed revoking the library regulations. From the number of votes, it seems that those who had favored revoking the regulations did not vote. However, two of the members who voted in favor of the revoking were elected: Guccia as secretary, and Porcelli as treasurer (he had been treasurer for a long time). However, Guccia and Porcelli resigned during another session held two weeks later, and a new secretary and treasurer were elected. Only thirteen of the thirty-three resident members attended the session, and only twelve members voted. The internal situation of the Circolo became quite unstable: one week later, the newly elected treasurer resigned, and one week after that a new treasurer was elected. This caused the post of secretary to be vacant, so, again one week later there were elections for secretary.[44]

In the session held on February 21, 1904, two important decisions were taken. The first came after the discussion of the yearly budget: there would be no more appointments for an employee in charge of the library. The second one was unexpected and far-reaching: it mandated that the Presidency seek another place in Palermo for the Circolo's library materials, or else leave them in a public library. Both proposals obtained sixteen favorable votes and five against.[45]

Meanwhile, the regular life of the Circolo seemed to have continued its normal pace, with sessions and mathematical communications. However, these changes in the library happened quickly: during March 29, 30 and 31, all library materials of the Circolo were taken out of the Palazzo Guccia on via Ruggiero Settimo, the seat of Circolo, and moved to the Institute for Physics of the University of Palermo. Not only were the books, memoirs and journals removed, but the shelves, furniture and all the documents were also taken away. Guccia had donated the shelves and the furniture when he founded the Circolo in 1884. The agreement that approved moving the library was signed one week later by the Presidency of the Circolo and by the director of the Institute for Physics. It is worth mentioning the persons involved in moving the library: the president of the Circolo was Francesco Caldarera, the member of the Steering Council that was not reelected ten years earlier when Segre and his colleagues left the Council; and the director of the Physics Institute was Damiano Macaluso, who at that time was also one of the secretaries of the Circolo.

Guccia had a strong reaction to the library relocation, and wrote a public memorandum on April 25 of that year to explain his view of the situation. This memorandum was widely distributed among mathematicians in Italy and abroad, and also outside the mathematical world, certainly all throughout Palermo. Unfortunately, we do not have a copy of the memorandum, we just have notice of its existence, and so we cannot read Guccia's arguments. The pressure over the group holding power in the Circolo became too strong, and finally the full Presidency resigned during an extraordinary session held on May 5. Five days later, in another extraordinary session with fifteen members attending, a new Presidency was elected. Not one of the previous officials was included in this new Presidency. Guccia was not elected for any position. Two weeks later, a new delegate for the library was appointed. By June 1, all books and other materials from the library were "reinstated in their former site and in perfect order as before."[46] In the meantime two members of the previous Presidency had resigned as members of the Circolo.

The minutes of the sessions of the Circolo provide a precise account of the sequence of events of this troubling episode, but give few hints about the reasons for

---

[44] Adunanze January 24, February 7, February 11, February 14, 1904. Rend. Circ. Mat. Palermo, 18 (1904), pp. 200–202.

[45] Adunanza February 21, 1904. Rend. Circ. Mat. Palermo, 18 (1904), p. 202.

[46] Adunanza May 5, 1904. Rend. Circ. Mat. Palermo, 18 (1904), pp. 221–222.

it. Two letters from Guccia provide invaluable information regarding this episode: one is to Michele de Franchis, dated May 16, 1905, and the other one to Ulisse Dini, dated January 9, 1906.

Guccia started his letter to De Franchis with a blunt statement: "With respect to ingratitude, I am a competent judge." He reprehended De Franchis for having sided with those who, loudly declaring that "professor Guccia is the public enemy", took over the management of the society with plans to reduce the print run and the number of pages of the volumes of the Rendiconti, as well as making other savings which would have been contrary to the interests of the non-resident members. All this, according to Guccia, was done so that three or four persons could make personal use of the library of the Circolo. There were plans to cancel some of the exchanges and to move the printing of the Rendiconti to Turin, with the excuse that it would be easier and provide a cheaper service for the authors. Guccia was clearly pointing to Segre, who was in Turin, as the person who had instigated the takeover of the management of the Circolo.[47]

More explicit was the letter to Dini written two years later, which shows that the episode was still fresh in Guccia's mind. Dini had been a student of Enrico Betti in Pisa and became a renowned analyst. He combined his research work with a deep involvement in academic affairs (he was Rector of the University of Pisa and Director of the Scuola Normale Superiore di Pisa) and in politics (he was deputy in parliament and a senator). He was a highly-respected figure inside and outside Italy. Guccia wanted Dini to join the Circolo for a long time. Once Dini became a member in 1900, he treated Dini with much care, placing him on the Steering Council but with a low workload, on account of his many other responsibilities. In December 1905, Dini wrote to Guccia politely requesting that Guccia send him, if allowed by the regulations of the Circolo, a series of volumes from the library of the Circolo for professor Onorato Nicoletti from Pisa. These books were not available in any other library in Italy. The answer had to be negative in accordance with the current regulations on the Circolo after the 1904 crisis. Despite this, Guccia had to answer with care, as Dini played an instrumental role in obtaining subsidies from the Ministry for the Rendiconti. On the other hand, his reply to Dini's request gave him a perfect opportunity to explain his viewpoint on the library crisis.

The response is dated January 9, 1906. Guccia started his reply by apologizing for the delay in answering, as the beginning of the year was quite busy for someone who carried on his shoulders the running of a society with 336 members. After legally justifying his proposal to revoke the library regulations, Guccia stated his objectives for the proposal. First was stopping the abuse of the library collection (for example, volumes of Acta Mathematica from the Circolo were seen in Rome). Second, he believed that devoting funds of the society to the salary of the library director in order to allow the borrowing of library material (which would only be available for members from Palermo or Turin, not from Australia, India or America)

---

[47]Letter from Guccia to De Franchis, May 5, 1905. Folder De Franchis, Archive of the Circolo Matematico di Palermo.

was a "fraud" to the foreign members, who could not make use of the library and should not have to pay for problems resulting from the lack of a university library in Palermo or elsewhere in Italy. The main objective, however, was related to the editorial work in the Rendiconti. To publish a high quality journal (as were many mathematical publications from Germany, France or England), it was necessary to have a full-time managing director, who needed all possible mathematical literature at hand to do the job properly. He contrasted these long-term goals with the petty-minded interests of his opponents, and concluded that the library crisis had been beneficial for the Circolo, as it had allowed the society to get rid of those members who were placing obstacles in the development of the Circolo (more from ineptitude and ignorance than from maliciousness, he specified). As a measure of the efficacy of this viewpoint, Guccia cited the figure of 159 new members of the Circolo since the crisis.

After the lengthy explanation (four full typed pages), Guccia excused himself for not being able to lend the volumes. In exchange, he offered Dini any volume from his personal private library (consisting of thirty complete journals). Unfortunately, he did not have any publication from Moscow or St Petersburg, which seemed to be what Nicoletti was looking for (the list of requests which accompanied Dini's letter is missing in the Archive of the Circolo). In passing, this correspondence shows the superb library that Guccia had been able to gather for the Circolo, mostly via exchanges with the Rendiconti. We will review the exchanges of the Rendiconti in detail in the next chapter; for now, here is the list of journals from the Russian Empire that could be found in the library of the Circolo in 1908:

Acta Societatis Scientiarum Fennicae
Öfversigt af Finska Vetenskaps-Societetens Förhandlingar
Bidrag till Kännedom af Finlands Natur och Folk
Observations metéorologiques de la Société des Sciences de Finlande
Meteorlogisches Jahrbuch für Finland
Bulletin de la Société Physico-Mathématique de Kasan
Annales de l'Université de Kharkov
Communications de la Société Mathématique de Kharkov
Bulletin de l'Université Impèriale de Kiew
Recueil Mathématique de Moscou
Mémoires de la Société des Naturalistes de la Nouvelle Russie (Odessa)
Bulletin de l'Académie Impériale des Sciences de St.-Pétersbourg
Mémoires de l'Académie Impériale des Sciences de St.-Pétersbourg

Guccia ended the letter by sharing with Dini his most precious project:

> To implement a program—which for a long time I have cherished—that seeks to give to our Society such progress that it will occupy, in a few years, the first place among the five major associations of mathematicians in the world (London, Paris, New York, Leipzig, Palermo)![48]

---

[48]Letter from Guccia to Dini, January 9, 1906. Folder Dini, Archive of the Circolo Matematico di Palermo.

By 1895, Guccia's scientific activity had almost ended. He published his last papers in the following years: in 1902, two notes entitled *Sulle curve algebriche piane. Sulle superficie algebriche* (items 40 and 41, Appendix 1), published in the Rendiconti del Circolo Matematico di Palermo, which continued the project of creating a geometric approach to the study of algebraic curves and surfaces from the viewpoint of projective geometry; in 1906, two notes published in the Comptes Rendus (and presented by Picard), entitled *Un théorème sur les courbes algébriques planes d'ordre n,* and *Un théorème sur les surfaces algébriques d'ordre n,* where the aim was to give a new geometrical meaning to the notion of order of a planar curve, and the corresponding extension to algebraic surfaces (items 42 and 43, Appendix 1). The last of Guccia's papers was *Sopra una nuova espressione dell'ordine e della classe di una curva gobba algebrica* (items 44, Appendix 1), where the results on planar curves were extended to twisted curves; it appeared in the Rendiconti del Circolo Matematico di Palermo in 1906.[49]

5.3 Towards the Constitution of an International Association

> As I already communicated to you, I have established a prize for an international competition aimed at encouraging some of the top and modern theories of Geometry. This prize, consisting of a gold medal and a sum of 3,000 lire in gold, would be conferred by the Circolo Matematico di Palermo in 1908, according to the judgment of an International Commission composed by three members, appointed by the President of our Society; this commission would be entrusted with the task of formulating, in all its details, the program of the competition.[50]

Guccia sent this letter to the president of the Circolo on July 23, 1904. In the session held the next day, the president of the society communicated the news, and announced that the prize would be given the name "Medaglia Guccia". He also announced that, in agreement with the founder of the prize, the international commission would be composed of Max Noether from the University of Erlangen, Henri Poincaré from the University of Paris, and Corrado Segre from the University of Turin.

Where did this proposal come from? In mid-July 1904, Guccia was informed that the mathematical section of the Accademia dei Lincei had recently discussed Vito Volterra's proposal for organizing the International Congress of Mathematicians of 1908 in Rome.[51] The Accademia dei Lincei was a legendary institution devoted to all branches of science, which had Galileo Galilei among its first members. The idea was well received in the Accademia, and it was agreed to seek advice and help from

---

[49] A. Barlotti, F. Bartolozzi, G. Zappa, *op. cit.,* pp. xx–xxii.

[50] Letter from Guccia to the President of the Circolo Matematico di Palermo, July 23, 1904. Folder Guccia, Archive of the Circolo Matematico di Palermo.

[51] Letter from Cerruti to Guccia, July 16, 1904. Folder Cerruti, Archive of the Circolo Matematico di Palermo.

Guccia, as it was widely known that he had many international connections that could be instrumental for the project to succeed. This was occurring just a few weeks after the resolution of the library crisis, and Guccia was exhausted from the battle over the control of the society. However, he realized that the proposal for organizing an International Congress in Rome was an opportunity to leave behind the crisis, and a perfect way to affirm the international role of the Circolo. Over all other considerations, celebrating an International Congress in Italy matched perfectly with his far-reaching plan to turn the Circolo Matematico di Palermo into the leading mathematical society in the world.

Immediately, Guccia wrote Volterra a tactful letter proposing to nominate Volterra as the representative of the Circolo to the next International Congress, which was going to be held in Heidelberg. Guccia sent Volterra this request in case Volterra planned to attend the Congress. Guccia also suggested that the two of them travel together from Rome to Heidelberg so that they could discuss a possible agreement with the Circolo.[52] Guccia also wrote to his friend in Paris, Georges Humbert, who had been one of the first non-Italian members of the Circolo.[53] Humbert was a respected and influential mathematician, who had won two prizes from the Académie des Sciences (one of them the Poncelet prize) and another prize from the Société Mathématique de France. He succeeded Hermite in his seat in the Académie and Jordan in his chair in the Collège de France. Guccia wanted Humbert to inquire about Poincaré's availability for the commission of the prize he was planning. Guccia's idea was to create a prize to be awarded for the first time at the International Congress in Rome, and intended to make it a permanent award associated with the international congresses. The topic of the prize would be the recent advances in Geometry, in particular, on the theory of algebraic twisted curves. Guccia was confident of Poincaré's backing ("[he] has never refused me his support"), and indeed, the response of Poincaré was positive. Max Noether was a clear candidate for membership of the commission given his previous works on the subject. The election of the third member, Corrado Segre, showed that beyond all the internal quarrels of the Circolo (where Segre had played a prominent role), Guccia wanted an international commission of first class to award the prize, and Segre was a natural candidate. Apart from having first class mathematicians, the committee reflected a balance of European nations, in line, for example, with the choice of plenary speakers in the International Congresses: one Frenchman, one German, and one Italian.

There were many awards at that time, granted by different learned institutions. The Medaglia Guccia could have followed as a model of the award established in 1889 by the King of Sweden and Norway, probably under the counseling of

---

[52]Letter from Guccia to Volterra, July 18, 1904. Folder Volterra, Archive of the Circolo Matematico di Palermo.

[53]Letter from Guccia to Humbert, July 21, 1904. Folder Humbert, Archive of the Circolo Matematico di Palermo.

Mittag-Leffler.[54] The Swedish prize was awarded for the first time to Poincaré for a note on the three-body problem. On that occasion, Charles Hermite, Gösta Mittag-Leffler and Karl Weierstrass formed the committee that awarded the prize. The award consisted of a gold medal engraved with an effigy of the King and 2,500 Swedish kronor in gold.

During the session of the Circolo where the prize was announced, it was also decided that the society would be represented by Guccia and Volterra in the coming Heidelberg International Congress of Mathematicians, which was going to be held from 8 to 13 August 1904. We do not know if Guccia and Volterra traveled together to Heidelberg (probably not, as Volterra was traveling with his wife Virginia to Cambridge, and Heidelberg was a stop in their trip), but both did participate in the Congress.

This was not Guccia's first encounter with an International Congress. In the first one, held in Zurich in 1897, there was a large group of Italian mathematicians attending, precisely twenty out of 204 participants. Among them there were Giuseppe Peano (one of the four plenary speakers), Volterra (secretary of the Congress), Brioschi (member of the presidency of the Congress and vice-president of the section on Analysis and Theory of Functions), and Segre (vice-president of the section on Geometry). But the only mathematician from Palermo was Francesco Gerbaldi,[55] who read a telegram of support from the Circolo Matematico di Palermo signed by Guccia. Something must have stopped Guccia from attending, since he had been involved in the organization of the Congress from the very beginning. Indeed, he was the person in charge of disseminating the announcements of the Congress in Italy (as Felix Klein was for Germany, Mittag-Leffler for Sweden and neighboring countries, Markov for the Russian Empire, and other mathematicians for other countries).[56] In those years Guccia also had to change his habit of travelling in August, as he was obliged to take care of the administration of the family's businesses, particularly the water supply business, due the old age of his father (who would pass away on February 1900). Guccia did participate in the second Congress, which was held in Paris in 1900. Italian participation increased moderately, up to twenty-three mathematicians out of 250 total participants. There, Volterra was vice-president of the Congress, Capelli was secretary of the Congress, and Tullio Levi-Civita was secretary of the section on Mechanics. This was the occasion when Volterra gave the plenary address *Betti, Brioschi, Casorati, three Italian Analysts, three ways to consider the question of Analysis* to which we have already referred.

The Heidelberg Congress, the third one in the series, was a success with respect to general participation, as 336 participants attended, but not for the Italians. This time,

---

[54] A. Guerraggio, P. Nastasi, *Roma 1908: il Congresso internazionale dei matematici*, Bollati Boringhieri, 2008, Torino, p. 115.

[55] On Francesco Gerbaldi, see M.R. Enea, *Francesco Gerbaldi e i matematici dell'Università di Palermo*, n. 34–35, Pristem/Storia, 2013, Milan.

[56] F. Rudio (ed.), *Verhandlungen des ersten Internationalen Mathematiker-Kongresses: in Zürich vom 9. bis 11. August 1897*, B.G. Teubner, 1898, Leipzig, p. 8.

there were only twelve of them. At the opening session, there was a remembrance to mathematicians who had passed away, in recent years, and Francesco Brioschi and Luigi Cremona (who died on June 1903, at the age of seventy-three) were mentioned. In contrast with the previous Congresses, there were no Italian officials, as this time all officials of the Congress (presidency and sections) were chosen by the German organizers to be Germans. For the plenary lectures, as in the previous Congresses, four mathematicians were invited. One of them was Corrado Segre, who spoke on "Today's Geometry and its links with Analysis." In his lecture, Segre exhibited his commitment to the necessity of rigor in Geometry: "With respect to the geometrical methods, allow me to express my thinking regarding an accusation that is sometimes made to them: that of little rigor. Already at the International Congress of Paris, Hilbert had vigorously protested against the opinion that only Analysis, and not Geometry, is susceptible to a fully rigorous treatment. [...] In general we can say that nowadays geometers aspire to rigor as much as analysts do!"[57]

On the last day of the Heidelberg Congress, the site of the following Congress was announced. The proceedings of the meeting described the occasion:

> Volterra from Rome transmits the invitation of the Accademia dei Lincei to celebrate the IV International Congress of Mathematicians in Rome in the spring of 1908, with the following words: "The members of the mathematics section of the Accademia dei Lincei, meeting last June, decided to propose to hold the next Congress of Mathematicians in Rome." After the assembly welcomes the invitation with lively applause, Volterra continues: "I appreciate the honor that you have given us by choosing Rome as host of the next congress. I propose to meet in the spring of 1908, leaving to the committee the decision of the precise date. At the same time I have the honor of informing this congress that M. Guccia has supplied the Circolo Matematico di Palermo the sum of 3,000 lire[58] to provide an international award, named the Guccia medal, to be granted during the next congress to a memoir which will provide substantial progress in the theory of inverse algebraic curves. The jury is composed of MM. Noether, Poincaré and Segre."[59]

The counterpoint of the scene came from Alfred Greenhill, from London, who stood up after the speech by Volterra and said:

> I left London under the impression that England was to be honored with the visit of the International Congress of Mathematicians on the next occasion after Germany; and I think this impression was shared by the other English present here. But we find now that Italy is the fortunate country, and is to receive the Congress in 1908. Disappointed in our expectation we must congratulate Italy and Rome on its good fortune, and we must content ourselves with the next best in our wish, and hope that England may be selected at this Assembly as the meeting place in 1911 or 12.[60]

---

[57]C. Segre, *La Geometria d' oggidí e i suoi legami coll' Analisi.* In: A. Krazer (ed.), *Verhandlungen des dritten Internationalen Mathematiker-Kongresses in Heidelberg*, B.G. Teubner, 1898, Leipzig, p. 111.

[58]The figure in the proceedings is 3,000 francs. Those were the times of the Latin Monetary Union, when French francs and Italian lire were equal in value.

[59]A. Krazer (ed.), *Verhandlungen des dritten Internationalen Mathematiker-Kongresses in Heidelberg*, B.G. Teubner, 1898, Leipzig, p. 53.

[60]*Ibid.*, pp. 53–54.

Greenhill's comment shows that Volterra and the Accademia dei Lincei, along with Guccia and the Circolo Matematico di Palermo, had quickly gathered enough support for their proposal for the fourth International Congress of Mathematicians to be held in Rome. The Rome proposal possibly gathered the support of many French and German mathematicians, and beat the English proposal.

Matters regarding the competition for the Guccia medal were sorted out quickly. During the session held on November 13, 1904, the president of the Circolo announced that the international commission in charge had developed the program for the competition. It appeared published on page 390 of volume XVIII of the Rendiconti. It established the following:

> The memoirs presented to the competition should be: unpublished, written in Italian, French, German, or English, typed (except for the formula) with typewriter. They should be sent (in three copies) to the President of the Circolo Matematico di Palermo, arriving before July 1, 1907, provided with an epigraph, and accompanied with a sealed envelope bearing on the cover the epigraph and inside the name and address of the author. The memoir receiving the award will be included in the Rendiconti, or in other publication of the Circolo Matematico di Palermo. The author will receive 200 reprints.
>
> In case none of the memoirs submitted to the competition was deemed worthy of the prize, this could be awarded to a memoir on the aforementioned theories, published after the publication of this program and before July 1, 1907.
>
> The prize will be conferred by the Circolo Matematico di Palermo in accordance with the judgment of an international commission of three members, composed by
>
> > Prof. Max Noether, from University of Erlangen,
> > Prof. Henri Poincaré, from the University of Paris,
> > Prof. Corrado Segre, from the University of Turin.
>
> The reading of the report of the commission, with the proclamation of the name of the author awarded and the presentation of the prize will be done in Rome, in 1908 in a session of the IV International Congress of Mathematicians. [61]

Some time later, Guccia sent 3,000 lire for the medal to the Circolo via a bank transfer.

An important issue for Guccia was the artistic side of the prize: the medal itself. Guccia approached this task with his usual thoroughness. He needed a designer, a sculptor, and a smelter. He chose first class artisans for all three tasks. The design of the medal was entrusted to Ernesto Basile, son of the architect who designed the Teatro Massimo of Palermo. Basile himself was an architect, as well as an important exponent of the Liberty style (the name in Italy for the Art Nouveau style), and avant-garde furniture designer. Guccia selected Antonio Ugo as his sculptor. Ugo was a renowned artist from Palermo who had made many busts and medals. Ugo collaborated with Basile in the design of the Liberty style furniture. The coining of the medal was done in the Florence workshop of the Palermitan sculptor Domenico Trentacoste, widely known for his marble and bronze statues.[62]

---

[61] Adunanza August 27, 1905. Rend. Circ. Mat. Palermo, 20 (1905), p. 377.

[62] Adunanza June 19, 1908. Suppl. Rend. Circ. Mat. Palermo, 3 (1908), p. 28.

Guccia described the medal in a letter to Segre.[63] On the reverse side of the medal, there was a bust of Archimedes sculpted in the style of the many classical images of the genius of Syracuse. Next to Archimedes there was a blank space that would be engraved with the name of the winner (Figs. 5.5 and 5.6).[64] It is interesting to note that many years later in the 1930s, John Charles Fields had the idea of establishing the "International Medal for Outstanding Discoveries in Mathematics"—nowadays known as the Fields Medal—, and also chose a bust of Archimedes for the medal design.[65] The front of the medal showed the logotype of the Circolo Matematico di Palermo (Fig. 5.7).[66] In the beginning, the Circolo had no identifying symbol. When the second statute was approved in 1888, the Circolo was no longer a local society. Guccia may have requested a logotype design from his personal friend Ernesto Basile to represent the local identity of a then international society. Basile did the design of the logotype, and afterwards he became member number 128 of the Circolo. From that moment on, the logotype was used in all publications and letterheads of the society. The motif chosen by Basile was the trinacria. Choosing the trinacria for the logotype was natural since the mathematical society was located in Sicily, and also because the founder of the society was Sicilian born. The origins of the trinacria are not clearly determined, and it is the subject of several legends. Related representations can be found in ancient archaeological remains around the Mediterranean Sea. The trinacria is composed of three equally distributed legs, bent at the knee, which radiate from the head of a gorgon, with wings as ears and snakes entwined in her hair. It is said to be related to the shape of Sicily: the three legs representing the three capes around the island. Throughout history, the trinacria has been associated with the national identity and the independence of Sicily.

Guccia was busy preparing for the International Congress from the moment that it was scheduled in Rome until the day that it took place. He had to balance these activities with his always-increasing dedication to the daily running of the Circolo. The Circolo was closed during Guccia's usual vacation period between September 1 and October 31. His custom was to stay for some time in the Swiss Alps and later go to Paris. In the summer of 1905, however, he changed his plans: he first stayed in the town of Rigi Kaltbad, in Switzerland, and from there he traveled through Germany. His first stop was Munich; he wanted to attend the performance of the third act of the Ring of the Nibelung. After attending, he believed that Richard Wagner's opera in Munich was as satisfactory as it could have been in Bayreuth ("though may think

---

[63] Letter from Guccia to Segre, February 28, 1908. Folder Segre, Archive of the Circolo Matematico di Palermo. See also: Adunanza (Straordinaria) June 19, 1908. Suppl. Rend. Circ. Mat. Palermo 3 (1908), p. 28, where it is explained that the President of the Circolo exhibited in the session a silver copy of the medal.

[64] Adunanza June 19, 1908. Suppl. Rend. Circ. Mat. Palermo, 3 (1908), p. 28.

[65] G.P. Curbera, *op. cit.,* p. 110.

[66] Adunanza June 19, 1908. Suppl. Rend. Circ. Mat. Palermo, 3 (1908), p. 28.

Fig. 5.5 Medaglia Guccia awarded to Francesco Severi at the 1908 Rome International Congress of Mathematicians. Courtesy of Melchiorre Di Carlo

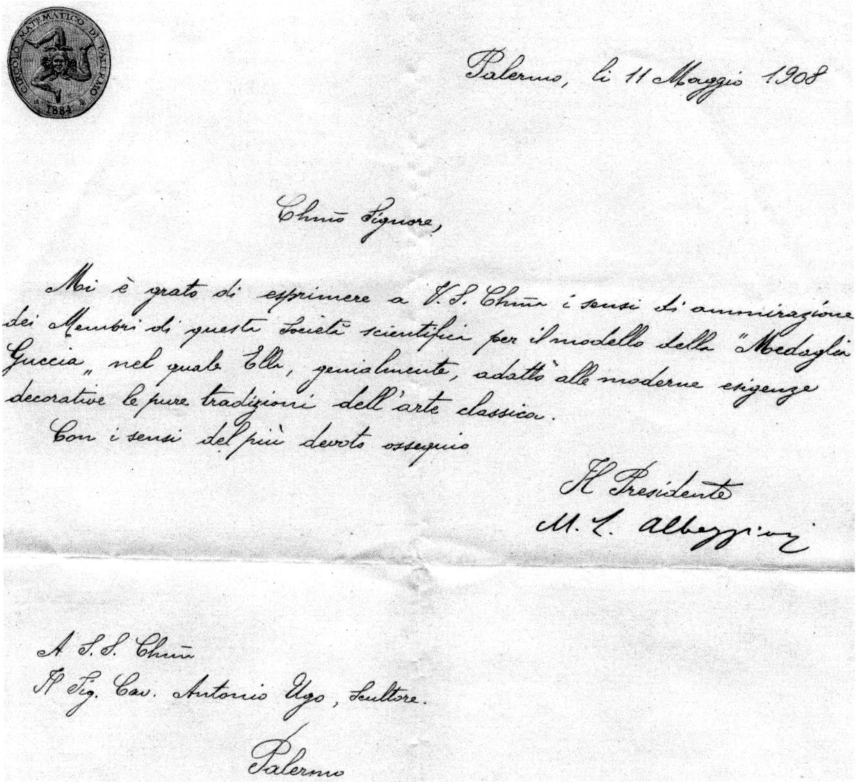

Palermo, li 11 Maggio 1908

Chmo Signore,

Mi è grato di esprimere a V. S. Chma i sensi di ammirazione dei Membri di questa Società scientifica per il modello della "Medaglia Guccia" nel quale Ella, genialmente, adattò alle moderne esigenze decorative le pure tradizioni dell'arte classica.

Con i sensi del più devoto ossequio

Il Presidente
M. L. Albeggiani

A S. S. Chma
Il Sig. Cav. Antonio Ugo, Scultore.
Palermo

Fig. 5.6 Letter from M. L. Albeggiani, president of the Circolo Matematico di Palermo, to the sculptor of the Medaglia Guccia, Antonio Ugo, transmitting the society's compliments for his fine work. Courtesy of Melchiorre Di Carlo

Donna Cosima"[67], Wagner's second wife and widow, to whom Guccia had planned to pay a visit). While in Munich he met with Alfred Pringsheim, who, apart from being a mathematician, was an excellent pianist and friend of Richard Wagner. From Munich Guccia went to Dresden, Leipzig and Berlin.

During that trip, he was in contact with Mittag-Leffler, and arranged a visit to Stockholm. The intense work associated with the Circolo had prevented Guccia and Mittag-Leffler meeting in June while Mittag-Leffler was staying in the Italian lakes. Guccia wrote Mittag-Leffler from Berlin to share the details of his arrival in Stockholm. Guccia also shared with Mittag-Leffler his worries about the troubles in Scandinavia, including the possibility of a cholera epidemic in Sweden, and the

[67]Letter from Guccia to Mittag-Leffler, September 11, 1905. Folder Mittag-Leffler, Archive of the Circolo Matematico di Palermo.

Fig. 5.7 Logo of the Circolo Matematico di Palermo. Courtesy of the Circolo Matematico di Palermo

political situation in Norway, where independence from Sweden was being crafted after a parliament vote and a plebiscite.

The newspaper Dagens Nyheter, probably through connections of Mittag-Leffler, reported on September 23, 1905, the visit of Guccia:

An eminent Italian mathematician visits Stockholm.
For the last days, our country and our mathematicians in Stockholm have had the eminent Italian mathematician G.B. Guccia, Marquis of Ganzaria, professor at the University of Palermo visiting Stockholm. The Marquis of Ganzaria is the founder and leader of Circolo Matematico di Palermo, highly regarded in the mathematical world, for whose magazine, edited by him, he has set up his own printing press in his palace. He also founded a prize for mathematics, which will be awarded at the congress of mathematics in Rome in 1908.[68]

Let us note that Guccia was not *Marchese di Ganzaria* but *Nobile dei Marchesi di Ganzaria;* see Chapter 7 for more details. In Stockholm, Guccia also met some of the collaborators of Mittag-Leffler, such as Edvard Phragmén, Ivar Bendixson, Ivar Fredholm and Helge von Koch. Guccia's intention for his visit to Mittag-Leffler

---

[68]*Dagens Nyheter*, September 23, 1905. Folder Mittag-Leffler, Archive of the Circolo Matematico di Palermo.

was clear, as he explained in one of his letters:

> My desire above all is to see you and talk calmly on some of my projects that aim to
> internationalize, spread and disseminate mathematical production worldwide, making use
> of the progress achieved by modern civilization in the field of international relations. I have
> great confidence in the advice you can give me on this matter.[69]

It is clear that Guccia had far-reaching projects in mind for the Circolo Matematico di Palermo that could profit from the International Congress being organized in Rome. After leaving Stockholm, Guccia went to Paris and met with Poincaré. There he received Poincaré's pioneering paper on relativity, *Sur la dynamique de l'electron.*[70] We have no record of what they discussed regarding the International Congress, but it is highly probable that Guccia wanted Poincaré's opinion on his project for the Circolo.

On his return trip from Paris, Guccia stopped in Rome and met with Guido Castelnuovo and Vito Volterra. These three mathematicians became the promoters and organizers of the Rome International Congress. Castelnuovo was an algebraic geometer from the University of Rome, very close to Segre. He became deeply involved in the preparations for the International Congress as secretary general. The topic of the meeting was the organization of the International Congress. Castelnuovo and Volterra had met before to prepare a common position for their meeting with Guccia. Indeed, Castelnuovo wrote to Volterra "I find most opportune the idea of Guccia to join us in Rome in November to agree on the organization of the Congress. Meanwhile, we can meet before the coming of Guccia and arrange on what we will have to decide in that session."[71] During the meeting, Guccia told them about his "project" for the Congress. It is not clear whether they discussed organizational matters related to the Congress or the details of Guccia's long-term project. However, it is unlikely that he discussed his long term project during this meeting. He may have discussed these matters with Mittag-Leffler and Poincaré, but he had to be more cautious about these plans with his Italian colleagues, given the growing institutional rivalry between the Accademia dei Lincei (and the University of Rome) and the Circolo Matematico di Palermo. The discussion of the arrangements for the Congress continued for several months through letters.[72]

Guccia proposed to distribute the work between Palermo and Rome in the following way: the Circolo would open a special office for the Congress in Palermo, which would take care of all the announcements of the Congress, the circular letters, the direct communication with individual participants, etc. The Circolo would

---

[69]Letter from Guccia to Mittag-Leffler, September 11, 1905. Folder Mittag-Leffler, Archive of the Circolo Matematico di Palermo.

[70]Letter from Poincaré to Guccia, September 24, 1905. Folder Poincaré, Archive of the Circolo Matematico di Palermo.

[71]Letter from Castelnuovo to Volterra, July 29, 1905. Vito Volterra Collection, Archive of the Accademia Nazionale dei Lincei.

[72]Letters from Guccia to Castelnuovo, May 28 and 31, 1905. Guido Castelnuovo Collection, Archive of the Accademia Nazionale dei Lincei.

publish the proceedings, and send a copy to each participant as well as one hundred reprints to every person delivering a lecture. The Circolo would also receive the payments from the participants of the Congress, and would retain a portion (eight out of twenty-five lire) of the payments before transferring them to the committee in Rome. It would also send to the inscribed participants the receipts, railway discount cards, brochures, etc. One month before the opening of the Congress, the office would be transferred from Palermo to Rome. This proposal was very attractive to Castelnuovo and Volterra because the Circolo would assume the heavy burden of the organizational and financial matters of the Congress, and they accepted it with no difficulties. They also saw another advantage: the committee in Rome would assume the scientific organization in full. This was a basic and fundamental agreement which guaranteed the holding of the Congress.

While the preparations for the International Congress were proceeding, a disagreement arose between Guccia and Volterra. Inspired by the prestigious British Association for the Advancement of Science, Volterra decided to launch a similar association in Italy. There was the precedent of the Società Italiana per il Progresso delle Scienze, whose last meeting had taken place in Palermo in 1875 (where young Giovanni Battista Guccia had met Luigi Cremona). The meetings of that association had been a success in part due to the nationalistic movement towards the unification of Italy, and once the unification was achieved, its drive ceased. Volterra considered those meetings "scientifically aristocratic", and favored a new trend of "scientifically democratic" meetings. The idea was counterbalancing the excessive specialization in science encouraging the exchange of ideas between the different branches of science, offering opportunities for young scientist and students, and setting the popularization of science as a key goal. For this task, a federation of scientific associations "with a broad base" could be the right tool. Volterra requested Guccia's help and the participation of the Circolo Matematico di Palermo. Guccia was aware of the idea's potential, but was skeptical about its possibilities for succeeding in Italy. In a letter to Volterra from June 30, 1906, he expressed very clearly, as usual with him, the two major difficulties for the project that he saw. First was the need for the right person with sufficient energy to carry out the project. The lack of such a person would render the project dead. He personally knew much about the tremendous effort involved in creating and sustaining a scientific organization. The second difficulty was more general:

> Politics: a microbe that in Italy enters throughout and... kills everything, in a particular way science! Politics! Here is the great enemy of Science in Italy! Here is why institutions that prosper and flourish in other countries cannot take root in Italy.[73]

Despite Guccia's views, Volterra went on with his plan and in September of 1907, the first congress of the new Società Italiana per il Progresso delle Scienze took place in Parma. Volterra had the scientific prestige and the political backing to support its creation, as he had been named senator of the Kingdom of Italy in 1905.

---

[73]Letter from Guccia to Volterra, June 30, 1906. Folder Volterra, Archive of the Circolo Matematico di Palermo.

Thus, the society was launched under the patronage of the King and with many socially important members on its high committee, such as senators, industrials, and bankers, in addition to scientists. Volterra presided over the congress and was elected the first president of the society. Although he sought the involvement of the Circolo, Guccia was quite skeptical, as we have seen, and kept his collaboration to the minimum.

A difficult moment in Guccia and Volterra's relationship occurred when Volterra requested that Guccia publish the lectures presented in the mathematics section of the meeting. Other scientific societies were also going to publish these lectures. The proposal included some payment for the cost of the corresponding volume of the journal.

Guccia answered three days later. He explained, somewhat theatrically: "I assure you, my dear friend, that your letter, which was delivered to me this morning, saddened me deeply; consider that I neglected other urgent business to reply to you right away and open you my mind", but his answer was clear and direct: "I regret not being able to accept your proposal." Guccia complained about the fact that such a congress could even be organized, given the involvement of the high authorities of the state, and the lack of funds for publishing its own proceedings. The main argument that he used to oppose to the publishing of the proceedings was the international character of the Circolo and of the Rendiconti, together with the negative effect on the journal: "What would become our Rendiconti, for which we have made so many sacrifices?" A passage of the letter reveals his deep and strong feelings towards the "Rome mathematical establishment":

> Our society does not receive any support from the Italian government! Our society, created and supported by private initiative, **is not recognized in Rome!** The mathematical International public (looking to the facts and not to the speeches) begins to realize this! I am the first one to suffer this, as an Italian, since I have the consciousness of having done all that I could do to save my country from this negative image. I could not do more, because in Rome I had not, as I have not (and as I do not wish to have), any of the official investiture that confers the degree of authoritative person! In Rome I am considered as a printer!!! a typographer! and nothing else! This could explain to you, once and for all, the fact that I rarely stop at Rome and only for causes of absolute necessity, that is, changing from one train and another! [74]

Documents in the Archive of the Circolo Matematico di Palermo provide evidence of Guccia's international activity related to the International Congress in Rome. A celebrated event took place in November of 1907 in Paris. As usual, Guccia was staying at the Hotel Continental. On Sunday November 3, he organized a meal at the restaurant of the hotel, for the main French mathematicians of the moment: Desiré André, Paul Appell, Èmile Borel, Pierre Boutroux, Gaston Darboux, Jules Drach, Georges Fouret, Jacques Hadamard, Georges Humbert, Camile Jordan, Charles-Ange Laisant, Louis Olivier, Paul Painlevé, and Henri Poincaré. Emile Picard was also invited, but excused himself.

---

[74]Letter from Guccia to Volterra, November 6, 1907. Vito Volterra Collection, Archive of the Accademia Nazionale dei Lincei.

The menu, devised by the exquisite taste of Guccia, was excellent:

Puff pastry American style
Little pot Henry IV
Trout salmon-coloured Galliera
Hen chicken sauté Mexican style
Foie gras parfait with Oporto
Young partridges encircled with quails
Rachel Salad
Asparagus with Muslin sauce
Covered pears Oriental style
Sweets
Fruits—Desserts

The wines served were Chateau Jilhol 1899 and Pomard 1er 1895, the champagne was Veuve Cliquot Carte Jaune, and for desert, there was coffee and liquors. All these details are kept in the Archive of the Circolo, along with the invoice for the meal for 554 francs—which Guccia paid from his pocket—and a drawing of how the guests were seated around the table—Guccia between Darboux and Jordan, and opposite of Poincaré. [75] Volterra sent a telegram to all at the gathering as a reminder of their invitation to attend the International Congress in Rome. The meal was reported in the Cronique section of the journal L'enseignement mathématique, in an article entitled *Un diner mathématique*, where it was explained that

> M. Guccia, in a speech full of charm, recalled his constant efforts, rewarded with success, whose goal was to create an international mathematical group, the Circolo Matematico di Palermo, and an organ, the Rendiconti, which has contributed and will contribute to the progress of science, in a disinterested way, without distinction of borders or nations. He then described the preparations for the 1908 Rome congress.[76]

We see that Guccia focused more on advertising the Circolo and the international role that it could play than on advertising the Rome Congress itself. That is why the choice of guests for dinner was so thoughtful. Guccia made sure that the main French scientific societies and journals in mathematics were represented at the meal: the Académie des Sciences, the Société Mathématique de France, the Journal des mathématiques pures et appliqueés, the Bulletin des sciences mathématiques, the Nouvelles Annales des mathématiques, the Intermédiaire des mathématiciens, the Revue du Mois, L'Enseignement mathématique, and the Revue Générale des Sciences.

The efforts of all people and institutions involved in organizing the Congress culminated on the evening of Sunday, April 5, 1908, when the Rector of the University of Rome, the mathematician Alberto Tonelli, hosted a welcome reception for members of the International Congress of Mathematicians in the Aula Magna of the old building of the university.

---

[75] Folder Guccia, Archive of the Circolo Matematico di Palermo.
[76] *Un dîner mathématique*, L'Enseignement Mathématique, 9 (1907), pp. 491–492.

The official opening of the Congress took place the next day in the magnificent Sala degli Orazi e Curiazi of the Campidoglio. The King of Italy presided over the opening ceremonies, and sat next to a statue of Pope Innocent X. Among the ceremonial speeches, the one by the Minister of Public Education stands out. He referred to Luigi Cremona—who had passed away five years before—, and recalled:

> The Italy of the urban republics and the Renaissance, with the names of Fibonacci, Tartaglia, del Ferro, Ferrari and so many others who prepared for the historical ripening of new spiritual and social demands for the blossoming of science. After Fibonacci, two streams appeared; one was seen in the studies of pure theory and the other grounded in the studies applied to commerce in which Italy was finding its renewed fortune. Thus, the double accounting of Luca Paciolo and his flourishing school of commercial arithmetic arose.[77]

Pietro Blaserna, the president of the Organizing Committee, and president of the Accademia dei Lincei explained to the audience how the Congress had been organized:

> The first International Congress of Mathematicians took place in Zurich, later at regular four-year intervals, in Paris and in Heidelberg. In this last one, in 1904, the Reale Accademia dei Lincei, that I have the honor to preside over, proposed, through its very distinguished member professor Volterra and with the agreement of the Circolo Matematico di Palermo, that the next International Congress be held in Rome; the proposal was very well received. The Accademia constituted itself as Organizing Committee, through its Mathematics and Mechanics sections; to which it added the eminent Rector of the University of Rome and several distinguished professors, as well as the Presidency of the Circolo Matematico di Palermo, that under the powerful initiative of professor Guccia has acquired national and also international value.[78]

The formal ceremony ended with the lecture by Vito Volterra entitled *Le matematiche in Italia nella seconda metà del secolo XIX*, where he paid tribute to the generation of mathematicians who were responsible for the burst of mathematics in Italy in the second half of the nineteenth century.[79]

The first plenary session took place the same day at 3 pm, and hosted a large audience. The sessions were held in the Palazzo Corsini, the seat of the Accademia dei Lincei, in the Trastevere, which was connected to the center of the city by a free shuttle of omnibuses. Pietro Blaserna, in his capacity of president of the Organizing Committee, invited the assembly to nominate the President of the Congress. Blaserna was elected to the post by acclamation. He immediately

---

[77] G. Castelnuovo (ed.), *Atti del IV Congresso internazionale dei matematici: Roma 6–11 Aprile 1908*, Reale Accademia dei Lincei, 1909, Rome, p. 29.

[78] *Ibid.*, p. 27.

[79] V. Volterra, *Le matematiche in Italia nella seconda metà del secolo XIX*. In: G. Castelnuovo (ed.), *Atti del IV Congresso internazionale dei matematici: Roma 6–11 Aprile 1908*, Reale Accademia dei Lincei, 1909, Rome, vol. I, pp. 55–65.

proposed the appointment of:

Vice-Presidents: Valentino Cerruti from Rome; Enrico D'Ovidio from Turin; Andrew R. Forsyth from Cambridge; Paul Gordan from Erlangen; Camile Jordan from Paris; Hendrik Antoon Lorentz from Leiden; Franz Mertens from Vienna. Gösta Mittag-Leffler from Stockholm; Simon Newcomb from Washington, A.V. Vasilief from St Petersburgh; and Hieronymus Georg Zeuthen from Copenhagen.
Secretary General: Guido Castelnuovo from Rome.
Vice-Secretaries: Gino Fano from Turin; V. Reina from Rome.
Deputy Secretaries: Ernest William Barnes from Cambridge; Émile Borel from Paris; Jacques Hadamard from Paris; T.F. Holgate from Evanston; Adolf Krazer, from Karlsruhe; Lars Edward Phragmén from Stockholm; and L. Schlesinger from Kolozswar.

The assembly accepted all these proposals.

Next was the award ceremony of the Medaglia Guccia. Corrado Segre read the report prepared by the commission in charge of the prize. Three memoirs had been submitted. While the commission praised the submissions, it did not consider any one of them deserving of the award. In accordance with the regulations of the prize, the commission examined other works on the topic of the award that were published within the submission period. Upon this second evaluation, the commission unanimously decided to award the prize to the work *Geometry of Algebraic Surfaces* by Francesco Severi from Padua (Fig. 5.8). Segre presented the Medaglia to Severi. From a scientific standpoint, the prize was a success: it was awarded by the best possible commission to a first-class treatise written by a renowned mathematician. In terms of publicity, everything also went well: the Medaglia Guccia was the first scientific prize awarded at an International Congress, the award ceremony had a central role in the program of the Congress, and there was great suspense among the audience members, as the name of the awardee had been carefully kept secret (at the special insistence of Guccia). Unfortunately, this was the first but also the only occasion when the Medaglia Guccia was awarded.

The regular program of the Congress started with the plenary lecture by Mittag-Leffler. Mornings were devoted to communications delivered in the scientific sections of the Congress, and afternoons to general plenary lectures. There were ten plenary lectures:

"Le partage de l'énergie entre la matière pondérable et l'éther," by Hendrik Antoon Lorentz from Leiden.
"La théorie du mouvement de la lune: son histoire et son état actuel," by Simon Newcomb from Washington.
"Sur les trajectories des corpuscules électrisés dans le champs d'un aimant élémentaire avec applications aux aurores boréales," by Carl Størmer from Christiania.
"Les origines, les méthodes et les problèmes de la géométrie infinitésimale," by Gaston Darboux from Paris.
"Die Encyklopädie der mathematischen Wissenschaften," by Walther von Dyck from Munich.
"On the Present Condition of Partial Differential Equations of the Second Order, as Regards Formal Integration," by Andrew R. Forsyth from Cambridge.

Fig. 5.8 Francesco Severi

"La mathématique dans ses rapports avec la physique," by Émile Picard from Paris.
"L'avenir des mathématiques," by Henri Poincaré from Paris.
"Sur la représentation arithmétique des fonctions analytiques générales d'une variable complexe," by Gösta Mittag-Leffler from Stockholm.
"La geometria non-archimedea," by Giuseppe Veronese from Padua.

This was the initial plan, but the lecture by Poincaré had to be read by Darboux, due to Poincaré's health problems, which kept him in his hotel in Rome all week (taken care of by his wife, and his friends Guccia and Volterra). Veronese's lecture was also canceled at the last moment due to a mild illness. Von Dyck was invited to replace Felix Klein, who was not able to attend the Congress. Hilbert did not attend also due to health problems, but his late cancellation made it impossible to organize a replacement.

The Congress was organized into scientific sections that were more or less similar to those of the previous Congress. The section on Applied Mathematics, however, was split into two. The scientific sections at the Rome Congress were:

Section I: Arithmetic, Algebra, and Analysis.
Section II: Geometry.
Section III (a): Mechanics, Mathematical Physics, and Geodesy.
Section III (b): Various Applications of Mathematics.
Section IV: Philosophical, Historical, and Didactical issues.

As had occurred in previous Congresses, there were lively debates from which visionary proposals came out. During the Rome Congress, proposals emerged to create an international commission to study the unification of vectorial notation (suggested by Jacques Hadamard); to create an archive of mathematical sciences; to celebrate jointly the International Congresses of mathematics and physics (also suggested by Jacques Hadamard); and to encourage the publication of Euler's works (which had been already proposed in the previous Congress), among others.

The Congress's social program met high expectations after the superb programs of the 1904 International Congress in Heidelberg. The Congress participants and their companions visited the Musei Capitolini in the Piazza del Campidoglio, by invitation of the city of Rome. Refreshments were served, and it was reported that "the halls were not cleared until midnight." They also visited the Ancient Roman sites in the Palatine Hill, where a buffet was served. An orchestral concert was organized in the Mausoleum of Augustus, which, according to Poincaré, had "revealed that Italy is, as France, torn by the pro and anti-Debussy quarrel."[80] The social climax of the Congress took place the day after its closing on Sunday, April 12. A special train took some 600 Congress members and guests to Tivoli. The American mathematician C.L.E. Moore described the excursion for the Bulletin of the American Mathematical Society:

> The first stop was at Hadrian's Villa. Carriages were ready to take those who did not care to walk from the station to the villa. On entering the ruins, we found refreshments, provided by the municipality of Tivoli, ready and waiting to be served. After spending about two hours here, we proceeded to Tivoli, where a banquet awaited us. The banquet closed with toasts in Italian, French, German and Latin. The afternoon was spent in visiting the cascades and the Villa d'Este. The returning trains arrived in Rome about 8 p.m. This was the unofficial but real close of the Congress.[81]

The day before that fantastic excursion, the Congress was officially closed with an invitation by the Cambridge Philosophical Society, supported by the London Mathematical Society, to hold the next Congress in Cambridge in 1912. The invitation was accepted with a unanimous vote and a lively applause. Let us note in passing that Mittag-Leffler, with the "capacity of Swedish mathematician and chief

---

[80]H. Poincaré, *Le congrès des mathématiciens à Rome*, letter to the publisher of Le Temps, April 21, 1908. See also: Suppl. Rend. Circ. Mat. Palermo, 3 (1908), pp. 19–22.

[81]C.L.E. Moore, *The Fourth International Congress of Mathematicians*, Bull. Amer. Math. Soc., 14 (1908), pp. 481–498.

editor of [the journal] Acta Mathematica", invited the International Congress of Mathematicians to meet in Stockholm in 1916. The decision for the 1916 Congress was postponed until the 1912 Congress in Cambridge, where it was later approved. Unfortunately, the 1916 Congress was canceled because of World War I and Mittag-Leffler's dream of hosting the International Congress in Stockholm was not fulfilled until 1962.

The Rome Congress differed from previous Congresses in two important ways. The first was the strong presence of applied mathematics and applications of mathematics. This was a traditional tendency of Italian mathematics, which began with the links to commercial trade in the Middle Ages and continued with the interest in problems arising in physics in the second half of the nineteenth century. The desire to extend the scope of the Congress beyond pure mathematics is seen in the choice of plenary lectures: three out of ten openly dealt with applications of mathematics, including those by Hendrik Antoon Lorentz (who explained why a heated solid emits light), by Simon Newcomb (who reported on recent progress in the theory of the Moon), and by Carl Størmer. The organization of the scientific content of the Congress into sections also reflects the applied trend of the Congress: the former section on Applied Mathematics was expanded into two subsections, (a): Mechanics, Mathematical Physics, and Geodesy, and (b): Various Applications of Mathematics, with the explicit intention of embracing Actuarial Mathematics. In fact, four national associations of actuaries sent delegates to the Congress. Another sign of the focus on applications was the diversity in the sources of funding. Funding for the Congress came from ministries other than that of Public Education, such as the Ministries of Agriculture, Industry, and Commerce, the Ministry of Finances, and the Ministry of Public Development. Four Italian insurance companies also participated in the funding.

The other innovation of the Rome Congress came from an idea conceived by David Eugene Smith from New York, which was firstly discussed in Section IV, on Philosophical, Historical, and Didactical issues, when considering the issue of the teaching of mathematics in secondary schools. The idea was turned into a project and later presented to the full Congress, where it was approved:

> The Congress, recognizing the importance of a thorough examination of the programs and of the methods of teaching mathematics at secondary schools of different nations, charges Professors Klein, Greenhill, and Fehr to constitute an International Commission to study these questions and to report to the next Congress.[82]

This is the origin of the Commission Internationale de L'Enseignement Mathématique. The commission was created with a four-year mandate, renewed in 1912, 1928, 1932, and 1936. The commission was initially led by Felix Klein, and Jacques Hadamard was its last president before World War II. In 1954, the Commission became part of the International Mathematical Union (IMU), which was newly reconstituted after World War II based on the principle of unrestricted internationalism. The Commission Internationale de L'Enseignement

---

[82]G. Castelnuovo, *op. cit.*, p. 51.

Mathématique was then renamed as the International Commission on Mathematical Instruction (ICMI). From its beginning, the Commission was linked to the journal L'Enseignement Mathématique, founded in 1899, which since then has been its official organ.

Overall, the Rome Congress was a success. Let us compare some figures from the first four International Congresses. Mathematician participation greatly increased at the Rome Congress: there were 208 mathematicians at the 1897 Congress in Zurich, 250 at the 1900 Congress in Paris, 336 at the 1904 Congress in Heidelberg, and 535 at the 1908 Congress in Rome (and 165 companions). The increase in the number of talks (apart from plenary lectures) is even more apparent: there were 30 in 1897 in Zurich; 33 in 1900 in Paris; 78 in 1904 in Heidelberg; and there were 127 in Rome in 1908. Consequently, the proceedings of the Congress grew from one volume with 320 pages for the Zurich 1897 Congress, to three volumes with a total of 1,146 pages for the Rome 1908 Congress.

The contribution of the Circolo Matematico di Palermo to the organization of the Congress was handled with Guccia's usual thoroughness: 2,500 copies of the first circular letter were sent in January-February 1907 to the members of the main mathematical societies and experts in actuarial science; 4,000 copies of the second circular letter were sent in January 1908 to the recipients of the first one, as well as to universities and academies all over the world; and copies were even sent to all Italian secondary schools!

A problem, however, arose with the printing of the proceedings of the Congress. The initial plan was to use the printing house Tipografia Matematica di Palermo— where the Rendiconti was printed—for its credited guarantee of thorough work under the supervision of Giovanni Battista Guccia. At the closing ceremony, the president of the Congress explained that a strike of Sicilian typographers had stopped all printing in Palermo (Fig. 5.9). As a result, Guccia had to ask that the Circolo be released from its commitment.[83] The change in the printing of the proceedings was unexpected and disrupted the previous arrangements for the organization of the Congress. In light of this unfortunate circumstance, the Presidency accepted the Circolo's request with much regret. The way in which it was announced to the congress was at least unusual. In any case, the Organizing Committee took over the task from the Circolo, and eventually the printing house Tipografia dei Lincei printed the proceedings.[84]

That was the official explanation. In fact, on March 15, 1908, a strike of typographers had started in Palermo; a few days later the Rendiconti announced a delay in the publication of the society's annual directory. However, the truth behind the printing affair had to do with the disagreements between Guccia and the "Rome establishment" that went back several years. These tensions started to arise from the very beginning of the preparations for the Congress. Guccia did not agree with many of the decisions of the organizing committee, as he found that

---

[83] *Ibid.*, p. 5 and p. 34.

[84] *Annunzio,* Suppl. Rend. Circ. Mat. Palermo, 3 (1908), p. 22.

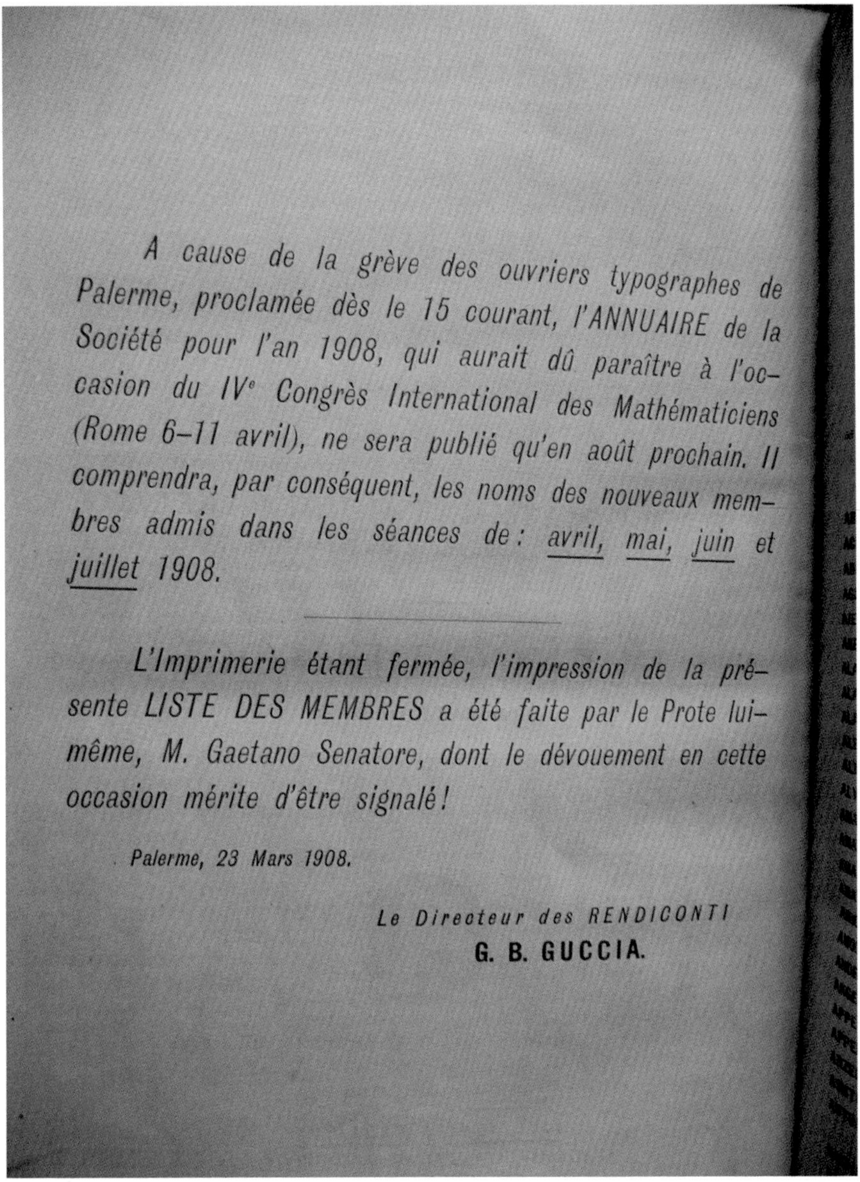

A cause de la grève des ouvriers typographes de Palerme, proclamée dès le 15 courant, l'ANNUAIRE de la Société pour l'an 1908, qui aurait dû paraître à l'occasion du IVᵉ Congrès International des Mathématiciens (Rome 6–11 avril), ne sera publié qu'en août prochain. Il comprendra, par conséquent, les noms des nouveaux membres admis dans les séances de : avril, mai, juin et juillet 1908.

L'Imprimerie étant fermée, l'impression de la présente LISTE DES MEMBRES a été faite par le Prote lui-même, M. Gaetano Senatore, dont le dévouement en cette occasion mérite d'être signalé !

Palerme, 23 Mars 1908.

Le Directeur des RENDICONTI
G. B. GUCCIA.

Fig. 5.9 Communication in the Rendiconti of the strike of Sicilian typographers in 1908. Courtesy of the Circolo Matematico di Palermo

they were against the character of the international Congresses ("In various other things they would not listen to me!").[85] For example, Guccia was frustrated that the actuaries were included in the Congress program, as well as by the removal of Spanish and Latin from the official languages of the Congress (although this had no effect on the running of the Congress). A deeper conflict concerned the publication of the proceedings of the Congress, as Guido Castelnuovo had been "continuously interfering" despite the fact that the publication task had been entrusted to the Circolo. By 1907, Guccia had commented that the interference was endangering the publication of the proceedings in a letter to Luigi Bianchi.[86]

The most serious disagreement was over the presidency of the International Congress. Guccia proposed that the election of the president and secretaries take place during the opening session, and that the candidates be selected by vote or by acclamation. From the very beginning, the people from Rome did not like the idea: in fact, Castelnuovo wrote to Volterra, "This issue needs to be discussed with calm".[87] The type of procedure proposed by Guccia had precedents. During the first International Congress, in Zurich in 1897, the Congress started by approving some Regulations, whose third article determined that the presidency of the Congress was to be elected during the first plenary session. This process was implemented immediately, and Carl Friedrich Geiser—who presided over the organizing committee—was elected president "by acclamation." In the next Congress, in Paris in 1900, it was the mathematician Jules Tannery, acting in his capacity of representative of the Ministry of Public Instruction and Fine Arts, who suggested to the Congress to elect Henri Poincaré for president, who was then named for the post "by acclamation." Poincaré in turn suggested naming Hermite as honorary president of the Congress, and "this proposal was received with prolonged applause."[88] The situation for the Heidelberg Congress of 1904 was different: at a meeting of the German Mathematical Society held in 1901, three years before the meeting of the Congress, Heinrich Weber was chosen to be the president of the Congress. Consequently, when the International Congress met in Heidelberg, no elections took place.

Guccia insisted on postponing the election of the president until the inaugural session of the Congress. This option, he argued, was advantageous because it allowed efforts to be concentrated on the organizing committee. Appointing the members of this committee would therefore cause fewer difficulties, he said, as no one would be interested in working for a committee that required a lot of work and yielded little public recognition. This was how Guccia presented his views to

---

[85]Letter from Guccia to Bianchi, December 30, 1907. Folder Bianchi, Archive of the Circolo Matematico di Palermo.

[86]*Ibid.*

[87]Letter from Castelnuovo to Volterra, February 2, 1906. Vito Volterra Collection, Archive of the Accademia Nazionale dei Lincei.

[88]E. Duporcq, *op. cit.*, p. 14.

Mittag-Leffler in 1906, and requested his friend's aid in convincing Volterra that his suggestion was the most convenient.

Despite Guccia's viewpoint, the president of the Congress was determined in advance. A natural candidate would have been Ulisse Dini, due to the national and international respect around his figure, but Dini refused the offer. The person then chosen was the president of the Accademia dei Lincei, Pietro Blaserna. He was then made president of the Organizing Committee so that his election as president of the Congress would follow naturally once the issue was presented at the inaugural session. Again, Guccia disagreed over both the procedure and the choice for president. Firstly, because Blaserna was not a mathematician, he was a physicist. Secondly, because of Blaserna's involvement in the legal case against Nunzio Nasi. A politician from Sicily, Nasi had been Minister of Public Education. He was once accused of misusing large amounts of public funds and official privileges. A mixture of political and judicial issues turned into a bizarre situation (Nasi escaping to France, street demonstrations in Sicily, Nasi's imprisonment and later release), which turned into a scandal that was thoroughly covered by the international press. Blaserna had briefly presided over the Senate court involved in the affair. Guccia argued that the choice of Blaserna as president of the Congress would damage the image of Italy in the eyes of the foreign mathematicians attending the Congress.[89]

The real conflict behind these quarrels was something else. From the very beginning Guccia saw holding the International Congress in Rome as a unique opportunity for transforming the Circolo Matematico di Palermo into the first, the leading (and perhaps eventually the only) mathematical society in the world. This was what he called in the letter to Mittag-Leffler "his project." It included "my dream (bold and ambitious), to have in Italy the most important journal of mathematics in the world."[90] Since then, he had been working hard to achieve this goal. The creation of the Medaglia Guccia was a step in that direction, as was the commitment to take responsibility for the full administrative organization of the International Congress. The same occurred with the publishing of Congress materials and the proceedings: it was another opportunity, this time for the Rendiconti, to advertise itself. By sending the circular letters and other communications of the Congress, Guccia was able to contact all participants directly, and benefit himself and the Circolo from those contacts. At that time, the Circolo already had more members than the London Mathematical Society and the Société Mathématique de France. In the next chapter, we will review in detail the growth of the society and the astonishing effect of the Rome International Congress on the Circolo's growth.

Guccia's interviews with influential mathematicians around the world had a double intention: gaging their reaction to his project and, when such reaction was positive, explaining the project in full. We don't know the responses of Mittag-Leffler, Poincaré or the group of French mathematicians present at the dinner in

---

[89]Letter from Guccia to Bianchi, December 30, 1907. Folder Bianchi, Archive of the Circolo Matematico di Palermo.

[90]*Ibid.*

Paris to Guccia's project, but we do know that Guccia did not lose his energy after those encounters.

Since Guccia and the Circolo's involvement and efforts towards the success of the International Congress had been so intense, Guccia expected that the rest of the organizers would recognize, and even share, the interest of launching from Rome the far-reaching project of an international association of mathematicians based on the Circolo Matematico di Palermo. Guccia believed that a way to express that recognition could be to give him the presidency of the International Congress, or at least the presidency of the Organizing Committee. From this post, he would be able to present "his project" to the assembly of the International Congress and maneuver for its approval. However, the "Rome mathematical establishment", consisting of the mathematicians from the Accademia dei Lincei and the University of Rome, did not agree with giving such a relevant role to Guccia and to the Circolo. He was only offered the Secretary of the Organizing Committee with Guido Castelnuovo. He declined with dissatisfaction, "I never asked anybody for anything for me and I do not accept any kind of decorations or honors."[91] Although it seemed purely institutional, naming Blaserna for the presidency of the Organizing Committee in fact allowed Vito Volterra, due to his close friendship with Blaserna, to have the actual control of the organization of the Congress. The last hope for Guccia was to maintain himself in a low-key position as "publisher of the International Congress," and wait until the opening of the International Congress, where he could try to be named for the presidency. From that position, he would then be able to propose to the assembly to give the Circolo the title of "International Association of Mathematicians."

The outcome was completely different from his expectations. Even though Guccia's activity and role was acknowledged all throughout the Congress—in Blaserna's opening speech, when the Medaglia was awarded, and in the closing ceremony when a letter addressed to Guccia from Georg Cantor was read, apologizing for not travelling to Rome—Guccia was extremely unhappy with the outcome of the Congress because of the failure of his plans for the Circolo Matematico di Palermo.

Two months later, in a letter to Mittag-Leffler that was labeled as "Private and Confidential", Guccia explained his sentiments on the events of the International Congress in Rome:

> Concerning the Congress of Rome, as you saw there our Society was removed from the scene as soon as the President of the Accademia dei Lincei announced at the opening session that he would take the presidency of the Congress. This circumstance, unexpected and bizarre, was in disagreement with the previous announcements and somehow surprised the audience. It was therefore necessary, without making noise, to save the dignity of our great international Society (600 members), which has, from all points of view, absolute independence, and to which belong the greatest mathematicians of the world. I do not think that I did wrong, as delegate of Circolo at the Congress, in suspending with a likely and convincing pretext (and, therefore, without any gossiping) our agreement with the Organizing Committee. [. . .]

---

[91] *Ibid.*

I received a month and a half ago from Rome a long letter with excuses. But since this letter cannot change the facts or the impression received in Rome by the mathematical community, I replied by telegram, "I will not respond!" That's all.

The only thing that really troubles me in this affair is that among the members of the Organizing Committee (the only one responsible to the public) there were three members of the Steering Council of the Circolo, which are my good and excellent friends, but who in those circumstances have forgotten completely about our Society. Between the Accademia dei Lincei, illustrious and famous, which pays attention to everything, even the diseases of the "gorgonzola" cheese (Rendiconti dei Lincei, April 26, 1908, page 568), and our modest society, that deals only with mathematics, on the occasion of the congress they chose without doubt: they turned to the Academy, to augment its glory. They may have been right, in their view. But I think that in their place, I would have previously presented my resignations from the Steering Council of the Circolo.[92]

The members of the Steering Council that Guccia refers to were Vittorio Cerruti, Alberto Tonelli and Vito Volterra. A succinct report on the Rome Congress was included in the minutes of the meeting of the Circolo from June 19, 1908.[93] The Supplemento of the Rendiconti included an "announcement" explaining the situation of the publication of the proceedings of the Rome Congress.[94]

What were the possibilities of "Guccia's project" succeeding and being approved by the International Congress if the Rome establishment had not blocked it? It is difficult to speculate on this issue. The unique contribution (though humble) supporting Guccia's project during the Rome Congress was made by Alberto Conti, from the University of Bologna, member of the Circolo and friend of Guccia. Conti presented at the end of the Congress the following proposal, which was approved by the assembly:

The Congress agrees that in the agenda of the next Congress is included the establishment of an International Association of Mathematicians.[95]

---

[92]Letter from Guccia to Mittag-Leffler, July 12, 1908. Folder Mittag-Leffler, Archive of the Circolo Matematico di Palermo.

[93]Adunanza June 19, 1908. Suppl. Rend. Circ. Mat. Palermo, 3 (1908), pp. 27–28.

[94]Suppl. Rend. Circ. Mat. Palermo, 3 (1908), p. 22.

[95]G. Castelnuovo, *op. cit.,* p. 33.

Chapter 6
Maximum Splendor

We review the period from 1908 to 1914 when the Circolo and the Rendiconti reached their maximum splendor. The society grew at a strong, steady rate, and the journal strengthened its international scientific prestige, playing an important role in the publication of first-line results. To illustrate the quality reached by the Rendiconti we comment on several mathematical papers published in the journal. The international nature and the internal organization of the society played a crucial role in such success, and we analyze their causes and effects in detail. Finally, the celebration in 1914 of the thirtieth anniversary of the foundation of the society was the highlight of the Circolo Matematico di Palermo, of the Rendiconti and of Guccia's projects.

6.1 The Rendiconti

Giovanni Battista Guccia was deeply disappointed with the outcome of the International Congress held in Rome in 1908. His longstanding plan for transforming the Circolo Matematico di Palermo into the International Association of Mathematicians had failed dramatically. This was attributed to the carefully planned attitude of the Italian mathematical establishment, led by the Accademia dei Lincei and the University of Rome, which ultimately meant that the person to be blamed was Vito Volterra. Despite all these well-established facts, there are still questions to be asked. Had Guccia been so naive as to set such a large-scale plan, requiring enormous amounts of effort, work and money, solely based on unfounded speculations? Were there any reasonable grounds for expecting the success of that project? A careful analysis of the development of the mathematical society that Guccia created and, particularly, of the mathematical journal that Guccia edited reveal that the true grounds for his expectations were based on the successful enterprise that Guccia was able to create around the Rendiconti.

© Springer International Publishing AG, part of Springer Nature 2018
B. Bongiorno, G. P. Curbera, *Giovanni Battista Guccia*,
https://doi.org/10.1007/978-3-319-78667-4_6

We have seen that the beginnings in 1884 of the Circolo Matematico di Palermo were humble: just twenty-seven members, all of them from Palermo. During its first year, eight new members joined the society. The number of new members entering the Circolo increased to fourteen in the second year, forty-one in the third year, and forty-six in the fourth year. However, there were also members who resigned from the society, passed away, or did not pay the fees and were expelled from the society. By March 1888, four years after its foundation, the Circolo had 102 members, an average of twenty-five new members per year. The place of residence of the members of Circolo will be a relevant criterion for understanding how the society evolved. The distribution of members according to their residency was as follows: forty-six resided in Palermo, fifty in the rest of Italy, and six were from outside Italy (precisely, two were from Paris, one from Liège, one from London, and two were from Prague and its vicinity).[1]

March 1888 was the moment when Guccia proposed to change the statute of the society in order to support its internationalization. There was a rather favorable indication that sustained that proposal.

We have already explained that the society had published each year a booklet with the minutes of the scientific sessions (there was still no official journal at that time). The first booklet corresponded to the period March 1884 to March 1885 and had forty-four pages. After the proceedings of the sessions of the society, the booklet ended with a section labeled "Library of the Circolo Matematico di Palermo: Publications sent by the authors." It consisted of a detailed list of all the articles (and some books) that the Circolo received during that year, specifying the author, the title and other bibliographical details. The authors were mathematicians that had responded positively to the invitation to donate copies of their publications in order to build up the mathematical library of the Circolo. The list was sixteen pages long; in total, there were some 370 publications from thirty-five authors. The contributors were mathematicians from Palermo, from the rest of Italy (Milano, Naples, Pavia, Pisa, Rome), and also from outside the country, such as De Boer from Leiden; Gordan and Noether from Erlangen; Hirst from London; Halphen, De Jonquières, Jordan and Mannheim from Paris; Klein from Leipzig; Mittag-Leffler from Stockholm; Schoute from Groningen; and Stephanos from Athens.[2]

The second booklet of the publication of the Circolo corresponded to the period April 1885 to April 1886 and had forty-eight pages. This time, the section with the contributions was labeled "Mathematical Library: Non-periodical publications." The list now occupied twenty-five pages, and contained more than 600 publications. The number of contributing mathematicians went up to ninety. Among them were the leading Italian mathematicians. However, more than half of the donors were from outside Italy, from Paris: Chasles, Hermite, Poincaré, and Fouret, who donated

---

[1] Rend. Circ. Mat. Palermo, 19 (1905), Note statistiche, p. xxx.

[2] Rend. Circ. Mat. Palermo, 1 (1887), Biblioteca del CMP, Pubblicazioni inviati dagli autori, pp. 29–44.

twenty-five articles; from Berlin: Weierstrass and Kronecker, who donated fifty articles; from Odessa: Starkoff, who donated fourteen articles; from Liège: Eugène Catalan, who contributed with ninety-six articles; and other donors from many other cities.[3]

This second booklet included a new section labeled "Mathematical Library: Periodicals." It listed the journals that had been received in exchange with the first booklet published by the Circolo. The journals in the list were:

American Journal of Mathematics, vol. 7 (4), vol. 8 (1, 2)
Annali della R. Scuola Normale superiore di Pisa, vol. 1–3
Bibliotheca Mathematica (Stockholm), 1884, 1885
Communication of the Mathematical Society of Kharkov, 1897–1884
Proceedings of the Moscow Mathematical Society, vol. 12
Memoirs of the Mathematical Section of the Odessa Natural Society, vol. 1–4
Bulletin de la Société Mathématique de France, vol. 13
Proceedings of the Canadian Institute, vol. 3
Jornal de Sciencias Mathematicas e Astronomicas (Coimbra), vol. 2–6
Casopis (Prague), vol. 14, 15

The list occupied five pages because it included very useful information for the readers: the detailed table of contents of each of the volumes received from those journals, with authors and titles of all the articles.[4]

The third and fourth booklets corresponded to the periods May 1886 to June 1886, and November 1886 to July 1887, respectively. They already constituted proper journal volumes, with 314 pages in total. The section on Periodicals contained seventeen journals (with Annals of Mathematics from Princeton, and Nieuw Archief voor Wiskunde from Amsterdam, among them) and the tables of contents occupied seven pages. The section on non-periodicals was not present this time (they appeared in the second volume of the Rendiconti published the following year). There were two other sections of practical character: an author index and a list of the members of the society with short curriculum vitae of each one of them.

The large number of scientific publications donated and the rapid exchanges with other journals can be interpreted as a sign of confidence and support of the international mathematical research community for Guccia's efforts to build the Circolo Matematico di Palermo. The support that the young scientific society was receiving was based on that confidence. At the initiative of Guccia, the Circolo changed its statute in 1888 and officially launched the Rendiconti. In what follows we will analyze the development of the journal and its role in Guccia's far-reaching project for the Circolo.

---

[3]Rend. Circ. Mat. Palermo, 1 (1887), Mathematical Library: Non-periodical publications, pp. 94–118.

[4]Rend. Circ. Mat. Palermo, 1 (1887), Pubblicazioni non periodiche, pp. 89–93.

However, the positive attitude towards the Circolo alluded to above did not reflect in the membership of the society: in the sixteen-year period from 1888 to 1904 the number of members of the society went from 102 to 195. The society almost doubled its size, but the annual average increase was low: six new members per year. This was the net variation of the number of members. If we look at the gross data, in the twenty years since its creation in 1884 to 1904 the number of persons who had joined the Circolo was 326; from that number, 131 had to be deducted, due to deaths, withdrawals, and expulsions due to lack of payment. We can see that the "failure rate" was high. The result was a low average increase in the membership of the society.[5]

This was the situation regarding membership in August 1904 when a double announcement was made at the International Congress held in Heidelberg: the following Congress was to be held in Rome in 1908, and a new prize, the Medaglia Guccia, would be presented at the Rome Congress. As a result of these announcements the Circolo's membership immediately increased. By March 1905 there was a net increase of sixty members. The total number of members was then 255, which were distributed according to their residency as follows:

Members residing in Palermo: 36.
Members residing in the rest of Italy: 138.
Members residing abroad: 81.[6]

The next year, in February 1906, the number of members had gone up to 362, and one year later, in February 1907, the number was 448. In two years, there had been a gross increase of 199 members, with only six dropouts, which resulted in a net increase of 193 members.

This rapid increase caused a deep change in the distribution of the members of the society according to their place of residency:

Members residing in Palermo: 44.
Members residing in the rest of Italy: 172.
Members residing abroad: 232.[7]

It is quite remarkable that within two years, the number of foreign members of the Circolo went from 81 to 232; it had multiplied almost threefold. Foreign members already constituted the majority of the Circolo. By the end of January 1908, membership continued to increase in a similar way and the society had then 550 members that were distributed as follows:

Members residing in Palermo: 57.
Members residing in the rest of Italy: 192.
Members residing abroad: 301.

---

[5]Rend. Circ. Mat. Palermo, 19 (1905), Note statistiche, p. xxx.
[6]*Ibid.*
[7]Rend. Circ. Mat. Palermo, 22 (1906), Note statistiche, pp. xi–xii.

It is enlightening to look at the distribution by country of the foreign members of the Circolo. They came from twenty-six different countries, and twenty of those countries had each fewer than ten members. The other six countries were:

The United States of America: 85 members.
Germany: 61 members.
France: 41 members.
Austria-Hungary: 24 members.
Great Britain and Ireland: 13 members.
Sweden: 10 members.[8]

The strong presence of German and French mathematicians was expected, as they were the countries with the largest mathematical communities. The novelty was the high number of mathematicians from "the New World": the United States of America.

The positive international reception was the fact upon which Guccia had founded his expectations for the future of the Circolo Matematico di Palermo, as well as his goal to transform his society into the "International Association of Mathematicians" at the Rome Congress.

The astonishing increase in the membership of the Circolo had a clear explanation: Guccia's management of the society's journal, the Rendiconti del Circolo Matematico di Palermo. Let us review how the Rendiconti was run.

The first 1884 statute did not allow the possibility of publishing a research journal. This was in line with Guccia's critical views on the misuse of publications for career purposes. The positive experience with the booklets giving accounts of the scientific sessions of the Circolo, however, changed his views on this matter, and in the 1888 statute the Rendiconti was created.

The Editorial Board of the Rendiconti was the Steering Council of the society, which was in charge of the scientific direction of the society. This allowed the separation of the administrative management of the society—assigned to the newly created Presidency—from the scientific matters, in particular, from the acceptance and rejection of articles. The Steering Council was designed to pursue articles of the highest quality for publication, and as a result, the internationalization of the journal. These were two aspects that reinforced each other: high quality articles are read by mathematicians worldwide, and authors are interested in submitting their best papers to widely read journals. To this end, the Council was large and diverse. It consisted of twenty members, and fifteen of these members had to reside outside Palermo. The election of the Steering Council was to take place every three years on the third Sunday of January. With the aim of involving the full membership of the society in the elections, voting by mail was allowed, regardless of the member's place of residency.

---

[8]Rend. Circ. Mat. Palermo, 25 (1908), Note statistiche, pp. xiii–xiv.

The first Steering Council, elected for the period 1888–90, had the following members residing outside of Palermo:

Battaglini from Naples	Del Pezzo from Naples
Beltrami from Rome	De Paolis from Pisa
Bertini from Pavia	D'Ovidio from Turin
Betti from Pisa	Jung from Milan
Brioschi from Milan	Pincherle from Bologna
Casorati from Pavia	Segre from Turin
Cerruti from Rome	Volterra from Pisa
Cremona from Rome	

The election took place just after the approval of the new statute, so the participation in the voting was limited: only twenty-seven out the 102 members of the Circolo voted.[9] For the 1891–1893 term, the only change in the Steering Council was the entrance of Poincaré who (as we already mentioned in Chapter 5) replaced Casorati, who passed away in 1890. For this election, the participation increased substantially: 114 of the 151 members voted.[10]

During the next three-year period, 1894–96, the Steering Council had a high number of changes, there were six. Three of them were natural and corresponded to members passing away (Battaglini, Betti and De Paolis). The other three changes (Bertini, D'Ovidio and Segre) were a consequence of the conflict over the promotion of Guccia to "professore ordinario", which we reviewed in Chapter 5. These vacancies resulted in the election to the council of six new members, in particular, of the second foreign member, Gösta Mittag-Leffler. Thus, the non-resident members of the council were

Beltrami from Rome	Loria from Genoa
Bianchi from Pisa	Maisano from Messina
Brioschi from Milan	Mittag-Leffler from Stockholm
Capelli from Naples	Peano from Turin
Cerruti from Rome	Pincherle from Bologna
Cremona from Rome	Poincaré from Paris
Del Pezzo from Naples	Volterra from Pisa
Jung from Milan	

This time, 140 out of 157 members voted. Participation was thus almost ninety percent, a dream for any scientific society.[11] On this occasion, the high participation can be interpreted as a sign of support for Guccia at the time when Segre and other mathematicians were attempting to take over control of the society.

---

[9]Rend. Circ. Mat. Palermo, 2 (1888), pp. 88–89.

[10]Rend. Circ. Mat. Palermo, 5 (1891), pp. 284–285.

[11]Rend. Circ. Mat. Palermo, 8 (1894), pp. 167–168.

There were no changes to the Steering Council elected for the 1897–1899 term except for an empty seat left after the death of Brioschi in 1897, and 99 out of the 170 members voted.[12] For the 1900–1902 term, there were two further changes among the Italian non-resident members: Del Re (Naples), Pascal (Pavia) and Tonelli (Rome) replaced Brioschi, Jung and Maisano (who became a resident member after moving to Palermo). Again, the council had one member fewer than the required fifteen non-resident members, due to the passing away of Beltrami just one month after the election took place. Voters this time were 82 of out the 183 members.[13] For the 1903–1905 term, the council remained the same and Dini took the empty seat left by Beltrami. On this occasion, 94 of the 189 members voted.[14] The non-resident members of the council were:

Bianchi from Pisa	Mittag-Leffler from Stockholm
Capelli from Naples	Pascal from Milano
Cerruti from Rome	Peano from Turin
Cremona from Rome	Pincherle from Bologna
Del Pezzo from Naples	Poincaré from Paris
Del Re from Naples	Tonelli from Rome
Dini from Pisa	Volterra from Pisa
Loria from Genoa	

We end this detailed listing with the Steering Council for the 1906–1908 term, which had only one change: after Cremona's death in 1903, D'Ovidio replaced his seat in the Council. This occurred twelve years after D'Ovidio's "expulsion" from the Council after the conflict over the promotion of Guccia. The voting on this occasion was noticeably high given the size of the society: of the 346 society members 165 sent their votes.[15]

We see that these councils were of a very high scientific quality: they had most of the best Italians mathematicians of the time, as well as, since 1894, two first class and influential foreign mathematicians: Mittag-Leffler and Poincaré.

The high participation in the elections for the Steering Council, together with the small dispersion of the voting, are clear signs of Guccia's control over the society. This control was based on the following procedure. When elections were coming up, Guccia would prepare a proposal for the composition of the Steering Council, carefully gauging all the different aspects of the candidates (scientific area and school, nationality, prestige and influence, etc.), and personally suggest his choices to the most relevant members of the society. Guccia played a crucial role in managing the journal; for this, he was always elected as a member of the Steering Council in the group of five members from Palermo and, after the election

---

[12]Rend. Circ. Mat. Palermo, 11 (1897), pp. 184–185.

[13]Rend. Circ. Mat. Palermo, 14 (1900), pp. 306–307.

[14]Rend. Circ. Mat. Palermo, 17 (1903), pp. 169–170.

[15]Rend. Circ. Mat. Palermo, 21 (1906), pp. 218–220.

had taken place, he was appointed by the council as a "delegate of the Steering Council for managing the publication of the Rendiconti", following the terms of the statute (which he insisted on recalling). Guccia's control over the society was also based on the general appreciation for his management.

Under the guidance of Guccia, and with his devoted and intense work, the Rendiconti flourished, and in thirty years became one of the most prominent mathematical research journals in the world.

Let us recall that the starting point of the the Rendiconti was the modest booklet published in 1885, containing twenty-eight pages devoted to the mathematical and internal proceedings of the Circolo, and sixteen pages listing the publications received, all wrapped in a humble paper cover. The second volume covered the year 1888 and had 320 pages. It was divided into two issues. The first issue contained thirty research articles, the proceedings of the meetings of the society, and a detailed list of members of the Circolo. Among the authors of the articles were Betti, Jordan, Peano, Poincaré, Segre and Volterra. The second issue listed additions to the so-called "Mathematical library", continuing what had started in the booklets: there were twenty-nine pages devoted to the "Non-periodical publications", and forty-seven pages devoted to the "Periodicals", with the table of contents of the journals which were received in exchange. The collection of journals received in exchange was now larger with the addition of Acta Mathematica (all published volumes since the first one in 1882), the Bulletin de Sciences Mathématiques, the Bulletin de la Société Mathématique de France, the Comptes Rendus de l'Académie des Sciences de Paris, the Journal für die reine und angewandte Mathematik, the Proceedings of the London Mathematical Society, the Philosophical Transactions Royal Society, the Proceedings of the Royal Society, the Zeitschrift für Mathematik und Physik, and many others.

The volumes of the Rendiconti had six installments that were sent to society members and subscribers with total regularity every two months, although occasionally two installments were published together. Each installment consisted of the first and second issues of the volume. With this structure, the journal continued its publication for twenty years until 1904. From 1887 to 1905, the Rendiconti had published 530 papers and communication from 170 authors.[16] Later, we will compare these figures with those of 1914, the year when the Rendiconti achieved the peak in the number of papers published.

As we have said, 1904 was a crucial year for the Circolo, as its membership grew substantially after the announcement made during the Heidelberg Congress regarding the 1908 Rome Congress and the Medaglia Guccia. The growth in membership had an impact on the Rendiconti: there was a great increase in the number and in the quality of manuscripts submitted for publication. This growth caused a huge scientific and management challenge for the Circolo. Guccia made several important technical decisions to face it. Starting the following year, 1905, the Rendiconti would be published annually in two volumes instead of one. This

---

[16]Suppl. Rend. Circ. Mat. Palermo, 1 (1906), p. 16.

implied that the number of pages and articles to be published each year would double. In 1904, volume XVIII of the Rendiconti had twenty-two articles over 381 pages, while the 1906 volumes XXI and XXI together added up to forty articles and 749 pages. From then on, the average number of pages published annually was 800. A more technical decision (which he announced in the Rendiconti)[17] was enlarging the size of the printed part of the page from 190×113 mm to 205×135 mm, without changing the size of the paper page. This amounted to a 29 percent increase in the printed surface of each page, which resulted in sensible savings in production costs.

Guccia also decided to limit the volumes of the Rendiconti exclusively to research articles, and created a new publication, the "Supplemento ai Rendiconti del Circolo Matematico di Palermo," for all other material. This publication, in a smaller format, had one volume per year that was sent to the subscribers in bimonthly installments. It contained the proceedings of the sessions of the society along with the list and table of contents of publications received by the society in exchange (all this information had been removed from the Rendiconti since 1894, volume VIII, due to its length). Also included were the indexes of publications and authors in the Rendiconti, statistical information on the society, and miscellaneous mathematical information (prizes, congresses, announcements, celebrations). The first volume of the Supplemento appeared in the second semester of 1906.

The Circolo had two further publications for its subscribers. One was the Annual Directory, "Annuario del Circolo Matematico di Palermo", which was at first published biannually and then annually since 1903 (with some exceptions). Most of the issues included biographical information on the members of the society. The first one, from 1884, had twelve pages, while the 1914 one had 220 pages. The other publication was an accumulative index of articles, ordered by author, that started to be published separately in 1908.

Regarding the scientific managing, Guccia was personally in charge of all the editorial work of the Rendiconti. He would receive the manuscripts submitted and read them carefully. After this first screening, a number of things could occur. The manuscript might not be accepted, and in this case, Guccia would personally communicate this to the author. If Guccia considered that the article was of interest for the journal but needed further work, he would communicate this to the author in a detailed letter, or the article might be sent to the appropriate member of the Editorial Board of the journal (which, recall, was the Steering Council of the society) for revision.

The further editorial process of articles in the Rendiconti was also under the complete control of Guccia. It was a thorough process guided by quality, precision and promptness. Guccia personally supervised the smallest details in the galley proofs. In the quest for a high-quality publication, Guccia intervened with many and substantial suggestions to the author (Figs. 6.1, 6.2, 6.3, 6.4, 6.5, 6.6, 6.7, 6.8, 6.9, 6.10, 6.11). He was especially insistent on the precision and adequacy of references. Guccia was well known for his masterly work in the preparation

---

[17] *Ibid.*

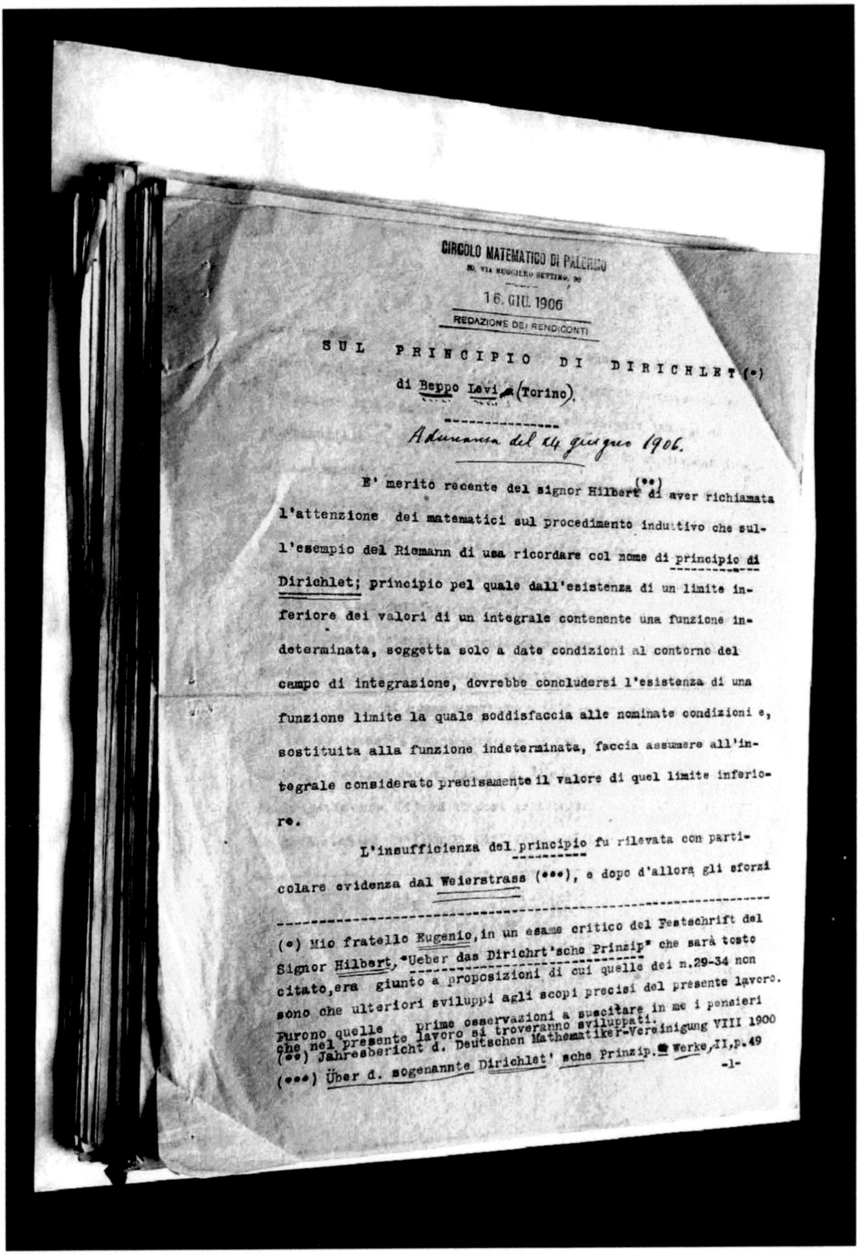
CIRCOLO MATEMATICO DI PALERMO
30, VIA RUGGIERO SETTIMO, 30
16. GIU. 1906
REDAZIONE DEI RENDICONTI

SUL PRINCIPIO DI DIRICHLET (*)

di Beppo Levi (Torino).

Adunanza del 24 giugno 1906.

E' merito recente del signor Hilbert (**) di aver richiamata l'attenzione dei matematici sul procedimento induttivo che sull'esempio del Riemann di usa ricordare col nome di principio di Dirichlet; principio pel quale dall'esistenza di un limite inferiore dei valori di un integrale contenente una funzione indeterminata, soggetta solo a date condizioni al contorno del campo di integrazione, dovrebbe concludersi l'esistenza di una funzione limite la quale soddisfaccia alle nominate condizioni e, sostituita alla funzione indeterminata, faccia assumere all'integrale considerato precisamente il valore di quel limite inferiore.

L'insufficienza del principio fu rilevata con particolare evidenza dal Weierstrass (***), e dopo d'allora gli sforzi

(*) Mio fratello Eugenio, in un esame critico del Festschrift del Signor Hilbert, "Ueber das Dirichrt'sche Prinzip" che sarà tosto citato, era giunto a proposizioni di cui quelle dei n.29-34 non sono che ulteriori sviluppi agli scopi precisi del presente lavoro. Furono quelle prime osservazioni a suscitare in me i pensieri che nel presente lavoro si troveranno sviluppati.

(**) Jahresbericht d. Deutschen Mathematiker-Vereinigung VIII 1900

(***) Über d. sogenannte Dirichlet'sche Prinzip. Werke, II, p.49

-1-

Fig. 6.1 Manuscript of a paper by Beppo Levi submitted to the Rendiconti del Circolo Matematico di Palermo. Courtesy of the Circolo Matematico di Palermo

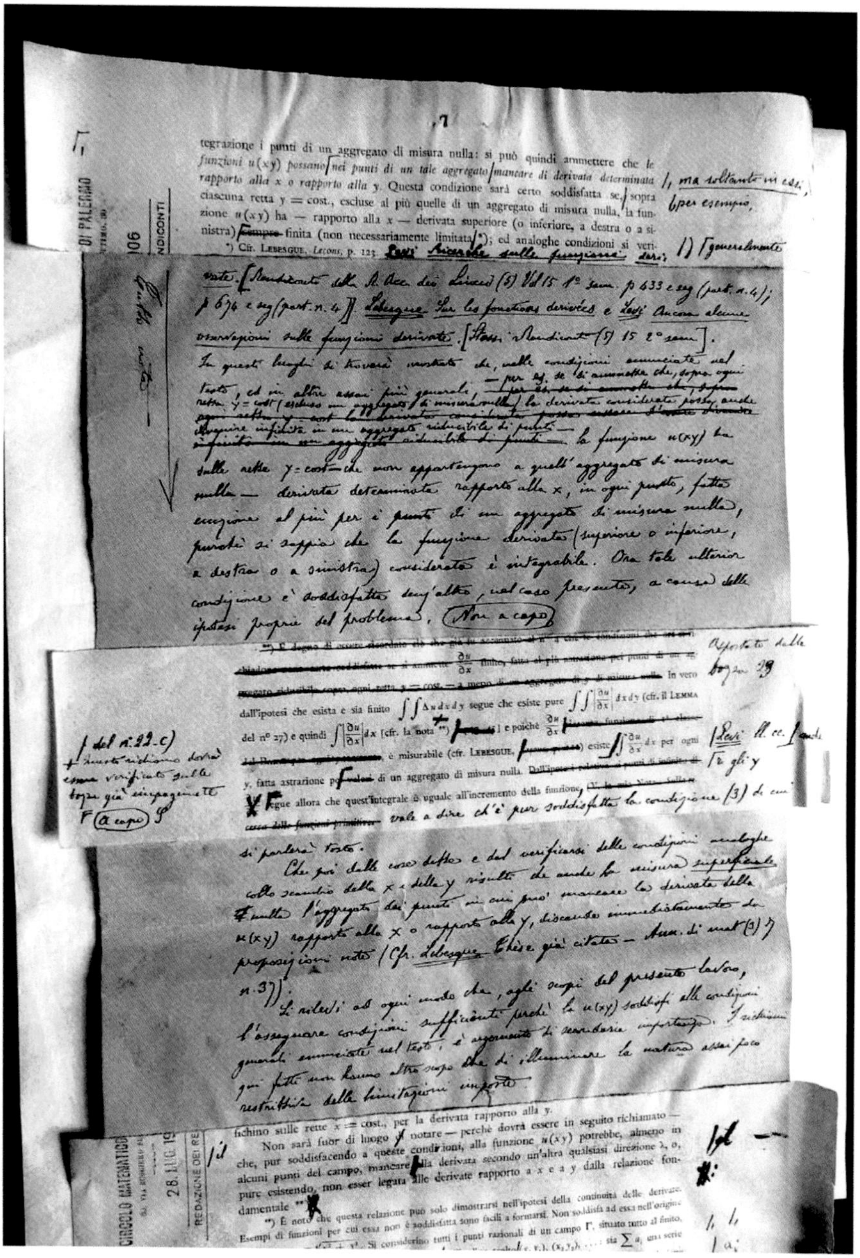

Fig. 6.2 Galley proofs of the paper by Beppo Levi with corrections by Guccia. Courtesy of the Circolo Matematico di Palermo

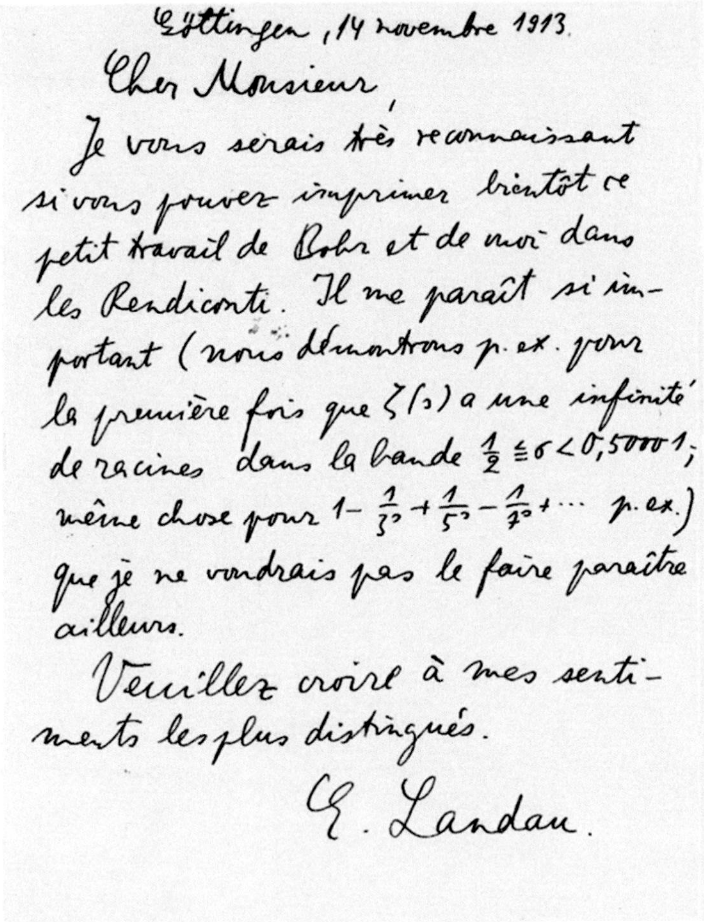

Göttingen, 14 novembre 1913.

Cher Monsieur,

Je vous serais très reconnaissant si vous pouvez imprimer bientôt ce petit travail de Bohr et de moi dans les Rendiconti. Il me paraît si important (nous démontrons p. ex. pour la première fois que $\zeta(s)$ a une infinité de racines dans la bande $\frac{1}{2} \leqq \sigma < 0,50001$; même chose pour $1 - \frac{1}{3^s} + \frac{1}{5^s} - \frac{1}{7^s} + \cdots$ p. ex.) que je ne voudrais pas le faire paraître ailleurs.

Veuillez croire à mes sentiments les plus distingués.

E. Landau.

Fig. 6.3 Letter November 14, 1913, from Edmund Landau to Guccia submitting a paper (joint with H. Bohr) to the Rendiconti del Circolo Matematico di Palermo. Courtesy of the Circolo Matematico di Palermo

of perfect bibliographic notes for articles published in the Rendiconti. His broad mathematical knowledge allowed him to be meticulously accurate regarding priorities, development of mathematical concepts and proof of results. This effort was publicly acknowledged by Émile Picard:

I just want to point out how difficult it is today to make a bibliography of some historical value. It can be accurate to say that half of the attributions are false and often even the first inventor is not cited. The history of science will become increasingly difficult to write; it is not difficult to find Encyclopedias where the historian risks drowning in a flood of citations among which the one who had the first idea disappears. You do, my dear friend, a great service through the Rendiconti in revising and often completing the citations by the authors,

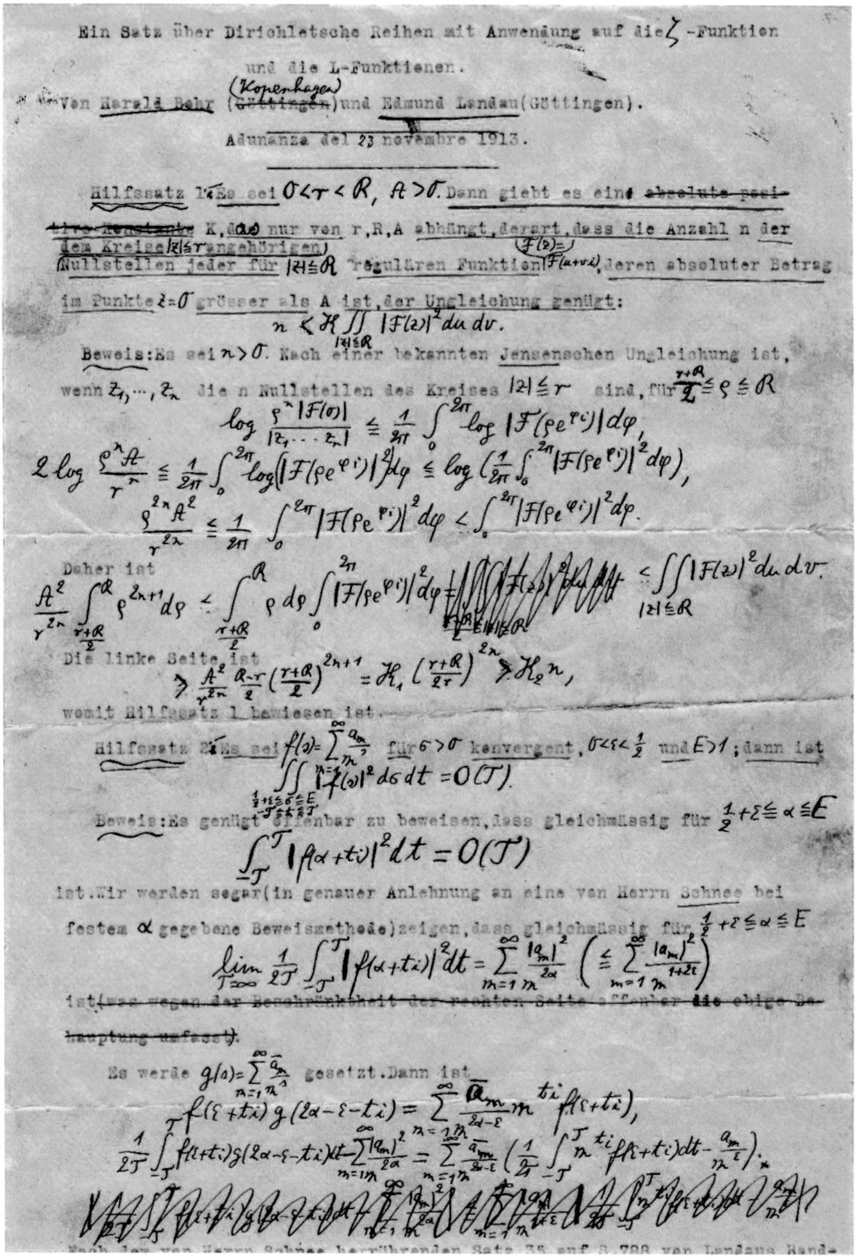

Ein Satz über Dirichletsche Reihen mit Anwendung auf die ζ-Funktion und die L-Funktionen.

Von Harald Bohr (Kopenhagen) und Edmund Landau (Göttingen).

Adunanza del 23 novembre 1913.

Hilfssatz 1: Es sei $0 < r < R$, $A > 0$. Dann giebt es eine Konstante K, die nur von r, R, A abhängt, derart, dass die Anzahl n der dem Kreise $|z| \leq r$ angehörigen Nullstellen jeder für $|z| \leq R$ regulären Funktion $F(z)$, deren absoluter Betrag im Punkte $z = 0$ grösser als A ist, der Ungleichung genügt:

$$n < K \iint_{|z| \leq R} |F(z)|^2 \, du \, dv.$$

Beweis: Es sei $n > 0$. Nach einer bekannten Jensenschen Ungleichung ist, wenn $z_1, \ldots, z_n$ die n Nullstellen des Kreises $|z| \leq r$ sind, für $\frac{r+R}{2} \leq \varrho \leq R$

$$\log \frac{\varrho^n |F(0)|}{|z_1 \cdots z_n|} \leq \frac{1}{2\pi} \int_0^{2\pi} \log |F(\varrho e^{i\varphi})| \, d\varphi,$$

$$2 \log \frac{\varrho^n A}{r^n} \leq \frac{1}{2\pi} \int_0^{2\pi} \log \left(|F(\varrho e^{i\varphi})|^2\right) d\varphi \leq \log \left(\frac{1}{2\pi} \int_0^{2\pi} |F(\varrho e^{i\varphi})|^2 d\varphi \right),$$

$$\frac{\varrho^{2n} A^2}{r^{2n}} \leq \frac{1}{2\pi} \int_0^{2\pi} |F(\varrho e^{i\varphi})|^2 d\varphi < \int_0^{2\pi} |F(\varrho e^{i\varphi})|^2 d\varphi.$$

Daher ist

$$\frac{A^2}{r^{2n}} \int_{\frac{r+R}{2}}^{R} \varrho^{2n+1} d\varphi < \int_{\frac{r+R}{2}}^{R} \varrho \, d\varphi \int_0^{2\pi} |F(\varrho e^{i\varphi})|^2 d\varphi \leq \iint_{|z| \leq R} |F(z)|^2 \, du \, dv.$$

Die linke Seite ist

$$\geq \frac{A^2}{r^{2n}} \frac{R-r}{2} \left(\frac{r+R}{2}\right)^{2n+1} = K_1 \left(\frac{r+R}{2r}\right)^{2n} \geq K_2 n,$$

womit Hilfssatz 1 bewiesen ist.

Hilfssatz 2: Es sei $f(s) = \sum_{n=1}^{\infty} \frac{a_n}{n^s}$ für $\sigma > 0$ konvergent, $0 < \varepsilon < \frac{1}{2}$ und $E > 1$; dann ist

$$\iint_{\substack{\frac{1}{2}+\varepsilon \leq \sigma \leq E \\ T \leq t \leq 2T}} |f(s)|^2 \, d\sigma \, dt = O(T).$$

Beweis: Es genügt offenbar zu beweisen, dass gleichmässig für $\frac{1}{2}+\varepsilon \leq \alpha \leq E$

$$\int_T^{2T} |f(\alpha+ti)|^2 dt = O(T)$$

ist. Wir werden sogar (in genauer Anlehnung an eine von Herrn Schnee bei festem α gegebene Beweismethode) zeigen, dass gleichmässig für $\frac{1}{2}+\varepsilon \leq \alpha \leq E$

$$\lim_{T \to \infty} \frac{1}{2T} \int_{-T}^{T} |f(\alpha+ti)|^2 dt = \sum_{m=1}^{\infty} \frac{|a_m|^2}{m^{2\alpha}} \left(\leq \sum_{m=1}^{\infty} \frac{|a_m|^2}{m^{1+2\varepsilon}} \right)$$

ist (was wegen der Beschränktheit der rechten Seite offenbar die obige Behauptung umfasst).

Es werde $g(s) = \sum_{n=1}^{\infty} \frac{\overline{a_n}}{n^s}$ gesetzt. Dann ist

$$f(\varepsilon+ti) \, g(2\alpha-\varepsilon-ti) = \sum_{m=1}^{\infty} \frac{\overline{a_m}}{m^{2\alpha-\varepsilon}} \, m^{ti} f(\varepsilon+ti),$$

$$\frac{1}{2T} \int_{-T}^{T} f(\varepsilon+ti) g(2\alpha-\varepsilon-ti) dt = \sum_{m=1}^{\infty} \frac{|a_m|^2}{m^{2\alpha}} = \sum_{m=1}^{\infty} \frac{\overline{a_m}}{m^{2\alpha-\varepsilon}} \left(\frac{1}{2T} \int_{-T}^{T} m^{ti} f(\varepsilon+ti) dt - \frac{a_m}{m^\varepsilon} \right).$$

Nach dem von Herrn Schnee herrührenden Satze 35 auf S. 788 von Landaus Hand-

Fig. 6.4 Manuscript of the paper by Edmund Landau and Harald Bohr. Courtesy of the Circolo Matematico di Palermo

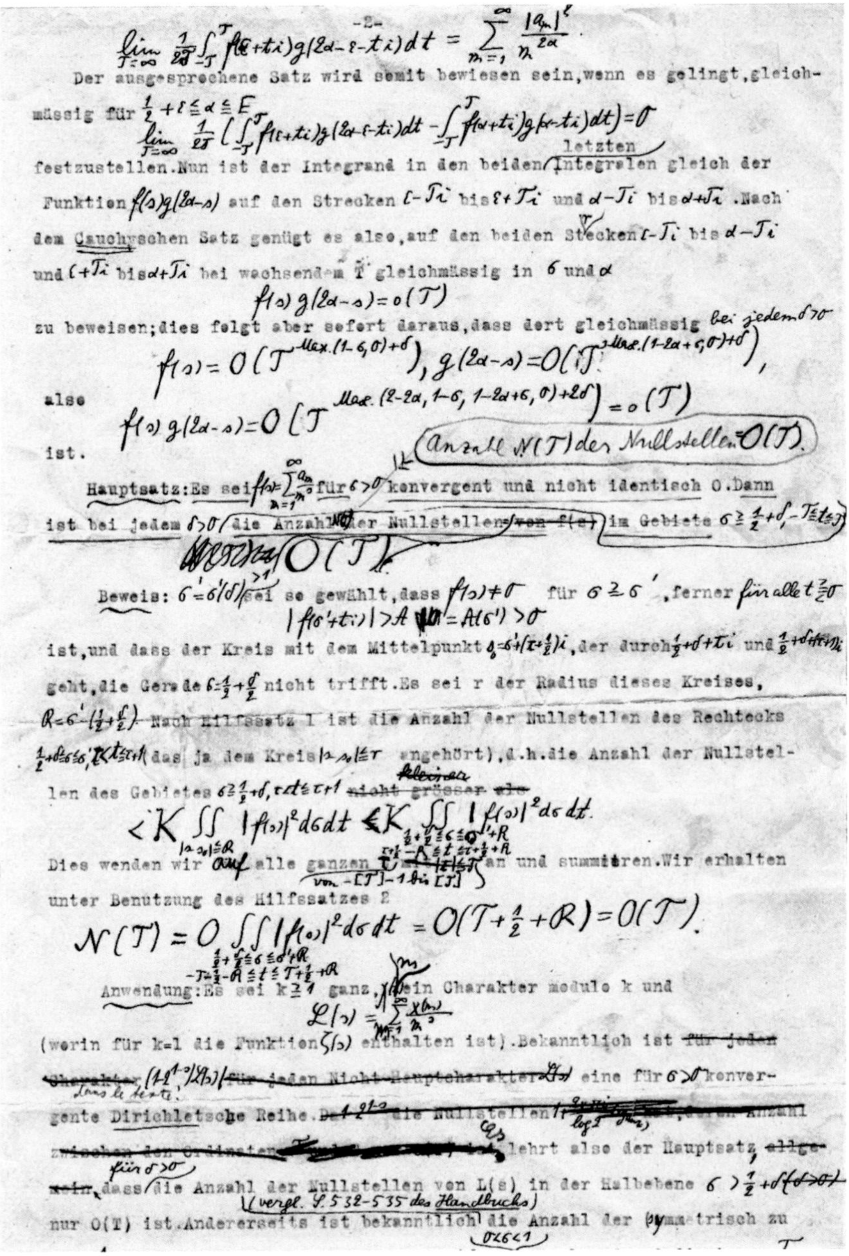

Fig. 6.5 Manuscript of the paper by Edmund Landau and Harald Bohr. Courtesy of the Circolo Matematico di Palermo

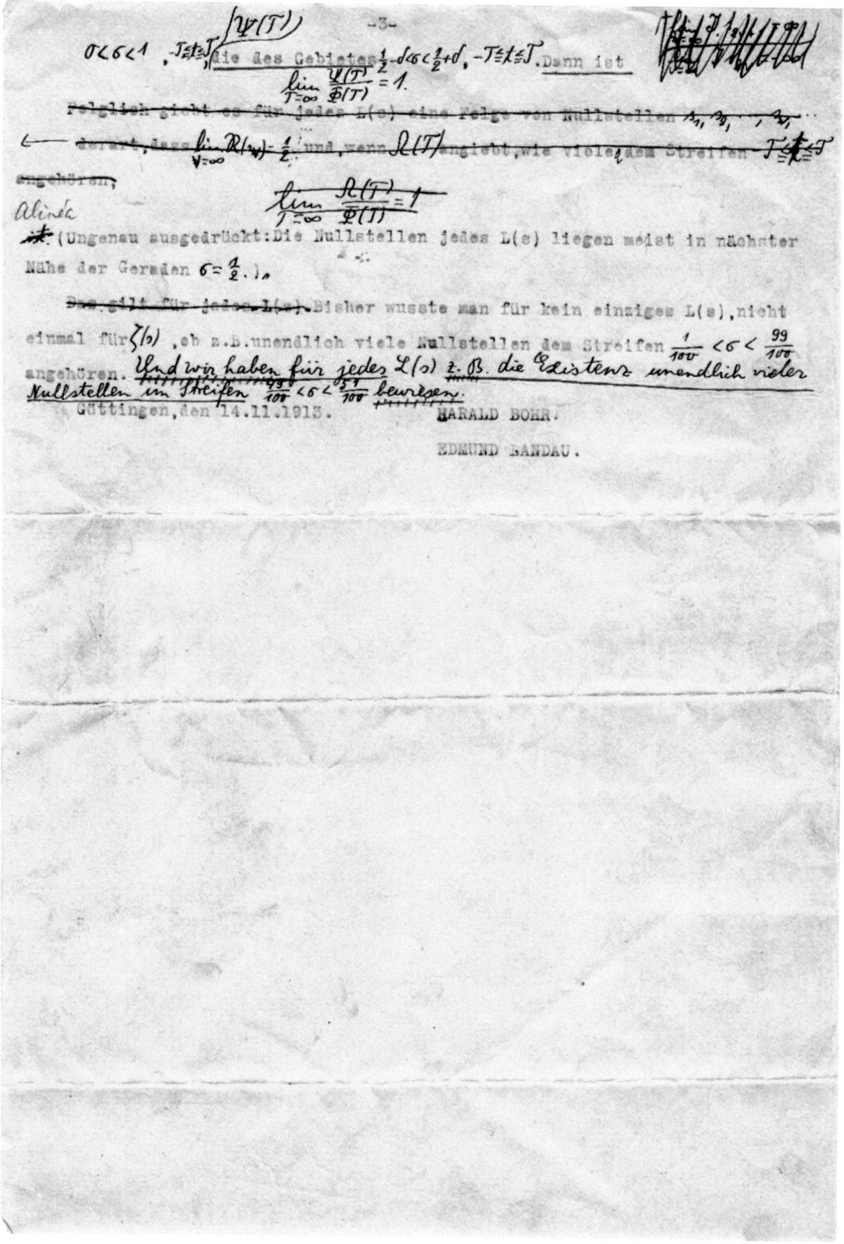

-3-

$0<\sigma<1$, $-T\le t\le T$, $\Psi(T)$ die des Gebietes $\frac{1}{2}-\delta<\sigma<\frac{1}{2}+\delta$, $-T\le t\le T$. Dann ist

$$\lim_{T=\infty}\frac{\Psi(T)}{\Phi(T)}=1.$$

~~Folglich giebt es für jedes L(s) eine Folge von Nullstellen~~ $\beta_1, \beta_2, \ldots, \beta_\nu, \ldots$

~~derart, dass $\lim_{\nu=\infty}\Re(\beta_\nu)=\frac{1}{2}$ und, wenn $R(T)$ angiebt, wie viele dem Streifen $-T\le t\le T$ angehören,~~

Alinea

$$\lim_{T=\infty}\frac{R(T)}{\Phi(T)}=1$$

(Ungenau ausgedrückt: Die Nullstellen jedes L(s) liegen meist in nächster Nähe der Geraden $\sigma=\frac{1}{2}$.).

~~Das gilt für jedes L(s).~~ Bisher wusste man für kein einziges L(s), nicht einmal für $\zeta(s)$, ob z.B. unendlich viele Nullstellen dem Streifen $\frac{1}{100}<\sigma<\frac{99}{100}$ angehören. Und wir haben für jedes $L(s)$ z.B. die Existenz unendlich vieler Nullstellen im Streifen $\frac{49}{100}<\sigma<\frac{51}{100}$ bewiesen.

Göttingen, den 14.11.1913. HARALD BOHR.

EDMUND LANDAU.

Fig. 6.6 Manuscript of the paper by Edmund Landau and Harald Bohr. Courtesy of the Circolo Matematico di Palermo

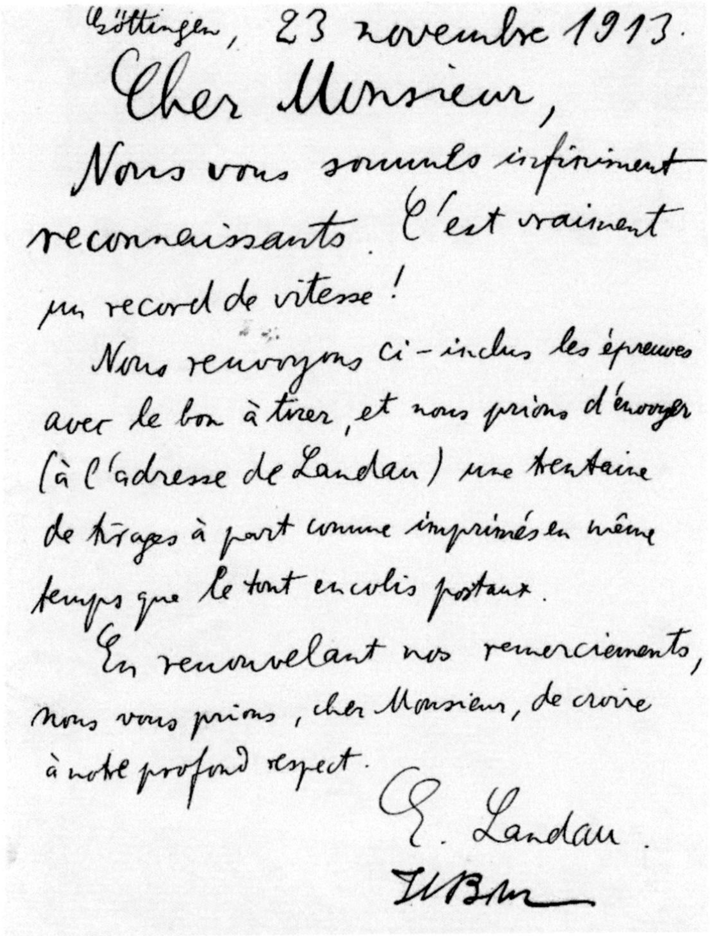

Göttingen, 23 novembre 1913.

Cher Monsieur,

Nous vous sommes infiniment reconnaissants. C'est vraiment un record de vitesse!

Nous renvoyons ci-inclus les épreuves avec le bon à tirer, et nous prions d'envoyer (à l'adresse de Landau) une trentaine de tirages à part comme imprimés en même temps que le tout en colis postaux.

En renouvelant nos remerciements, nous vous prions, cher Monsieur, de croire à notre profond respect.

E. Landau.

H Bohr

Fig. 6.7 Letter November 23, 1913, from Edmund Landau to Guccia regarding the paper submitted to the Rendiconti del Circolo Matematico di Palermo. Courtesy of the Circolo Matematico di Palermo

in part responsible for this state of affairs due to the little care placed on bibliographic information.[18]

The Archive of the Circolo Matematico di Palermo contains folders with the complete editorial process of many of the papers published in the Rendiconti when Guccia was managing the journal (Figs. 6.1–6.11). Each paper reflects Guccias great care and attention. As part of this careful process, Guccia used the library materials, such as books and articles, from both the library of the Circolo and his

[18]Letter from Picard to Guccia, Rend. Circ. Mat. Palermo, 36 (1913), pp. 277–278.

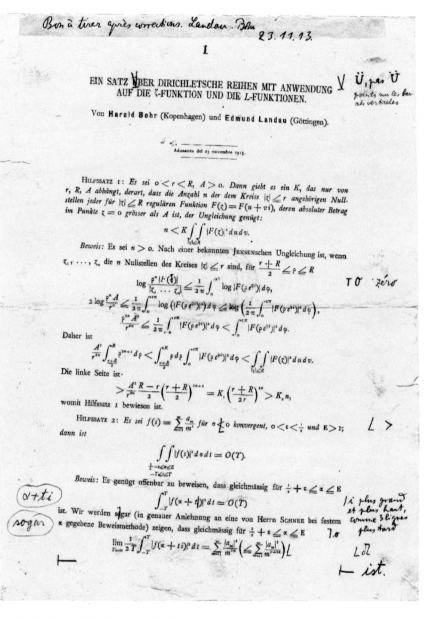

Fig. 6.8 Galley proofs of the paper by Edmund Landau and Harald Bohr with corrections by Guccia. Courtesy of the Circolo Matematico di Palermo

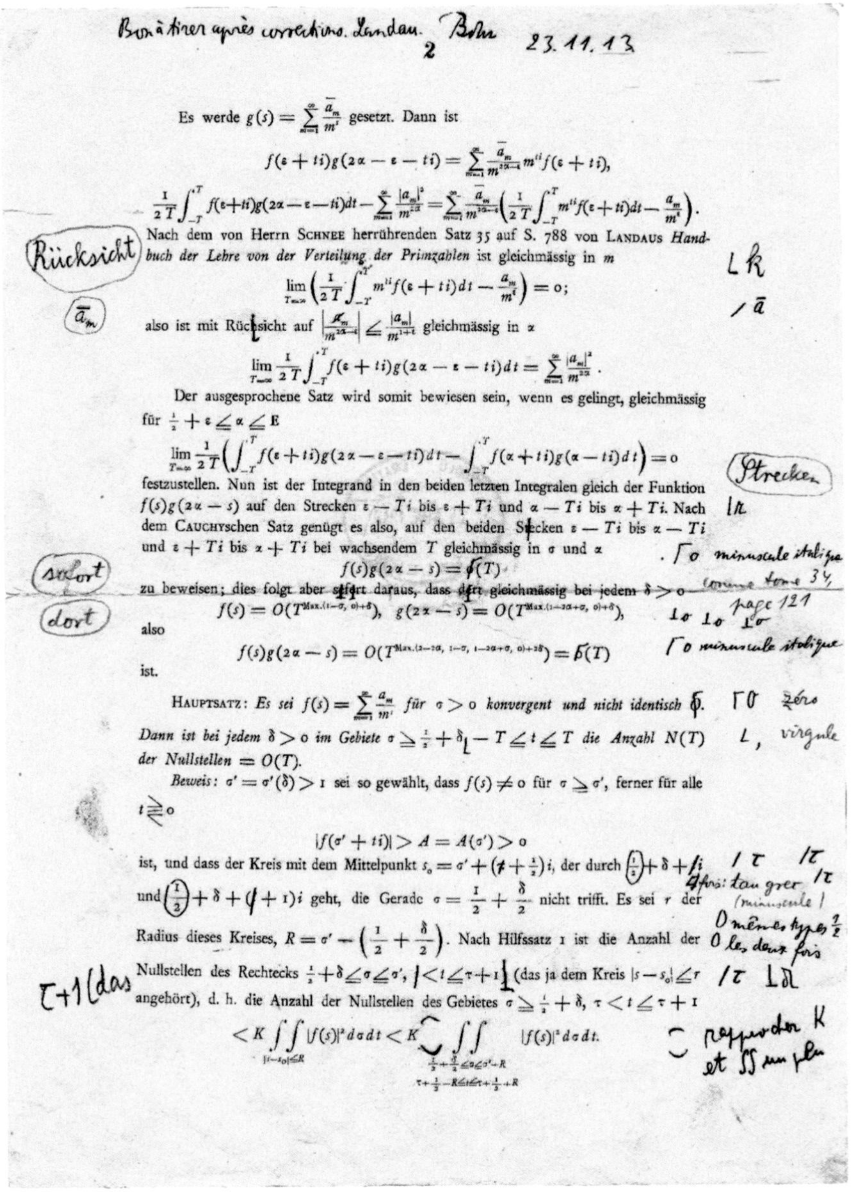
Bon à tirer après corrections. Landau. Bohr 23.11.13

2

Es werde $g(s)=\sum_{m=1}^{\infty}\frac{\bar a_m}{m^s}$ gesetzt. Dann ist

$$f(\varepsilon+ti)g(2\alpha-\varepsilon-ti)=\sum_{m=1}^{\infty}\frac{\bar a_m}{m^{2\alpha-\varepsilon}}m^{ti}f(\varepsilon+ti),$$

$$\frac{1}{2T}\int_{-T}^{T}f(\varepsilon+ti)g(2\alpha-\varepsilon-ti)dt-\sum_{m=1}^{\infty}\frac{|a_m|^2}{m^{2\alpha}}=\sum_{m=1}^{\infty}\frac{\bar a_m}{m^{2\alpha-\varepsilon}}\left(\frac{1}{2T}\int_{-T}^{T}m^{ti}f(\varepsilon+ti)dt-\frac{a_m}{m^{\varepsilon}}\right).$$

Nach dem von Herrn SCHNEE herrührenden Satz 35 auf S. 788 von LANDAUS *Handbuch der Lehre von der Verteilung der Primzahlen* ist gleichmässig in m

$$\lim_{T=\infty}\left(\frac{1}{2T}\int_{-T}^{T}m^{ti}f(\varepsilon+ti)dt-\frac{a_m}{m^{\varepsilon}}\right)=0;$$

also ist mit Rücksicht auf $\left|\frac{\bar a_m}{m^{2\alpha-\varepsilon}}\right|\leqq\frac{|a_m|}{m^{1+\varepsilon}}$ gleichmässig in α

$$\lim_{T=\infty}\frac{1}{2T}\int_{-T}^{T}f(\varepsilon+ti)g(2\alpha-\varepsilon-ti)dt=\sum_{m=1}^{\infty}\frac{|a_m|^2}{m^{2\alpha}}.$$

Der ausgesprochene Satz wird somit bewiesen sein, wenn es gelingt, gleichmässig für $\frac{1}{2}+\varepsilon\leqq\alpha\leqq E$

$$\lim_{T=\infty}\frac{1}{2T}\left(\int_{-T}^{T}f(\varepsilon+ti)g(2\alpha-\varepsilon-ti)dt-\int_{-T}^{T}f(\alpha+ti)g(\alpha-ti)dt\right)=0$$

festzustellen. Nun ist der Integrand in den beiden letzten Integralen gleich der Funktion $f(s)g(2\alpha-s)$ auf den Strecken $\varepsilon-Ti$ bis $\varepsilon+Ti$ und $\alpha-Ti$ bis $\alpha+Ti$. Nach dem CAUCHYschen Satz genügt es also, auf den beiden Strecken $\varepsilon-Ti$ bis $\alpha-Ti$ und $\varepsilon+Ti$ bis $\alpha+Ti$ bei wachsendem T gleichmässig in σ und α

$$f(s)g(2\alpha-s)=o(T)$$

zu beweisen; dies folgt aber sofort daraus, dass dort gleichmässig bei jedem $\delta>0$

$$f(s)=O(T^{\text{Max.}(1-\sigma,\,0)+\delta}),\quad g(2\alpha-s)=O(T^{\text{Max.}(1-2\alpha+\sigma,\,0)+\delta}),$$

also

$$f(s)g(2\alpha-s)=O(T^{\text{Max.}(2-2\alpha,\,1-\sigma,\,1-2\alpha+\sigma,\,0)+2\delta})=o(T)$$

ist.

HAUPTSATZ: *Es sei $f(s)=\sum_{m=1}^{\infty}\frac{a_m}{m^s}$ für $\sigma>0$ konvergent und nicht identisch 0. Dann ist bei jedem $\delta>0$ im Gebiete $\sigma\geqq\frac{1}{2}+\delta$, $-T\leqq t\leqq T$ die Anzahl $N(T)$ der Nullstellen $=O(T)$.*

Beweis: $\sigma'=\sigma'(\delta)>1$ sei so gewählt, dass $f(s)\neq 0$ für $\sigma\geqq\sigma'$, ferner für alle $t\gtreqless 0$

$$|f(\sigma'+ti)|>A=A(\sigma')>0$$

ist, und dass der Kreis mit dem Mittelpunkt $s_0=\sigma'+(\tau+\frac{1}{2})i$, der durch $\frac{1}{2}+\delta+\tau i$ und $\frac{1}{2}+\delta+(\tau+1)i$ geht, die Gerade $\sigma=\frac{1}{2}+\frac{\delta}{2}$ nicht trifft. Es sei r der Radius dieses Kreises, $R=\sigma'-\left(\frac{1}{2}+\frac{\delta}{2}\right)$. Nach Hilfssatz 1 ist die Anzahl der Nullstellen des Rechtecks $\frac{1}{2}+\delta\leqq\sigma\leqq\sigma'$, $\tau\leqq t\leqq\tau+1$ (das ja dem Kreis $|s-s_0|\leqq r$ angehört), d. h. die Anzahl der Nullstellen des Gebietes $\sigma\geqq\frac{1}{2}+\delta$, $\tau<t\leqq\tau+1$

$$<K\iint_{|s-s_0|\leqq R}|f(s)|^2d\sigma dt<K\iint_{\substack{\frac{1}{2}+\frac{\delta}{2}\leqq\sigma\leqq\sigma'+R\\ \tau+\frac{1}{2}-R\leqq t\leqq\tau+\frac{1}{2}+R}}|f(s)|^2d\sigma dt.$$

Fig. 6.9 Galley proofs of the paper by Edmund Landau and Harald Bohr with corrections by Guccia. Courtesy of the Circolo Matematico di Palermo

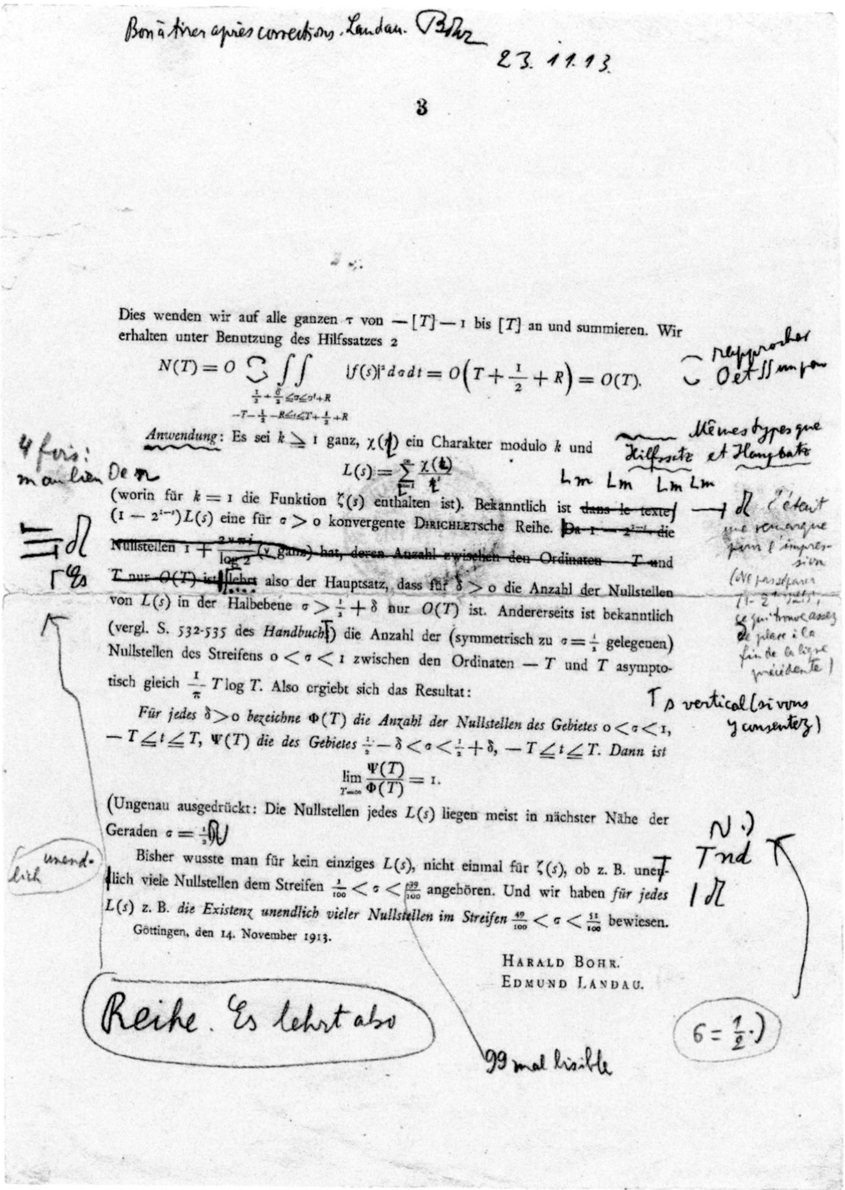

Bon à tirer après corrections. Landau. Bohr 23. 11. 13

3

Dies wenden wir auf alle ganzen τ von $-[T]-1$ bis $[T]$ an und summieren. Wir erhalten unter Benutzung des Hilfssatzes 2

$$N(T) = O \iint_{\substack{\frac{1}{2}+\frac{\delta}{2} \leq \sigma \leq \sigma' + R \\ -T-\frac{1}{2}-R \leq t \leq T+\frac{1}{2}+R}} |f(s)|^2 \, d\sigma \, dt = O\left(T + \frac{1}{2} + R\right) = O(T).$$

Anwendung: Es sei $k \geq 1$ ganz, $\chi(n)$ ein Charakter modulo k und

$$L(s) = \sum_{n=1}^{\infty} \frac{\chi(n)}{n^s}$$

(worin für $k = 1$ die Funktion $\zeta(s)$ enthalten ist). Bekanntlich ist dans le texte $(1 - 2^{1-s})L(s)$ eine für $\sigma > 0$ konvergente DIRICHLETsche Reihe. Da $1 - 2^{1-s}$ die Nullstellen $1 + \frac{2\nu\pi i}{\log 2}$ (ν ganz) hat, deren Anzahl zwischen den Ordinaten $-T$ und T nur $O(T)$ ist, lehrt also der Hauptsatz, dass für $\delta > 0$ die Anzahl der Nullstellen von $L(s)$ in der Halbebene $\sigma > \frac{1}{2} + \delta$ nur $O(T)$ ist. Andererseits ist bekanntlich (vergl. S. 532-535 des *Handbuchs*) die Anzahl der (symmetrisch zu $\sigma = \frac{1}{2}$ gelegenen) Nullstellen des Streifens $0 < \sigma < 1$ zwischen den Ordinaten $-T$ und T asymptotisch gleich $\frac{1}{\pi} T \log T$. Also ergiebt sich das Resultat:

Für jedes $\delta > 0$ bezeichne $\Phi(T)$ die Anzahl der Nullstellen des Gebietes $0 < \sigma < 1$, $-T \leq t \leq T$, $\Psi(T)$ die des Gebietes $\frac{1}{2} - \delta < \sigma < \frac{1}{2} + \delta$, $-T \leq t \leq T$. Dann ist

$$\lim_{T=\infty} \frac{\Psi(T)}{\Phi(T)} = 1.$$

(Ungenau ausgedrückt: Die Nullstellen jedes $L(s)$ liegen meist in nächster Nähe der Geraden $\sigma = \frac{1}{2}$.)

Bisher wusste man für kein einziges $L(s)$, nicht einmal für $\zeta(s)$, ob z. B. unendlich viele Nullstellen dem Streifen $\frac{1}{100} < \sigma < \frac{99}{100}$ angehören. Und wir haben *für jedes* $L(s)$ z. B. *die Existenz unendlich vieler Nullstellen im Streifen* $\frac{49}{100} < \sigma < \frac{51}{100}$ bewiesen.

Göttingen, den 14. November 1913.

HARALD BOHR.
EDMUND LANDAU.

Reihe. Es lehrt also

99 mal lisible

Fig. 6.10 Galley proofs of the paper by Edmund Landau and Harald Bohr with corrections by Guccia. Courtesy of the Circolo Matematico di Palermo

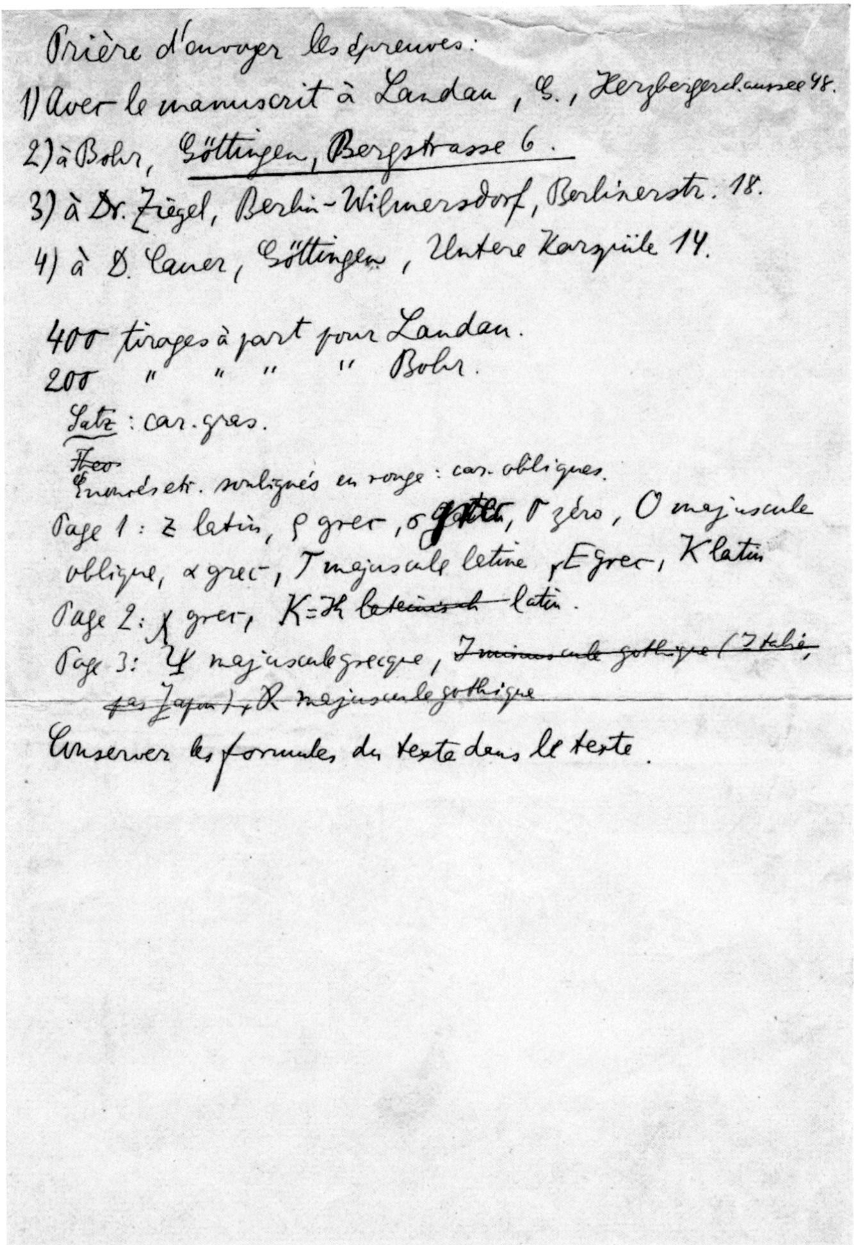

Prière d'envoyer les épreuves:

1) Avec le manuscrit à Landau, G., Herzbergerchaussee 48.

2) à Bohr, Göttingen, Bergstrasse 6.

3) à Dr. Ziegel, Berlin-Wilmersdorf, Berlinerstr. 18.

4) à D. Cauer, Göttingen, Untere Karspüle 14.

400 tirages à part pour Landau.

200 " " " " Bohr.

Satz: car. gras.

~~Theo~~ Énoncés etc. soulignés en rouge: car. obliques.

Page 1: z latin, ρ grec, σ ~~grec~~, 0 zéro, O majuscule oblique, α grec, T majuscule latine, E grec, K latin

Page 2: χ grec, K=K ~~latin~~ latin.

Page 3: Ψ majuscule grecque, ~~I majuscule gothique (Italie, pas Japon)~~, R majuscule gothique

Conserver les formules du texte dans le texte.

Fig. 6.11 Editorial instruction for the paper by Edmund Landau and Harald Bohr. Courtesy of the Circolo Matematico di Palermo personal library (which according to reports was superb). His collaborator Michele de Franchis wrote:

> Guccia worked each hour of the day and sometimes several hours of the night to prepare these publications [...]. No one who has not attended his daily work could imagine how much it mattered to him. The citations were reviewed and modified in an essential way; the shape and sometimes even the substance of the paper were often modified. A continued recourse to the Editorial Board [...] ensured the importance of the papers presented, so that for authors publishing in the Rendiconti was, apart from an advantage due to its wide dissemination, a moral satisfaction, as it certified the quality of their papers. This, together with the advantage of receiving one hundred free reprints.
> One of the merits of Rendiconti was not distinguishing between authors already famous and novice authors: the Rendiconti has baptized many illustrious mathematicians nowadays.[19]

It is important to note that De Franchis's observation points to one of the particular characteristics of the Rendiconti during its first thirty years of existence, which was the special attention it paid to young mathematicians. Guccia's broad knowledge of mathematics, along with his fine intuition and the extended network of relations within the research community, allowed him to detect quite early strongly talented young mathematicians. In these cases, it was well known that he was willing to speed up publication and reprint mailing to comply with the needs of young mathematicians in competitions for positions (obviously, provided the quality of the articles involved).

We halt our story to present some influential papers published in the Rendiconti at the time when the authors were young mathematicians. The test for the quality of a mathematical journal is the test of time over the articles that it published. It should be noted that the selection is based solely on the mathematical taste of the authors of this book, and not on a thorough study of the mathematical contributions published by the Rendiconti. This study has not yet been done, and we do not attempt to it do here.

Sur quelques points du calcul fonctionnel, M. Fréchet, Rend. Circ. Mat. Palermo, 22 (1906), pp. 1–74.

This is Maurice Fréchet's doctoral thesis, completed under Jacques Hadamard's supervision. Fréchet was twenty-eight years old when the paper was published. The seventy-four pages of the article occupied a full monthly installment of the journal. According to Dieudonné, "The extensions of the ideas of limit and continuity which had been formulated always were relative to special objects such as curves, surfaces or functions. The possibility of defining such notions in an arbitrary set is an idea which undoubtedly was first put forward by Fréchet in 1904 and developed by him in his famous thesis of 1906."[20]

Sul principio di Dirichlet, Rend. Circ. Mat. Palermo, B. Levi, 22 (1906), pp. 293–360.

---

[19]M. De Franchis, *Il Circolo Matematico di Palermo dalla sua fondazione ad oggi*, In: *Atti della Società italiana per il progresso delle scienze, XVIII Riunione, 1929*, Rome, 1930, 358–364, p. 362.

[20]J. Dieudonné, *History of functional analysis*, North-Holland, 1981, Amsterdam, p. 116.

Beppo Levi was a student of Corrado Segre with a thesis in algebraic geometry.[21] In March 1906, Levi sent Guccia an 88-page-long manuscript (which still exists in the Archive of the Circolo Matematico di Palermo, Fig. 6.1) requesting its fast publication, as he needed the reprints for a competition for a position at the University of Cagliari. The way that Levi presented his request did not please Guccia. It is not known who the reviewer of the manuscript was, but his recommendation must have been extremely positive, since on June 18, 1906, the paper was accepted and a few months later it was published. Levi obtained the position in Cagliari. The spaces of type Beppo Levi, related to Sobolev spaces and studied by Deny and Lions go back to this paper.[22]

Contribution à l'étude de la représentation d'une fonction arbitraire par les intégrales définies, M. Plancherel, Rend. Circ. Mat. Palermo, 30 (1910), pp. 289–335.

After obtaining his doctoral degree from the University of Fribourg, Michel Plancherel followed courses in Göttingen and in Paris. He was twenty-five years old when the paper was published. Dieudonné wrote, "In 1940 A. Weil showed how most results concerning Fourier series and integrals could be generalized to all locally compact commutative groups; the central theorem was the generalization of the Parseval relation [...] which had been proved by M. Plancherel in 1910."[23]

Ein Satz über Dirichletsche Reihen mit Anwendung auf die Z-Funktion und die L-Funktionen, H. Bohr, E. Landau, Rend. Circ. Mat. Palermo, 37 (1914), pp. 269–272.

Harald Bohr was twenty-seven years old when this paper was published (though Landau was thirty-seven). Heath-Brown commented, "The theory of the Riemann zeta function developed a great deal during Landau's lifetime, and he contributed much to the process. One of his best known results, in joint work with Harald Bohr, concerns the number of zeros of..."[24]

Some historians have interpreted Landau's speech, which we included in the Introduction, quite literally and thus exclusively attribute the success of the Rendiconti to "its great speed in publishing papers."[25] There is ample evidence that this was not the case. A pioneering study of mathematical journals based on their citations was published in the scientific journal *Science* in 1929, and determined that the Rendiconti ranked fifth in the world from 1909 to 1919. It was preceded only by Mathematische Annalen, the Transactions of the American Mathematical

---

[21] N. Schappacher, R. Schoof, *Beppo Levi and the Arithmetic of Elliptic Curves,* Math. Intelligencer 18 (1996), pp. 57–69.

[22] J. Deny, J.L. Lions, *Les espaces du type de Beppo Levi*, Ann. Inst. Fourier, Grenoble 5 (1953–54), pp. 305–370.

[23] J. Dieudonné, *op. cit.,* p. 204.

[24] D.R. Heath-Brown, *Edmund Landau: Collected Works*, Bull. London Math. Soc. 21 (1989), 342–350, pp. 347–348.

[25] U. Bottazzini, *Review of "Il Circolo Matematico di Palermo" by A. Brigaglia and G. Masotto,* Mathematical reviews, MR658129 (84d:01077), 1984.

Society, the Proceedings of the London Mathematical Society and the Comptes Rendus.[26] The society's large and extended membership shows that a large number of mathematicians were willing to pay the annual fee just to receive the Rendiconti. Poincaré held the Rendiconti in high regard: in a newspaper article published after the Rome 1908 International Congress he praised the work of Guccia for having founded "one of the most widespread mathematical journals in the world."[27] André Lichnerowicz recalled a mathematical discussion that he witnessed in 1936 where Paul Montel said to young Pierre Lelong, "your first results are interesting; write an article that is clear and sufficiently long to be clear; if it is worthy of that journal, we will present it to the Circolo di Palermo."[28] Let us end citing another French mathematician, Jean Dieudonné, in his praise of the Rendiconti: "It is well known that between 1900 and 1910, the Rendiconti published some of the most important papers of that time."[29]

Returning to the management of the Rendiconti, there is an important issue to be noted regarding the publication of research articles in the journal. Article 32.2 of the 1888 statute of the society established that the Rendiconti del Circolo Matematico di Palermo would compulsorily publish "the original notes and memoirs communicated by the members [of the society] and accepted by the Editorial Board." Therefore, in order to have a paper published in the Rendiconti, authors had to be members of the society. Obviously, this requirement had no effect on those authors who were already members of the Circolo. However, how did non-members publish their manuscripts? The editorial process was organized in such a way that when an author who was not a member of the Circolo submitted a manuscript, he would receive a letter explaining the society's membership requirements, which included a necessary annual fee (for non-resident members, there was no admission fee). Couldn't this be seen as some form of extortion? Two concerns arise from this question. First, once the international prestige of the journal was established, it is reasonable to think that authors would be willing to make the first payment; they could even know of the requirement in advance from colleagues or advisors. Second, once a paper was published, why would the author remain as member of the Circolo and keep paying the annual fee?

It is possible that the Circolo retained its members by providing them with access to publications and other information that was useful for the working mathematician. A mathematician who subscribed to the Rendiconti would receive twelve booklets per year, each with sixty to seventy pages of mathematical articles.

---

[26]E.S. Allen, *Periodicals for mathematicians,* Science, N.S. 70, n. 1825 (1929), pp. 592–594, Table II, p. 593.

[27]H. Poincaré, *Le congrès des mathématiciens à Rome,* letter to the publisher of Le Temps, April 21, 1908.

[28]A. Lichnerowicz, *Saluto in nome della comunità matematica internazionale,* Atti del Convegno celebrativo del 1° centenario del Circolo Matematico di Palermo, Suppl. Rend. Circ. Mat. Palermo, 8 (1985), p. 17.

[29]J. Dieudonné, *review of "The Mathematical Circle of Palermo" by A. Brigaglia,* Mathematical reviews, MR863153 (88m:01104), 1988.

He would also receive every year a volume of the Supplement, which contained the table of contents of most of the mathematical journals in the world (the ones which were received in the Circolo in exchange for the Rendiconti), and other miscellaneous information, such as congresses, prizes, awards and various announcements of interest to mathematicians. Every year he would also receive a Biographical Directory for the members of the Circolo and a cumulative Index of all articles published in the Rendiconti. Apart from the mathematical content of the research articles, the collection of information received was close to fulfilling the projects envisioned at the first International Congresses of Mathematicians for structuring the international mathematical community. All this was received in exchange for the annual fee of fifteen lire, the same amount as in 1884. Guccia, who had been raised in a family devoted to business, believed that this was a fair deal. Moreover, it was a service close to what an international association of mathematicians should provide to its members.

We can say that overall, the international success of the Circolo Matematico di Palermo was mostly due to the quality of its journal, the Rendiconti, and to the professional benefits and services that the members of the society received from their membership.

To develop in full his project of a mathematical journal, Guccia needed two conditions: high quality printing and complete control over the printing work. The first condition was in accordance with his standards and with his criticism of the low quality of many publications at the time. The second condition came from the importance that he gave to the speed of the process of printing. It was a way of paying attention to the needs and interest of authors, many of them, as we have already commented, were competing for academic positions. Both conditions were related to the printing of the journal.

Since the founding of the Circolo in 1884 the Rendiconti was printed in the printing house Tipografia Michele Amenta, one of the many printing houses of Palermo; precisely, volumes I to VI and some parts of volume VII. In 1893, Guccia created his own printing house, the Tipografia Matematica di Palermo. It was independent of the Circolo, but it worked exclusively for the Circolo (at least at first).[30] With his usual thoroughness, Guccia bought the necessary machinery and types from professional suppliers at a national and international level[31] and he set up a facility in his own palace, on via Ruggiero Settimo 28, just next to the seat of the Circolo on via Ruggiero Settimo 30 (where sessions were held and the library was situated). A further element was needed. Guccia knew throughout his whole life that the success of any enterprise depended upon finding the most appropriate and skilled person to oversee it. In this case, he needed a skillful and committed

---

[30]*Discorso del Prof. Ing. Michele Albeggiani*, XXX anniversario della fondazione del Circolo Matematico di Palermo, Adunanza solenne del 14 aprile 1914, Suppl. Rend. Circ. Mat. Palermo, 9 (1914), p. 14.

[31]The printing machine was Maroni type, from the firm Augusto Dell'Orto from Milan. The Archive of the Circolo contains many documents and invoices related to the printing business.

typographer. Thus, he hired a young typesetter from the Tipografia Michele Amenta named Gaetano Senatore.

The history of the Tipografia Matematica di Palermo is closely tied to that of the Circolo Matematico di Palermo, especially from 1894 until 1914 (indeed, the Rendiconti was printed by the Tipografia Matematica until 1942!). In 1899, the Tipografia moved its premises to a nearby property on Via Villareale 11, which Guccia's father had bought some years before and donated to his son. That same year, Gaetano Senatore obtained the degree of master typesetter (in Italian, "proto-compositore"), as it was reflected in the printed issues of the Rendiconti. The premises of the Tipografia changed on several occasions to different properties of Guccia, all in the vicinity of his palace, until 1911 when it settled in Vicolo Guccia 11, on the street just at the rear of the palace.

Guccia highly valued Gaetano Senatore's commitment to the Rendiconti. On the occasion of the Rome International Congress in 1908, the Rendiconti had planned to publish the Annual Directory. A strike of Palermo's typographers on March 1908 prevented the publication on time. In the next volume of the Rendiconti, a full-page was inserted explaining of the situation and announcing the inclusion of a list of members of the Circolo. The announcement ended with the following paragraph:

> The printing press is closed, the printing of this LIST OF MEMBERS was made by the master typesetter Mr. Gaetano Senatore himself, whose dedication on this occasion deserves note![32]

This appreciation and confidence in Senatore appears twice in Guccia's will. First, he gave

> To Gaetano Senatore, typographer of the Circolo, the ownership of the building located in Vicolo Guccia, no. 11, no. 13, no. 15, and all the goods existing in that building [the typography].[33]

Following Guccia's death, the Rendiconti listed Gaetano Senatore as Director-owner of the Tipografia Matematica. Guccia's will also included the following recommendation:

> To my secretary engineer Ester Paolo Guerra, treasurer of the Circolo Matematico di Palermo, and to my typographer Gaetano Senatore, my dear friends and co-workers, I strongly recommend this international scientific society, whose development and rapid increase is largely due to their loving care.[34]

He also gave Ester Paolo Guerra, his engineer and personal secretary, an annual income of one thousand lire.[35]

---

[32]Suppl. Rend. Circ. Mat. Palermo, 3 (1908).

[33]Notar Ferdinando Lionti, October 31, 1914. Archivio Notarile Distrettuale di Palermo.

[34]*Ibid.*

[35]*Ibid.*

The failure of Guccia's plan to have the Circolo Matematico di Palermo recognized as the "International Association of Mathematicians" at the Rome Congress in 1908 did not discourage him. He immediately started to plan a new strategy for obtaining that recognition in the International Congress of Mathematicians to be held in Cambridge in 1912. In that Congress, according to the proposal approved in Rome, "the establishment of an International Association of Mathematicians [would] be discussed."

With the Rendiconti as Guccia's main tool for his plan, he proposed a change to the 1888 statute of the Circolo. Thus, in the session held on June 14, 1908, thirty members of the Circolo presented a proposal where the Steering Council would have forty non-resident members instead of the fifteen that it had at that time (it would continue to have five resident members). The proposal was justified by explaining that

> The amendment, designed to ensure for the Editorial Board of the Rendiconti a wider and more effective international cooperation, was required by the increased number of members from various countries.[36]

The final aim was to enhance the international character of the society via greater international participation in the Editorial Board. The voting on the proposal was set for August 20. Including the votes received by mail, the result was 341 votes in favor of the proposal and none against.[37] These results indicated a strong-backing for the proposal, since the society had at that moment some 544 members. Guccia was overjoyed by both the participation and the result. He wrote to Mittag-Leffler, "Everybody finds necessary and useful to emphasize, in our society, the international imprint."[38] With this modification, forty-five members, five residents and forty non-residents formed the Steering Council.[39]

The next step was proposing a Steering Council that would be prestigious, effective as the Editorial board of the Rendiconti, and have the approval of the Society. Guccia planned this, and informed Mittag-Leffler of his plans in detail:

> I give you some explanations concerning the list that we recommend for next Steering Council of the Circolo for the years 1909–10–11.
> The voting will take place on the third Sunday of January; but the circular letters for the most remote countries (Australia, Japan, etc.) are leaving now. Concerning the 40 non-resident members of the Steering Council, 15 will be Italians and 25 from abroad (in proportion with the 197 Italian non-resident members and the 347 foreign members, Annuario 1908). Here are the 25 foreigners, France: Poincaré, Picard, Humbert, Hadamard, Borel; Germany: Klein, Hilbert, Noether, Stäckel, Landau, Carathéodory (he should to be considered as a representative of Germany, due to his nationality); Sweden: Mittag-Leffler, Fredholm; Austria-Hungary: Mertens, Wirtinger, Féjér; England: Forsyth, Love (A.E.H.);

---

[36] Suppl. Rend. Circ. Mat. Palermo, 3 (1908), pp. 26–27.

[37] Suppl. Rend. Circ. Mat. Palermo, 3 (1908), pp. 43–46.

[38] Letter from Guccia to Mittag-Leffler, July 12, 1908. Folder Mittag-Leffler, Archive of the Circolo Matematico di Palermo.

[39] For the modification of the statute, the favorable voting of 2/3 of the voters was required (Article 44 of the Statute).

Russia: Liapounoff, Stekloff; Denmark: Zeuthen; Belgium: De la Vallée Poussin; Greece: Stéphanos; U.S.A.: Moore (E.H.), Osgood.

They have all agreed, and agreed knowing that it is not an honorary committee but an Editorial Board where the members by their fine advice, sending papers, and revising manuscripts, assist the editorial office of Rendiconti in the difficult task of maintaining this international journal at the highest level of scientific quality.[40]

On the third Sunday of January 1909 the elections took place. The full Steering Council for the period 1909–1910 was:

Albeggiani	Hilbert	Pascal
Bagnera	Humbert	Picard
Bertini	Klein	Pincherle
Bianchi	Landau	Poincaré
Borel	Levi-Civita	Scorza
Carathéodory	Liapounoff	Segre
De Franchis	Loria	Severi
Dini	Love	Somigliana
Enriques	Marcolongo	Stäckel
Fejér	Mertens	Stekloff
Forsyth	Mittag-Leffler	Stéphanos
Fredholm	Moore, E.H.	Vallée Poussin, de la
Gebbia	Noether, M.	Vivanti
Guccia	Osgood	Wirtinger
Hadamard	Ovazza	Zeuthen

More than half of the members of the council were foreigners. Guccia's objective had been achieved. The voters were 397 (plus one blank vote) out of the 643 members that the society had at the time. Hilbert received the most votes with 396 votes, followed by Poincaré, Guccia and Albeggiani, who had 395 votes each. Of the elected members, the ones with fewer votes were Carathéodory and Gebbia with 383 votes each. There was little dispersion of the votes.[41] Let us mention in passing that the following Steering Council, for the 1912–1914 term, was elected on January 21, 1912, and was similar to the previous one with the only change of Ovazza with Castelnuovo. The voters in this case were 353 out of 821 members.[42]

---

[40]Letter from Guccia to Mittag-Leffler, October 25, 1908. Center for History of Science of the Royal Swedish Academy of Science.

[41]Suppl. Rend. Circ. Mat. Palermo, 4 (1909), pp. 3–6.

[42]Suppl. Rend. Circ. Mat. Palermo, 7 (1912), pp. 4-6.

In preparation for the Cambridge Congress, Guccia had another initiative also aimed at supporting the international character of the Circolo. On January 21, 1910, Giuseppe Bagnera, professor of Analysis and member of the Steering Council of the Circolo, argued the following during a speech he delivered at the meeting of the Faculty of Science of the University of Palermo:

> Who will replace in the future Professor Guccia in his work of administering and managing the Rendiconti? All Faculties of Science in Italy must consider this question. In particular, our Faculty since the Circolo Matematico di Palermo currently gives invaluable services to the Mathematical Sciences. Consequently, it would be a great loss for Italian Science if the future of the society was endangered or if it was forced to move its seat abroad.[43]

Bagnera's objective was to obtain, before the 1912 Congress in Cambridge, the support of all the Faculties of Science in Italy for Guccia's project of internationalizing the Circolo. The pretext was the weak health of Guccia, which indeed was a real worry; in fact, on September 22, 1911, Guccia decided to prepare a personally handwritten will.[44] Guccia's plan was a risky enterprise, however, as it had to be played in the arena of the Italian academic world, where his project had already faced strong opposition at the Rome Congress. The Faculty of Science of the University of Palermo, his own university, waited several months to approve a vague statement requesting financial support for the Circolo from the Government and inviting all other Faculties of Science in Italy to take common actions to support the Circolo and assure that its seat was kept within Italy (not even in Palermo). The weakness of the support went to the point that the voting in the Faculty of Science of the University of Palermo was not even unanimous.[45] Old scores, from the library crisis in 1904, were settled on that occasion. The result was the failure of Guccia's strategy to obtain the support of the Italian Faculties of Science and hence, the support of the Italian mathematical community.

There was a distinct contrast between the national and the international positioning towards the Circolo Matematico di Palermo. Foreigners would happily support and contribute to the development of the Circolo, located in the distant and neutral city of Palermo. On the contrary, the Italian mathematical establishment, mainly coming from Northern Italy, was permanently mortified by the existence and the success of a strong mathematical society situated in a city that was located on Italy's geographic and cultural peripheries. Even worse, the existence of a strong Circolo was seen as a permanent obstacle for the creation of a "genuine" Italian mathematical society.

The fifth International Congress was held in Cambridge in August 1912. Guccia attended, as well as thirty-four other Italian mathematicians and twenty other members of the Steering Council of the Circolo. At the closing ceremony, the

---

[43]Minutes of the meeting held on January 21, 1910. Archive of the Faculty of Science of the University of Palermo.

[44]Notar Ferdinando Lionti, October 31, 1914. Archivio Notarile Distrettuale di Palermo.

[45]Minutes of the meeting held on June 7, 1910. Archive of the Faculty of Science of the University of Palermo.

President of the Congress, the astronomer Sir George H. Darwin, reviewed the resolutions adopted at the Rome 1908 Congress. He officiated at the passing of Guccia's dream when he cuttingly said:

> It was proposed at Rome that a constitution should be formed for an International Association of Mathematicians. I have not heard that any proposal will be made to-night and I do not hesitate to express my own opinion that our existing arrangements for periodical Congresses meet the requirements of the case better than would a permanent organisation of the kind suggested.[46]

Politically molded nationalistic sentiments soon dominated the international scene, and invaded areas such as scientific thought and scientific cooperation which up to that time had been held apart from such conflicts. This situation eventually led to the outbreak of World War I. The paradox to Sir George Darwin's statement was that the International Congresses were also about to be in serious danger.

6.2 The 1914 Celebration

After the Rome Congress in 1908, the relationship between Guccia and Volterra ceased. This was because Guccia blamed Volterra for blocking the Circolo's internationalization project. The disagreement was much deeper than the one caused by Volterra's project of the renewed Società Italiana per il Progresso delle Scienze in Parma the year before. A longstanding friendship that started in 1885 was seriously damaged. As a result, Volterra was not on the ticket for the Steering Council elections of 1909 and 1912. It was the first time that this had occurred since 1888. After the Cambridge Congress in the summer of 1912, it seems that Volterra was willing to soften Guccia's feelings of personal defeat. On November and December 1912, there was a polite exchange of letters between Guccia and Volterra dealing with various professional issues (papers submitted to the Rendiconti; the passing away of Poincaré; Guccia's ideas for homage to his French friend and supporter of the Circolo).[47]

In mid-December 1912, Volterra officially informed Guccia that, together with Guido Castelnuovo, he had started to collect money from the international members of the Circolo in order to give Guccia a gold medal to recognize him as the founder of the society. Guccia was taken by surprise by this gesture, and was deeply moved. He waited two weeks to reply. In his ornate Palermitan style, he politely protested

> If you had allowed me to speak I would have said this: I do not deserve such an honor, because even if my work has been of any use, it was quite modest and excluded any high reward. It would have sufficed for me to know (but not with a medal, which is too

---

[46]E.W. Hobson, E.A.H. Love (eds.), *Proceedings of the Fifth International Congress of Mathematicians*, Cambridge University Press, 1913, Cambridge, p. 40.

[47]Letter from Guccia to Volterra, December 3, 1912. Vito Volterra Collection, Archive of the Accademia Nazionale dei Lincei.

much!) that Italian mathematicians are satisfied with the work done to place in Italy the
seat of the largest international mathematical society and one of the most disseminated
(and perhaps one of the most important, if one believes the opinion of some illustrious
foreigners) international journals in higher mathematics. But a medal is too much! This
initiative (dear and kind) gives an excessive honor to my humble name, and places me at a
great embarrassment, because I know that I do not deserve so much.[48]

The initiative went on, and a committee was formed, consisting of twenty-
seven of the most important Italian mathematicians of the time; among them was
Bertini, Bianchi, Dini, D'Ovidio, Enriques, Levi-Civita, Scorza, Segre, Severi,
and Veronose. The committee decided to organize a double ceremony on March
2, 1914, combining the presentation of the medal to Guccia with the celebration of the thirtieth anniversary of the foundation of the Circolo Matematico
di Palermo. The latter was appropriate since the celebration in 1909 of the
twenty-fifth anniversary was cancelled due to the consequences of the devastating earthquake that had occurred in the strait of Messina on December 28,
1908.[49]

When returning from his summer traveling in 1913, Guccia paid a visit to
Volterra in Rome. Personal relations had been already restored. Some days later,
when he was back in Palermo, he sent Volterra a postcard (the one which has a
photograph of the *Biblioteca Guccia* on the front, Fig. 4.3). He explained, "As soon
as I arrived, I went to bed; I hope that in a few days I can get back to work, which
is waiting for me and it is huge."[50] On December 30, 1913, Guccia sent Volterra a
New Year greetings postcard and explained: "I have been ill for two months, since
we met in Rome. Since last week I am convalescent, gradually resuming my duties,
but I feel very weak."[51]

The next letter to Volterra, from January 7, 1914, was labeled as "reserved".
Guccia explained that even though he was recovering, he was still very weak,
writing, "my weight has dropped from 65 to 59 kilos." He made several comments
regarding the ceremony and made the following "warmest pleas" to Volterra:

1st The ceremony should remain limited and confined to the mathematical family.
    [...] I accept everything from the mathematicians; nothing from the others.
2nd In case someone has the idea of proposing any decoration for me, you must
    stop that from occurring.[52]

The ceremony was postponed for two reasons. The main one was that Guccia's
health was improving very slowly and he could not have attended the celebration in

---

[48] *Ibid.,* December 29, 1912.

[49] Adunanza, January 10, 1909. Suppl. Rend. Circ. Mat. Palermo, 4 (1909), pp. 2-3.

[50] Letter from Guccia to Volterra, October 30, 1913. Vito Volterra Collection, Archive of the
Accademia Nazionale dei Lincei.

[51] *Ibid.,* December 30, 1913.

[52] *Ibid.,* January 7, 1914.

the scheduled time. The other reason was that the Library of the Circolo was being reorganized, and this occupied much of Guccia's energy and attention.[53]

The ceremony was finally held on April 14, 1914 in the Aula Magna of the University of Palermo (Fig. 6.12). At the presidency table were academic authorities of the university, officers of the Circolo, the Mayor of Palermo, Vito Volterra, Edmund Landau and Giovanni Battista Guccia. Apart from the more ceremonial addresses, the president of the Circolo, Michele Albeggiani, delivered an interesting account of the development of the society; Edmund Landau, as representative of the Göttingen Mathematical Society, delivered the address that we included in the Introduction of the book. The other noteworthy address was that of Vito Volterra; in Chapter 5, we saw its beginning. It was a long address where he also said:

> He [Guccia] gave to his journal a genuine international imprint, thus placing it on secure and solid grounds. Due to esteem and sympathy, he acquired for the Rendiconti works by mathematicians from all over the world. With assiduous, persevering and ingenious work, with fine discernment, he was able to choose among authors, among schools, among directions, so that the Circolo has published works that have made history in the scientific world and will remain classic and fundamental.
> Today the Circolo di Palermo is the premier mathematical society in the world in membership, which includes some of the brightest names in science. The thirty-seven volumes that it has published contain memoirs on Higher Analysis, on Geometry, on Mechanics and on Mathematical Physics through which the course of the admirable progress of mathematics in the last three decades can be traced. The most distinguished mathematicians are honored to publish their works; the Rendiconti has penetrated wherever Geometry is cultivated and rapidly disseminates the most recent discoveries.[54]

Thirty-eight institutions sent or nominated a representative to the celebration; nine of those institutions were foreign, five of them from the United States. Letters and telegrams of solidarity came from eighty-seven institutions, scientific societies and academies, from twenty different countries. Sixty-six mathematicians sent letters or telegrams, among them Camille Jordan and Émile Picard.[55]

The sculptor Antonio Ugo, who had sculpted the Guccia medal for the Rome International Congress, was in charge of coining the medal. The front of the medal showed the logotype of the Circolo (based on the Sicilian trinacria) and the reverse a bust of Guccia. The medal was 116 grams of solid gold and 62 mm in diameter.[56] The collection for the medal was a success. There were 233 contributors and the total amount received was 3,187.91 lire. The largest personal contributions were

---

[53] *Ibid.* See also: Suppl. Rend. Circ. Mat. Palermo, 9 (1914), p. 6.

[54] *Discorso del Senatore Prof. Dr. Vito Volterra,* XXX anniversario della fondazione del Circolo Matematico di Palermo, Adunanza solenne del 14 aprile 1914, Suppl. Rend. Circ. Mat. Palermo, 9 (1914), pp. 18–19.

[55] Suppl. Rend. Circ. Mat. Palermo, 9 (1914), pp. 21–65.

[56] M. Di Carlo, M.A. Spadaro, *Commemorare a Palermo. Le Medaglie di Antonio Ugo,* Kalós Edizioni d'arte, 2014, Palermo.

Fig. 6.12 Newspaper reporting the celebration of XXX anniversary of the Circolo Matematico di Palermo. Courtesy of the Biblioteca Centrale della Regione Siciliana

from Jordan, Mittag-Leffler, Picard and Volterra (together with two other non-mathematicians) who contributed 50 lire each. Carathéodory and André contributed 40 lire each and Landau 36. The largest donation, 200 lire, came, surprisingly, from the Romanian Ministry of Labor.[57] Those who donated ten or more lire received from the organizers a bronze copy of the medal.[58]

There was a large attendance at the ceremony: mathematicians from Palermo, Italy and abroad, Palermo's nobility, relatives of Guccia, and others. Among them, let us point out two young participants. One was Giuseppe Tomasi di Lampedusa, great-grandson of Giulio Fabrizio Tomasi di Lampedusa, the uncle astronomer of Guccia. He was eighteen years old at the time. Forty-three years later, he wrote *The Leopard*. The other one was Pietro Tortorici, a twenty-three year old mathematician who had just graduated from the University of Pisa and contributed ten lire to the collection of the medal. For his donation, he received a bronze copy of the gold medal. His copy of the medal is the one reproduced in this book (Fig. 6.13).

The ceremony ended with a short address by Guccia. He was deeply moved and excused himself, due to his health condition, for not being able to express his profound gratitude as he would have liked to do.

Guccia was touched by the celebration. His life time of effort was highly valued by the international and national communities, the local community, the professional community, and his social community. He could be honestly proud of his achievements. The situation of the Circolo on April 1914 was superb. The society had 924 members that according to residency were distributed as follows:

Members residing in Palermo: 72.
Members residing in the rest of Italy: 234.
Members residing abroad: 618.

Regarding the distribution by country of the foreign members, the Circolo exhibited the true international nature of the society: foreign members came from twenty-six different countries (the same figure as in 1908). The largest figures were from the following countries:

Germany: 140.
United States of America: 140.
Austria-Hungary: 77.
France: 67.
Russia: 44.
Great Britain and Ireland: 29.
Sweden: 21.

---

[57] *Ibid.*, pp. 66–71.

[58] *Sottoscrizione per la medaglia d'oro offerta al prof. G.B. Guccia*, Archive of the Circolo Matematico di Palermo.

Fig. 6.13 Medal offered to Giovanni Battista Guccia on the occasion of the XXX anniversary of the Circolo Matematico di Palermo. Courtesy of Maurizio Urbano Tortorici

The remaining countries had twelve or fewer members.[59] Guccia was proud of these figures. Indeed, the statistical section of the 1914 Annual Directory where

---

[59] Annuario Circ. Mat. Palermo 1914, Note statistiche, p. 3.

those figures were given also contained a comparison with the data of national versus foreign members for other mathematical societies[60]

	founded	national	foreign	total	%
London Mathematical Society	1865	260	60	320	19
Société Mathématique de France	1872	189	109	298	37
Edinburgh Mathematical Society	1883	199	20	219	9
Circolo Matematico di Palermo	1884	306	618	924	67
American Mathematical Society	1888	648	55	703	8
Deutsche Mathematiker-Vereinigung	1890	480	289	769	36

With respect to the Rendiconti, the figures were also impressive: since 1887, it had published thirty-seven volumes, with 958 memoirs and communications from 327 authors. It had exchange arrangements with some 250 periodicals (scientific journals and proceedings of academies) coming from all parts of the world.[61] At that moment the Rendiconti was the mathematical journal with the largest print run in the world: it had reached the extraordinary figure of 1,200 copies.

6.3 Other Projects

Guccia had a clear picture of what was needed for the development of mathematics in the new scientific context of the twentieth century. His views were in accordance with the internationalization project that was recurrently presented at the International Congresses of Mathematicians. He tried to develop these other projects, but with no success. We briefly review them.

> But much, much more, remains to be done! I'll frankly tell you what is on my mind, as I am encouraged by the success that day by day our Society achieves abroad each day. As soon as we are well constituted, with 400 members from all countries, and when we feel strong enough [...] in my opinion we should take a big step ahead, a VERY DARING one. It is a difficult and daring step that should be considered carefully before taking it![62]

This is how Guccia prepared Valentino Cerruti on June 1905 for the announcement of his new project. The success of the Circolo after 1904 had encouraged Guccia to publicly propose other projects for the future development of the Circolo.

In this case, Guccia's reaction was motivated by Felix Klein's (and also other influential mathematicians, he claimed) movement towards applied mathematics.

---

[60] Annuario Circ. Mat. Palermo 1914, Note statistiche, pp. 7–8.

[61] *Discorso del Prof. Ing. Michele Albeggiani*, XXX anniversario della fondazione del Circolo Matematico di Palermo, Adunanza solenne del 14 aprile 1914, Suppl. Rend. Circ. Mat. Palermo, 9 (1914), p. 15.

[62] Letter from Guccia to Cerruti, June 15, 1905. Folder Cerruti, Archive of the Circolo Matematico di Palermo.

Klein had become a member of the editorial board of the journal "Zeitschrift für Mathematik und Physik", which Oskar Schlömilch founded in 1856. Some years later, the historian of mathematics Moritz Cantor joined Schlömilch on the editorial team of the journal. In 1901, there was a substantial change in the management of the journal: R. Mehmke and Carl Runge became editors-in-chief and a large editorial committee was set, with Felix Klein as a member. Simultaneously, a subtitle was added to the name of the journal highlighting its editorial line: "Organ für Angewandte Mathematik" (in English, "Publication for Applied Mathematics").

In view of these events taking place in the mathematical arena, Guccia argued:

> We cannot remain indifferent! The idea naturally arises that something should also be done in Italy for Applied Mathematics. [...] This should be done by the Circolo Matematico di Palermo, for various reasons:
>
> 1st Because (when we reach 400 members), our extensive membership abroad will help us, I am convinced, to successfully carry out this second program.
> 2nd Because our old Rendiconti (publication for pure mathematics), already so much credited, will help credit in a short time the new Rendiconti, that would be published at the same time with the name "Organo per le matematiche applicate" [in English, Publication for Applied Mathematics], or something like that.
>
> At the same time, the statute should be revised along the following lines:
>
> 1st The Circolo will publish two "Rendiconti": a Rendiconti for pure mathematics, and another Rendiconti for applied mathematics.
> 2nd Each member of the Society, by paying the 15 lire fee, will have the right to receive (at his choice) one or the other Rendiconti, or both of them for an additional fee. [...]
>
> Organizations and methods would be the same, but this is not enough to ensure the success of the enterprise. We need a man who has, knowledge of the wide field of applied mathematics, special and extended competence, a great working capability, and numerous personal relationships abroad, etc. This person should have the will and the time to entirely devote himself to this mission, without aiming any other reward than to render a great service to our country.[63]

Guccia had a clear view of the conditions needed for the success of this new project. Importantly, it required a mathematician, situated in the area of applied mathematics, who would devote his full life to the project, in a similar way as Guccia had done himself for the Rendiconti.

In his answer, Cerruti considered it highly desirable to promote the work in applied mathematics, and praised Guccia's project. However, he pointed out that "This is a delicate issue. As for today, the scholars in applied mathematics do not even reach mediocrity, in Italy I mean. Though in the past we have had distinguished ones."[64] Under these conditions, Cerruti did not feel like he was able to recommend a suitable person for that position.

---

[63] *Ibid.*

[64] Letter from Cerruti to Guccia, June 17, 1905. Folder Cerruti, Archive of the Circolo Matematico di Palermo.

We have found no other document where Guccia raised this topic. Maybe Cerruti's answer was devastating enough for Guccia that he left the project aside. This, however, does not match well with his tenacious personality. We are convinced that Guccia discussed this issue with other mathematicians and requested their opinion; possibly with Poincaré and Mittag-Leffler.[65] Since the Rendiconti for Applied Mathematics never appeared, it is certain that any responses to Guccia's inquiries were negative, and that he did not find the person that he thought that was needed for the success of the project.

Another issue that greatly interested Guccia was the review of mathematical literature. We have already commented on the creation in 1868 of the Jahrbuch über die Fortschritte der Mathematik and its important role despite the difficulties caused by the late appearance of the volumes (sometimes a three year delay). A related project, although with a completely different viewpoint, was that of the Répertoire Bibliographique des Sciences Mathématiques sponsored by Poincaré. Guccia was the person responsible for the Italian literature included in the Répertoire. He devoted a great amount of energy and work to this project, but the success of the project required a large number of mathematicians. Many of the candidates for collaborating in Italy were not willing to do so, and the ones who did, mostly under pressures from Guccia, did not comply with Guccia's accurate and meticulous working style.

In a letter to Cerruti, Guccia complained:

> I have to finalize the quite unfortunate "Repertorio bibliografico delle scienze matematiche in Italia", which I would like to publish in 1906 (the first volume, of course) in separate volumes. You certainly cannot imagine (I tell you this confidentially) how much nonsense our friends who collaborated with me have done. I am seriously embarrassed.[66]

As a consequence of the disagreements between Guccia and his collaborators, the "Repertorio bibliografico delle scienze matematiche in Italia" was never published. We have not found any of the large amounts of data which were collected in the process.

The failure of the Répertoire did not distract Guccia from his project of reviewing mathematical literature. Indeed, in the "Annuario del Circolo Matematico di Palermo" he created a section consisting of the indices of all journals received in exchange by the society, which by 1914 were some 250 journals. This enormous amount of information was distributed to the members of the Circolo, and it allowed them to be permanently informed of almost everything that was published in mathematics.

---

[65] No other trace of this topic has been found in the letters in the Archive of the Circolo Matematico di Palermo, the Archives Henri-Poincaré, and the Center for History of Science of the Royal Swedish Academy of Sciences (where the correspondence of Mittag-Leffler is now stored).

[66] Letter from Guccia to Cerruti, June 15, 1905. Folder Cerruti, Archive of the Circolo Matematico di Palermo.

However, Guccia's view went further. What was lacking was a systematic and objective service of reviews for published articles. This, he explained to Ulisse Dini:

> Reading the important preface as well as the introduction and a few chapters of your book made me think about the lack in our society of a service for reviews of periodical publications. I should confess to you that I have always been afraid to address this issue! [...] Every day we see works which are full of errors, naivety and plagiarism, praised beyond measure by indulgent reviewers, driven by interests that are not always those of science and education! Then, I say, which and how many will be the difficulties that will be encountered by the editor of a journal who intends to regulate and moralize such a service?[67]

The project of a review service within the Circolo did not go beyond that point.

Another one of Guccia's projects regarded an approach to the popularization of mathematics for the general public. He commented on this in a letter to Poincaré from December 1905:

> I have not abandoned the idea that I discussed with you recently at the Hotel Continental; that is, to publish every year in the Annuario (that we distribute for free in a very large number of copies) an article by a master of the Science on a topic which could interest even those who only have a modest understanding of science. In my opinion, this should give to the general public an idea of the great achievements of science in recent times, and increase the number of friends of mathematics. The links between pure and applied science escape the general public and, sometimes, even learned people!
> It would be useful, in my opinion, to properly highlight (when addressing the general public) everything that relies on high mathematics in the major modern discoveries of practical order. But, such a delicate and at the same time deep argument can only be treated by a scholar like you.[68]

Poincaré immediately welcomed the initiative.[69] On December 28, that same year, Guccia shared his project with Mittag-Leffler[70] who also was very positive about the idea.[71] Unfortunately, the project was never realized.

All these initiatives have two features in common: they arise around the same time, and they all share the same fate. They arise after the 1904 Heidelberg Congress when Guccia saw the number of possibilities opened up by the international success of the Circolo. All these initiatives failed. These failures have their roots in two unavoidable facts: the increasing weakness of Guccia's health and the structural weakness that the Circolo carried inside. We will expound these two facts in the next chapter.

---

[67]Letter from Guccia to Dini, August 4, 1907. Folder Dini, Archive of the Circolo Matematico di Palermo.

[68]Letter from Guccia to Poincaré, December 19, 1905. Folder Poincaré, Archive of the Circolo Matematico di Palermo.

[69]Letter from Poincaré to Guccia, December 24, 1905. Folder Poincaré, Archive of the Circolo Matematico di Palermo.

[70]Letter from Guccia to Mittag-Leffler, December 28, 1905. Center for History of Science of the Royal Swedish Academy of Science.

[71]*Ibid.*, February 10, 1906.

We end with one last project of Guccia that eventually succeeded. The aim was to reprint the complete works of renowned mathematicians. Volterra clearly stated the importance of understanding past research in mathematics:

> Men of science love to study the treasures gathered by others, in order to know well the value of their own. In the person dedicated to mathematical studies such curiosity is much greater than in those occupied with other disciplines.[72]

Guccia since long had the idea of reprinting of the complete works of Paolo Ruffini. Ettore Bortolotti,[73] a historian of mathematics from the University of Modena, had published a booklet on the correspondence of Paolo Ruffini. In December 1906, Guccia contacted Bortolotti and asked if he "would be willing to collect, sort out and ensure the reprint of the works of Ruffini in the Rendiconti?"[74] Guccia and Bortolotti exchanged letters, and a few months later, Guccia elaborated on fine details of the project:

> Unfortunately, our budget does not have any margin for fees or compensations to collaborators: it would then be working for the honor of ending the unfair oblivion in which, for a century, Italians have left the main geometer of Modena. [...]
> Keep in mind that the distribution of our Rendiconti exceeds that of any other reputed mathematical journal in the world! Therefore, there will be readers for whom the name of Ruffini is absolutely new. [...] You should inform, briefly but clearly, the readers of the Rendiconti of the events of the famous "impossibility"; of the role played by Ruffini; of the recent movement of recognition (Burkhardt, Bortolotti) in favor of Ruffini; and take the opportunity to exhibit to the readers work that no one had the possibility of reading, because it was not in the market, and that illuminated a very important point in the history of Algebra! But all this, should be done without inducing any intention of starting a campaign against Cauchy or others.[75]

In this letter, Guccia tactfully wanted to avoid mentioning the negative reaction at the time on behalf of the mathematical community (and of many French mathematicians in particular) to Ruffini's proof of the impossibility of solving the quintic equations with radicals.[76] The exception was precisely Cauchy, who in 1821 had written to Ruffini:

> Your memoir on the general resolution of equations has always seemed to me worthy of attracting the attention of geometers and, in my opinion, fully demonstrates the algebraic insolubility of general equations of degree greater than four. [...] Your work on the non-algebraic solvability of equations is precisely the memoir that I have recently mentioned

---

[72]E.S. Allen, *The Scientific Work of Vito Volterra,* The American Mathematical Monthly, 48 (1941), pp. 516–519, p. 516.

[73]U. Bottazzinni, *Italy. Ettore Bortolotti.* In: J.W. Dauben, C.J. Scriba, (eds.), *Writing the History of Mathematics: Its Historical Development,* Birkhäuser, 2002, Basel, pp. 90–91.

[74]Letter from Guccia to Bortolotti, December 3, 1906. Folder Bortolotti, Archive of the Circolo Matematico di Palermo.

[75]*Ibid.,* May 12, 1907.

[76]R.G. Ayoub, *Paolo Ruffini's Contributions to the Quintic,* Arch. Hist. Exact Sci., 23 (1980), pp. 253–277.

in your favor to some members of the Academy, on the occasion of the appointment of a corresponding member in the section Geometry.[77]

Bortolotti accepted the job and the official announcement was made, with due emphasis, at the meeting of the Società Italiana per il Progresso delle Scienze held in Parma in 1907. Guccia's plan evolved from the original idea of reprinting Ruffini's main work *Teoria Generale delle Equazioni, in cui si dimostra impossibile la soluzione algébraica delle equazioni generali di grado superiore al quarto* (in English, *General theory of equations, in which is proven the impossibility of solving by radicals general equations of degree greater than four*), published in Bologna in 1799 with 324 pages, to the reprinting of the complete mathematical works and the scientific correspondence of Ruffini.

In the session of the Circolo held on August 31, 1907, with the previous favorable vote of the Steering Council, the plan was approved. The Council detailed several provisions for the project:

1st   To publish in a separate volume, by the Circolo Matematico di Palermo and under the direction of Prof. Guccia, director of the Rendiconti, the *Opere matematiche* of Paolo Ruffini and his correspondence with the scientists of his time.

2nd   To accept, with applause, Prof. Guccia's offer to advance the necessary funds for such publication.

3rd   To reimburse Prof. Guccia from the sales of the *Opere* up to the cost of the publication, and, in accordance with Prof. Guccia's proposal, that any possible profit will benefit the Circolo and any possible loss would be on the side of Prof. Guccia.[78]

Guccia communicated to Bortolotti the approval of the project and started to buy memoirs and other works of Ruffini that the Circolo did not have in its library. Meanwhile, Bortolotti organized the photographing Ruffini's full scientific correspondences.[79]

Unfortunately, the project did not progress as planned. Volume I was not published by the Circolo until seven years later, in 1915. The delay could have been caused by Guccia's dislike of Bortolotti's critical stand on the attitude of French mathematicians towards Ruffini and his work. The unfavorable circumstances created by World War I affected the sales, and only a few copies were sold. At the International Congress of Mathematicians held in Bologna in 1928, Bortolotti

---

[77]Letter from Cauchy to Ruffini, September 20, 1821. In: E. Bortolotti, *Opere matematiche di Paolo Ruffini*, Vol. III (mathematical correspondence), Unione Matematica Italiana, Ed. Cremonese, 1954, Rome, pp. 88–89.

[78]Adunanza, August 31, 1907. Suppl. Rend. Circ. Mat. Palermo, 2 (1907), p. 41.

[79]Letter from Guccia to Bortolotti, December 30, 1907. Folder Bortolotti, Archive of the Circolo Matematico di Palermo.

reported on the status of the publication.[80] Volume II was published by the Rome publisher Edizioni Cremonese with the support of the Unione Matematica Italiana (which, as we will later see, had been created in the meantime) and printed in Palermo in 1943 by the Tipografia Matematica of Gaetano Senatore. Before the volumes were sent to Rome, however, the volumes were irreparably damaged in the bombing of Palermo during World War II. Only three or four copies remained undamaged. Bortolotti himself passed away in 1947. Volume II was recovered from the existing copies and a facsimile reproduction was published in 1953. A year later, in 1954, the third and last volume containing Ruffini's correspondence with his contemporary mathematicians was published, as Bortolotti had prepared it.[81]

---

[80]E. Bortolotti, *La pubblicazione delle opere matematiche di Paolo Ruffini*. In: Atti del Congresso Internazionale dei Matematici: Bologna, 3–10 Settembre 1928 (VI). Nicola Zanichelli, 1929–1932, Bologna, vol. 6, pp. 401–406.

[81]E. Bortolotti, *Opere matematiche di Paolo Ruffini*, Vol. I, Circolo Matematico di Palermo, 1915, Palermo; Vol. II, Unione Matematica Italiana, Ed. Cremonese, 1943 (facsimile edition, 1953), Rome; Vol. III (mathematical correspondence), Unione Matematica Italiana, Ed. Cremonese, 1954, Rome.

Chapter 7
Epilogue

Guccia's death occurred as World War I began. The dramatics of the war deteriorated his already poor health; as a true internationalist, he felt great pain at the spectacle of Europe destroying itself. The death of Guccia and the aftermath of the war had serious consequences on the Circolo and the Rendiconti. However, the society's members were able to overcome the difficulties and later to resist the pressure of Mussolini's regime until World War II, when the bombing of Palermo literally destroyed Guccia's project. We finish this story with an evaluation of the work, the efforts and the contributions of Guccia to the main aim of his life: the development of an international community of mathematicians.

7.1 The Death of Giovanni Battista Guccia

Family intermarriage has had a long tradition within European Royal families. It was a means of maintaining power and stability. The House of Habsburg, the House of Bourbon, and the House of Hanover engaged in these practices, but they also suffered from the health consequences of inbreeding. In many countries, this custom extended to nobility; this was the case in Sicily. In the third generation of the Guccia family lineage, founded by the patriarch Giovanni Battista Guccia e Bonomolo, there was a marriage between two cousins: Giovanni Battista Guccia e Di Maria married Carolina Cipponeri e Guccia. In the next generation, their daughter Chiara Guccia e Cipponeri married Giuseppe Maria Guccia e Vetrano, who was stepbrother of her grandfather. From this marriage was born the mathematician Giovanni Battista Guccia e Guccia. This family history could have been a health risk for Guccia. Indeed, ten children were born into the Guccia e Guccia family. Four of them died within the first year of life. August 1852 was a sad month for the family: on the same day, August 7, one child was born, Rosalia, and another one died at the age of eleven months; Rosalia survived only three days. Four family members died

© Springer International Publishing AG, part of Springer Nature 2018
B. Bongiorno, G. P. Curbera, *Giovanni Battista Guccia*,
https://doi.org/10.1007/978-3-319-78667-4_7

Fig. 7.1 Guccia mausoleum in the Cappuccini cemetery in Palermo

when they were precisely 59 years old, Concetta Maria, Carolina, Maria Stella and the mathematician Giovanni Battista (there is a mausoleum of the Guccia family in the cemetery of the Cappuccini in Palermo, Fig. 7.1).

Giovanni Battista Guccia suffered from poor health throughout his life. In his letters, he regularly discusses his illness or convalescence. He was constantly trying to avoid high temperatures and their effect on his health; this he commented to Cremona before the defense of his thesis.[1] His regular travels abroad also affected

---

[1]Letter from Guccia to Cremona, June 4, 1880. Folder Cremona, Archive of the Mazzini Institute of Genoa.

his health: in his first visit to Paris in 1880, he suffered a bronchopneumonia that required "ten days of fever and ten of convalescence."[2] It was also common that upon returning home he was not in good condition. He wrote, "Upon arriving in Palermo I went to bed with a gastric fever, which gave me a bit to do."[3] He often visited health resorts to rest and recover; just before the presentation of the medal and celebration of the anniversary of the Circolo in 1914 he stayed at the spa city of Karlsbad.[4]

His friend and collaborator Michele de Franchis recalled Guccia's health condition during the celebration in his honor and in the later months:

> I still remember those days. From autumn of 1913 Guccia had already been badly struck by the illness that so quickly extinguished his life. By April the disease had already progressed enormously. However, seeing the Circolo's importance for Science being universally recognized by the oldest and most notable scientific institutions and by the most distinguished scientists in the discipline infused him with new vigor. Ephemeral vigor! The day that the medal was presented to him he wanted to speak to thank the participants. A tremor of emotion went across the audience: his voice seemed to come from the hereafter and on his gaunt face inexorably wafted death.

> Everything was tried in vain to heal him, consulting distinguished doctors and hospitals. It was precisely when he was in Valmont's hospital that the horrendous war that still disrupts Europe started. He worried about the massacres and the unleashed hatred among nations which he very much admired and where he had beloved friends. He worried about the inevitable consequences for an international scientific society like ours.[5]

On October 9, Guccia was in Florence when he sent a telegram to his close acquaintance Fortunato Bellina, a Sicilian who worked in the Ministry of Post and Telegraph in Rome. In the telegram, he explained:

> With my brother and nurse I am leaving for Rome, arriving today at 18:55. We then go to the Hotel Quirinale.

Fortunato Bellina must have previously agreed to be Guccia's intermediary with Volterra, as he immediately informed Volterra of Guccia's telegram.[6] We do not have Volterra's response to Guccia, but from Bellina's next letter to Volterra we know that on October 14, Guccia had to go through "a surgical operation to extirpate a gland for its histological examination."[7] One week later, Guccia sent a handwritten note to Volterra:

---

[2] *Ibid.*, July 17, 1880.

[3] Letter from Guccia to Pasquale Villari, May 16, 1908. Folder Villari, Archive of the Circolo Matematico di Palermo.

[4] Letter from Volterra to Guccia, March 17, 1914. Folder Volterra, Archive of the Circolo Matematico di Palermo.

[5] M. De Franchis, *op. cit.,* pp. xi–x.

[6] Letter from Bellina to Volterra, October 9, 1914. Vito Volterra Collection, Archive of the Accademia Nazionale dei Lincei.

[7] *Ibid.*, October 14, 1914.

Dearest friend, in the last few days I have felt less well and my doctor has forbidden me to have visits. Please, excuse me if I have not contacted you before. We leave tomorrow morning for Naples to take in the evening (if there is no storm) the ship to Palermo.[8]

On October 24, Guccia sent a telegram to Bellina explaining that he had "arrived in moderate condition." Bellina sent the telegram to Volterra, adding in handwriting "As long as there is life there is hope (as we say in Sicily)."[9]

Michele de Franchis recalled the final moments of Guccia's life:

In the early hours of October 29, his great soul exhaled. Shortly before, he was thinking about the Circolo with phrases that no one, unfortunately, could understand.[10]

Fortunato Bellina wrote to Volterra that same day:

The catastrophe has occurred and, although it was predictable, I have learned it with the greatest pain. I comply with the sad duty of communicating it to you.[11]

Giovanni Battista Guccia had just turned 59 years old when he passed away (Figs. 7.2 and 7.3). According to the letter written by Michele de Franchis to Mittag-Leffler on December 4, 1914,[12] the cause of Guccia's death was what was known at the time as "Hodgkin's disease" which is an oncological illness of the lymphatic system (Fig. 7.4). The histological analysis of the glands carried in Rome must have shown such an advanced stage of the disease that the doctor suggested a fast return to his home in Palermo.

7.2   The Circolo After Guccia

The second half of 1914 was dramatic for the world: the outbreak of World War I was the starting point of four years of terrible destruction and deep changes. The situation was quite a threat to the Circolo Matematico di Palermo and to the Rendiconti as their activity was heavily based on international collaboration. The absence of the Circolo's founder and leading figure was an additional difficulty during these challenging times.

It appears that the scientific sessions of the Circolo continued regularly in Palermo during the war. We do not have direct evidence of this fact, since the "Supplemento ai Rendiconti", where the proceedings of the sessions were recorded since 1906, stopped its publication during the war. However, there is indirect evidence indicating that sessions were held. Namely, in the published volumes of

---

[8]*Ibid.*, October 21, 1914.

[9]*Ibid.*, October 24, 1914.

[10]M. De Franchis, *op. cit.*, p. x.

[11]Letter from Bellina to Volterra, October 29, 1914. Vito Volterra Collection, folder Guccia, Archive of the Accademia Nazionale dei Lincei.

[12]Letter from De Franchis to Mittag-Leffler, December 4, 1914. Center for History of Science of the Royal Swedish Academy of Science. The letter is not recorded in the catalog of the archive.

Con l'animo angosciato annunziamo la morte, avvenuta in Palermo, il 29 ottobre 1914, alle ore 3, del benemerito Fondatore del
Circolo Matematico di Palermo e Direttore delle pubblicazioni periodiche di questa Società: « *Rendiconti del Circolo Matematico di Palermo* »,
« *Supplemento ai Rendiconti del Circolo Matematico di Palermo* », « *Annuario del Circolo Matematico di Palermo* », « *Indici delle Pubblicazioni del
Circolo Matematico di Palermo* » :

D.ʳ GIOVANNI BATTISTA GUCCIA

Nobile dei Marchesi di Ganzaria,

Professore ordinario di Geometria Superiore nella R. Università di Palermo.

Il Circolo Matematico di Palermo tutto a Lui deve.

Immersi nel dolore del recente grave ed irreparabile lutto, non
possiamo per ora trovare parole per commemorare l'illustre Estinto.

Per quanto dubbiosi di poter dire degnamente di Lui *(tanto no-
mini nullum par elogium!)*, in principio del prossimo Tomo XXXIX,
ne tenteremo una biografia.

Per deliberazione dell'Assemblea dei Soci (22 novembre 1914), la
Società prende il lutto per un anno.

Inoltre, in apposita adunanza straordinaria, i Soci saranno convocati
per una solenne commemorazione.

Novembre 1914.

LA REDAZIONE.

Fig. 7.2 Notification of the death of Giovanni Battista Guccia in the Rendiconti. Courtesy of the
Circolo Matematico di Palermo

Belgio

Angelo Ta-
ondo elenco
a della Soc.
e L. 200 —
ista, L. 5 —
hiaramonte
hiaramonte
2, per spon-
rai dolcieri
pa Di Chia-
lattiolo Gio-
o l. 0,50; La
sina Salva-
,20; Ernesto
colò l. 0,50;
la Vincenzo
Notaro An-
l. 0,50; Ca-
esco l. 0,50;
sterer 0,50,
. 0,50, Blind
0,50; N. N.
13,45).
e, che desi-
ha inviato
ie completo
ie completo
ivali, calze,
stato.

Nazionale

mma da Ro-
nissione pa-
ll'on. Cava-
la quistione

è occupata
'Economista
e ci sembra
'importanza

a necessità,
opportunità

Per la morte del prof. G. G. Guccia

A pochi mesi di distanza del giubileo della gloriosa fondazione di G. B. Guccia, il
Circolo Matematico di Palermo, oramai
istituzione scientifica internazionale che co-
stituisce l'ammirazione di tutti i cultori di
scienze esatte, scende nel sepolcro il suo
fondatore.

Di G. B. Guccia, professore di matematiche superiori alla nostra R. Scuola d'ap-
plicazione per gl'ingegneri ed architetti,
studioso costante della difficile scienza di
Euclide, propagatore insigne di essa con
idee e metodi da vero Mecenate diranno degnamente gl'ingegni eletti che seguono lo
studio di quelle severe discipline.

L'opera del nostro scienziato oltrepassa i
confini della patria, essa per la sua bontà
per la sua efficacia scientifica è divenuta
mondiale.

Il *circolo* matematico di Palermo è l'istituzione più importante del genere, i reso-
conti da esso pubblicati bimestralmente,
sono la sintesi e l'esponente di ciò che di
meglio vi è nella produzione internazionale in materia, la tipografia matematica da
Lui istituita è un meraviglioso completa-
mento dell'azione scientifica esercitata con
la pubblicazione periodica.

Palermo intellettuale dovrà essere rico-
noscente a questo suo figlio, cui nel 14 dello
scorso aprile venne giustamente tributata
una medaglia d'oro commemorativa per
l'opera altamente civile da esso compiuta,
usando nella maniera più consona ai nuovi tempi delle ricchezze patrimoniali e delle
doti del suo ingegno.

Per conto nostro facciamo un voto augu-
rale, ed è che l'Istituzione Scientifica fon-
data e fatta prosperare dal prof. G. B. Guccia non scenda nel sepolcro con Lui; ma
gli sia *monumentum aere perennius*, pren-
dendo sempre maggiore incremento, per il
progresso delle scienze matematiche, e a
maggior decoro della nostra Città.

Il banchetto all'on. Di Cesarà

R. Istituto

Nella sessi
tennero la li
giovani:
Sezione ca
doardo. Di I
cat Emilio,
Simoncini Pi
lino.
Sezione mc
Fell Giovann
colò, Pecorai
Sezione cos
Mento Letter
Vincenzo, Ma
Il giorno 2
zioni.
Col giorno
rogabilmente

Per i

I garibaldi
per quanto r
Essi invoca
legge la qua
si provvederà
pagna e per
degli assegna
Concludono
garibaldini d
diritto per
2 giugno 186
da Milazzo

Liceo

Nel R. Lic
vranno princ
Alle lezion
coloro che e
alle varie cl
tasse se nuo
già interni
Le tasse e
cati nell'albo

R. Scuo

Nella R. S
avranno pri
ore 8,30, e l
istituti medi
prorogabilme
Le alunne
da altra scu
se presentera
di tassa di

Fig. 7.3 News of the death of Giovanni Battista Guccia in the Giornale di Sicilia. Courtesy of the
Biblioteca Centrale della Regione Siciliana

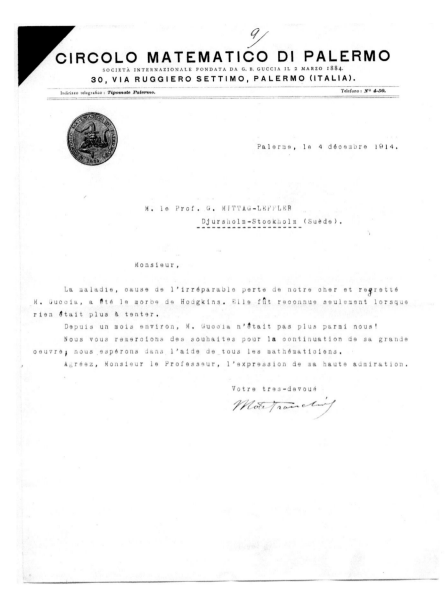

Fig. 7.4 Letter from Michele de Franchis to Gösta Mittag-Leffler regarding the cause of the death of Giovanni Battista Guccia. Courtesy of the Center for History of Science of the Royal Swedish Academy of Sciences

the Supplemento we see that before the war, session number 618 was held on March 15, 1914, and after the war, session number 717 was held on January 26, 1919. The Circolo's statute stipulated that sessions should be held every second and fourth Sunday of the month, except for the vacation months of September and October.

Fig. 7.5 Michele de Franchis. Courtesy of Aldo de Franchis

Accounting for both scientific sessions and other administrative sessions held during those years, an average of twenty sessions were held per year. The papers published in the Rendiconti during the war also carry the date of the session where they were accepted. The scientific sessions of the Circolo resumed after the war; session number 864 took place on March 28, 1926.[13]

The publication of the Rendiconti also continued during the war. The management of the Rendiconti was in the hands of the Editorial Board and the managing editor.

When Guccia died, Michele de Franchis became the managing editor of the Rendiconti. De Franchis (Fig. 7.5) was a Palermitan mathematician, twenty years younger than Guccia, who had been Guccia's student at the University of Palermo. He worked in Algebraic Geometry and held positions in Cagliari, Parma, and

---

[13] See Suppl. Rend. Circ. Mat. Palermo: vol. 9 (1914); vol. 11 (1919–1920); vol. 12 (1921); vol. 14 (1924–1925); vol. 15 (1926–1928); vol. 16 (1929–1930).

Catania, and succeeded Guccia in his professorship at the University of Palermo.[14] He was admitted to the Circolo Matematico di Palermo in 1895, when he was twenty years old. Since De Franchis joined the Circolo, he became very close to Guccia, who helped him on many occasions.

The Bordin prize of 1907 was one of these occasions when Guccia intervened in favor of De Franchis. For the prize, the Académie des Sciences de Paris proposed the problem of classifying a certain class of hyperelliptic surfaces.[15] The deadline for the submission of the papers was December 31, 1906. De Franchis worked with Giuseppe Bagnera (a mathematician from Palermo and also a member of the Circolo) but as the deadline was approaching, they realized that they could not finalize their paper on time. Meanwhile, Federigo Enriques and Francesco Severi had presented a solution to the problem on time. A few months later, De Franchis, speaking with Enriques and Severi, noted that there were several errors in their solution. This caused Enriques and Severi to submit a revised version after the due date. They submitted the revised paper, as Jeremy Gray explains, "even though they knew that Bagnera and De Franchis were also candidates for the same prize, and indeed had better results."[16] Guccia was in Paris when De Franchis explained the situation to him, including the precise mathematical details of the case, and requested his help on account of Guccia's friendship with many members of the Académie.[17] Responding to De Franchis, Guccia interceded before the Académie. He also helped De Franchis translate the summary of his paper into French for the Comptes Rendus.[18] Despite Guccia's efforts, however, the Académie awarded the 1907 Bordin prize to Enriques and Severi, and announced a new prize on the same subject for 1908. The fate of the prize for Bagnera and De Franchis still was undecided: their paper was sent off to the Académie on December 27, 1908, four days before the new deadline. The train carrying the parcel containing De Franchis and Bagnera's manuscript stopped in Messina just when the terrible earthquake— that we have already mentioned—destroyed the city. However, the parcel was found and arrived in Paris on January 15, 1909. Due to the exceptionality of the case, the work was accepted and the prize was awarded to Bagnera and De Franchis in 1909. A full account of the vicissitudes of the case was published in the Rendiconti.[19]

The Steering Council of the Circolo (and Editorial Board of the journal) was elected for the last time before the war in January 1912 for the 1912–1914 term. Twenty-five of its forty-five members were foreigners. This international diversity

---

[14]C. Ciliberto, E. Sernesi, *Some aspects of the scientific activity of Michele de Franchis,* In: *Opere di de Franchis,* Rend. Circ. Mat. Palermo (2) Suppl. No. 27 (1991), 3–33.

[15]*Prix Bordin,* Suppl. Rend. Circ. Mat. Palermo, 2 (1907), pp. 114–115.

[16]J. Gray, 2006, *op. cit.,* p. 16.

[17]Letters from De Franchis to Guccia, October 21, October 25, 1907. Folder De Franchis, Archive of the Circolo Matematico di Palermo.

[18]*Ibid.,* October 26, November 1, 1907.

[19]Suppl. Rend. Circ. Mat. Palermo, 4 (1909), pp. 34–36.

created a delicate situation when the war ended, as Guccia had foreseen in his last days of life.

After World War I, the governments of the Allied Powers launched a new policy for international cooperation in science. This policy was implemented by the International Research Council, which was created in Brussels in 1919. The declared objectives of the International Research Council were to coordinate international efforts in science and to foster the creation of international scientific associations. The Council also incorporated into its aims the content of some resolutions that had previously been approved in governmental conferences held in London and Paris in 1918. In these resolutions, the policy of excluding Germans and their allies in the war from scientific international cooperation was explicitly stated.[20] The hidden agenda of the International Research Council was to diminish the strength of German science.

In mathematics, the consequences were felt immediately. In accordance with the policies of the International Research Council, the creation of the International Mathematical Union was outlined at the 1919 Brussels conference. The decision to hold the next International Congress of Mathematicians in Stockholm was overruled. The new venue for the International Congress was the disputed city of Strasburg, which had just been recovered by France after its loss to Germany in the Franco-Prussian war of 1870–1871. In 1920, the Strasburg Congress was held, and the International Mathematical Union was officially created. A new era had started in international scientific cooperation: the members of the International Mathematical Union were not individual mathematicians, but countries. Germans, Austrians, Hungarians and Bulgarians were not invited to attend the Congress or to join the Union.[21]

Departing from the Rendiconti's previous aversion to political rhetoric, volume XI of the "Supplemento ai Rendiconti" corresponding to 1919–1920 opened with a text entitled "The Editorial office to the members of the society," which declared:

> Despite the fact that difficulties of all kinds, especially economic ones, have made our work more and more painful and despite the fact the effects of the great cataclysm, in which Europe almost annihilated itself, instead of fading away are growing in intensity, affecting alarmingly our international scientific societies, we want to demonstrate our good will by resuming the publication of the "Supplemento ai Rendiconti del Circolo Matematico di Palermo" that was interrupted in 1914. We hope with this to please our members. [...]

> We should express our wish that the Circolo Matematico di Palermo will be able to overcome the disadvantages of the current period, returning to the prosperous situation that it had during the life of its great founder G.B. Guccia so that it still may constitute one of the bodies for collaboration between the intellectuals from all countries, collaboration which is necessary for the peaceful and fraternal cooperation between people, for the benefit of the entire Humanity.[22]

---

[20] O. Lehto, *Mathematics without borders: A History of the International Mathematical Union*, Springer, 1998, Berlin, pp. 15–21.

[21] *Ibid.*, pp. 24–33.

[22] Suppl. Rend. Circ. Mat. Palermo, 11 (1919-1920), pp. 5–6.

This was a different style of addressing the members of the Circolo to the one that had been in use up to that moment. The world had changed after the war, and the nineteenth century style was part of the past. De Franchis' rhetoric was not Guccia's rhetoric.

In this context, the Circolo still seemed to be an isle of fraternal coexistence. However, several important mathematicians in the Circolo's Steering Council were also in charge of enforcing the exclusion policies in the new scientific-political structures of the time. Two notable examples of this were Émile Picard, who was president of the International Research Council from 1919 to 1931, and Charles de la Vallée Poussin, who was president of the International Mathematical Union from 1920 to 1924. Picard and De la Vallée Poussin, among others, pushed for the expulsion of Germans from the Circolo's Steering Council.[23] As a response, De Franchis strongly defended the legacy of Guccia: the Circolo was a purely scientific international society, where there should be no political interference. De Franchis won, and was able to maintain the Council as it was elected: with mathematicians from eleven different countries; among them France and Belgium as well as Germany, Austria and Hungary.[24]

The journal was able to continue its publication due to the work and efforts of two other persons in Palermo: the treasurer of the Circolo Ester Paolo Guerra and the typographer Gaetano Senatore. In the text "The Editorial office to the members of the society," it was explained that

> The editorial office feels obliged to publicly thank the treasurer E.P. Guerra, who has carefully done a large part of the work required for this publication, despite the great effort needed for the administration of the society, which relies entirely on him. Praise also goes to the owner-manager of the Tipografia Matematica of Palermo, Mr. Gaetano Senatore, who was able to overcome the difficulties that affect the printing industry and maintained for our publications, as much as possible, the traditional typographical precision.[25]

The other problem that the Circolo had to face after the war was that of its finances. For thirty years, the society had been financed by the fortune of its founder, who was willing to use his personal wealth to support the Circolo.

From the foundation of the Circolo in 1884 to the end of 1900, the financial balance was (rounding the quantities to hundreds):

1884–1900					
Income	Fees	Publications	Subsidies	Guccia	Total
	35,400	9,400	9,800	2,000	61,200
Expenses	Publications	Administration		Premises	Total
	39,500	13,200		4,400	61,200[26]

---

[23]Letter from R. Rothe to De Franchis, March 24, 1921. Folder De Franchis, Archive of the Circolo Matematico di Palermo.

[24]A. Guerraggio, P. Nastasi, *Italian Mathematics Between the Two World Wars*, Springer, 2006, Berlin, pp. 58–59.

[25]Suppl. Rend. Circ. Mat. Palermo, 11 (1919–1920), pp. 5–6.

[26]*Conti consuntivi 1884-1900*, Rend. Circ. Mat. Palermo, 15 (1901), pp. 156–157.

The main income came from the fees from members, selling the Rendiconti to institutions, and author's payments for reprints. The Ministry of Public Education and the City of Palermo had given the Circolo occasional subsidies, which altogether averaged a contribution of 500 lire per year. The personal contribution of Guccia in this period was of 2,000 lire. With respect to expenses, the main item was the printing and mailing of the Rendiconti and author's reprints. Administrative expenses consisted of the administrative functions of the editorial offices, as well as the cost of the Circolo's library.

The situation changes substantially when we look at the years of the international burst of the Circolo, after the Heidelberg 1904 and the Rome 1908 International Congresses:

		Income			
Year	Fees	Publications	Subsidies	Guccia	Total
1908	8,400	3,600		6,500	18,600[27]
1910	9,200	4,000	600	4,300	18,000[28]
1912	11,300	3,500	600	4,600	20,000[29]

	Expenses			
Year	Publications	Administration	Premises	Total
1908	12,600	1,800	2,800	18,600[30]
1910	14,000	1,100	2,800	18,000[31]
1912	16,700	1,100	2,200	20,000[32]

The increase in the number of members of the society had the negative effect of increasing the deficit resulting from the difference between the amount collected from the fees and the cost of publishing the Rendiconti (and the other publications of the Circolo). Since the member's fee was still set at the initial amount of fifteen lire (which was established in 1884), Guccia had to make a large personal contribution every year to achieve a financial balance.

The financial help and all his personal effort in the running of the society were not the only contributions that Guccia made to the Circolo. Since the founding of the Circolo, the seat of the society was located in the Palazzo Guccia, on Via Ruggiero Settimo. The library of the Circolo was also in the palace, comfortably furnished at Guccia's expense. When the printing house Tipografia Matematica was created,

[27] *Conto consuntivo 1908,* Supp. Rend. Circ. Mat. Palermo, 5 (1910), p. 3.

[28] *Conto consuntivo 1910,* Supp. Rend. Circ. Mat. Palermo, 6 (1911), p. 67.

[29] *Conto consuntivo 1912,* Suppl. Rend. Circ. Mat. Palermo, 9 (1914), p. 91.

[30] *Conto consuntivo 1908,* Supp. Rend. Circ. Mat. Palermo, 5 (1910), p. 3.

[31] *Conto consuntivo 1910,* Supp. Rend. Circ. Mat. Palermo, 6 (1911), p. 68.

[32] *Conto consuntivo 1912,* Suppl. Rend. Circ. Mat. Palermo, 9 (1914), p. 91.

it was also placed inside the palace. With time, the premises of both the Circolo and the Tipografia were moved into different rooms of the palace. This occurred because of changes in the use of the palace by the Guccia family, and also because the rooms in the ground floor of the palace facing via Ruggiero Settimo were rented to businesses. These rooms were very profitable since the street had become part of an important commercial area of the city.[33] We already commented that in 1899 the Tipografia was moved to a building owned by Guccia behind the palace on Via Villareale. In 1907, it was moved to Via Guccia at the rear of the palace. The final move for the seat and library of the Circolo was in 1914. They occupied three rooms in the building of Via Villareale. This allowed for a much better organization of the huge bibliographic materials. During the first thirteen years of the Circolo, the rooms for its seat and the library were free from rent; starting in 1897, the society had to pay rent. In his will, Guccia left his personal library to the Circolo, and established two years of free rent for the rooms where the Circolo was situated at that time, in the ground floor of Via Villareale.[34]

The economic situation of the Circolo during and after the war was extremely difficult, and this had a direct effect on the publication of the Rendiconti. Bianchi and other mathematicians from Pisa sent in 1919 the following letter to the Circolo

> The undersigned, while enclosing money orders to fulfill the request addressed by the cashier of Circolo Matematico concerning the payment of the fees of 1918 and 1919, would like to remark that the failure in sending these fees was justified by the suspension of all social activity [of the Circolo], and would have appreciated if the payment request had been accompanied by the explicit promise of the forthcoming publication of the delayed volumes of the Rendiconti.[35]

Guccia had prepared the first volume of 1914, which was number 37. The next volume, corresponding to the second half of 1914, was published on time even though the war had already started. On the front cover, the name of Giovanni Battista Guccia appeared as founder of the journal, and his collaborator Michele de Franchis as "Direttore". The two volumes for 1915 were also published on time. The first one had the usual number of pages, some 400, but the second one did not reach 350 pages. Starting in 1916, the Rendiconti went back to publishing only one volume per year, as had occurred until 1905.

De Franchis wrote to Bianchi explaining the difficulties which had accumulated as a consequence of the war:

> As you know, when the late Prof. Guccia was alive every year he arranged to cover the cash deficit with his personal means; in significant amounts. Those were normal times and the cost of labor and paper were not the current ones. My management commenced with the disaster that we all know. Now we also have to pay the rent for the premises and the cost of

---

[33] A. Chirco, M. Di Liberto, *op. cit.*, p. 96, pp. 165–166.

[34] Notar Ferdinando Lionti, October 31, 1914. Archivio Notarile Distrettuale di Palermo.

[35] Letter from Bianchi to De Franchis, June 9, 1919. Folder Bianchi, Archive of the Circolo Matematico di Palermo.

printing has tripled. After the outbreak of the war, a large number of members do not pay [the fee].[36]

The most difficult years for the journal were 1918 and 1919. The devaluation of many currencies was an extra problem for the finances of the Circolo. Eventually, one volume, number 43, covering the years 1918 and 1919 was published. The difficulties that De Franchis had to face were many; he ended his letter writing, "Summarizing, I have collected a lot of bitterness."

De Franchis did not mention in the letter to Bianchi that Guccia had established in his will an annual subsidy to the Circolo of 3,000 lire and two years of free rent for the premises that the Circolo occupied on Via Villareale.[37] In 1920, the rent for the premises was already larger than the annual subsidy.

In 1921, the Circolo launched an "extraordinary subscription" (a fundraiser), and requested a 25 lire donation from each member. The subscription went quite well: there were 192 donors, 97 of whom contributed double or more the suggested amount, and 49 of whom contributed one hundred or more lire. There were some extraordinary contributions: George Birkhoff sent 500 lire; H.D. Thompson 250 lire; and Henrik Block from Lund, Matsusaburo Fujiwara from Tohoku, and Harris Hancock from the U.S.A., sent 200 lire each. The total amount collected was some twelve thousand lire. There was a substantial contribution by a non-mathematician: Count Lanza di Mazzarino contributed 375 lire. He was the grandfather of Gioacchino Lanza Tomasi in whose personal library are mathematical books of Guccia's uncle, the Prince of Lampedusa.[38]

The financial balance for the years from 1917 to 1924 exhibits great instability. There were tremendous oscillations in the amounts received from the fees of members: from one year to the next, the total revenues could be halved or doubled, or even tripled. It was clearly difficult to manage the finances of the society under those circumstances. Eventually, the member fee was doubled and set at thirty lire.[39] Subsidies from the Ministry of Public Education resumed, but neither regularly nor in uniform amounts.[40]

In 1926, the International Education Board of the Rockefeller Foundation granted 900 dollars (which corresponded to 22,338 lire) to the Circolo Matematico di Palermo as a contribution for the Rendiconti. The objective was "to overcome the economic difficulties of the present time" so that the journal could gradually resume the full activity achieved in the years prior to the war. Tullio Levi-Civita, along with the mathematicians from the Accademia dei Lincei, had prompted the Accademia

---

[36]Letter from De Franchis to Bianchi, June 15, 1919. Folder Bianchi, Archive of the Circolo Matematico di Palermo.

[37]Notar Ferdinando Lionti, October 31, 1914. Archivio Notarile Distrettuale di Palermo.

[38]Suppl. Rend. Circ. Mat. Palermo, 12 (1921), pp. 7–11.

[39]Suppl. Rend. Circ. Mat. Palermo, 15 (1926–1928), p. 21.

[40]Suppl. Rend. Circ. Mat. Palermo, 15 (1926–1928), p. 3 and p. 12.

dei Lincei to request this subsidy from the Rockefeller Foundation.[41] The next year
the Foundation also gave a smaller subsidy. The list of all subsidies received by
the Circolo from 1884 until 1927 appears in the 1928 Annual Directory; the total
amount is 193,837 lire, from which 83,947 came from the founder Giovanni Battista
Guccia.[42]

The Archive of the Circolo holds a number of sad letters from mathematicians
resigning their membership due to economic reasons. The reports of the sessions
appearing in the Supplemento at that time regularly included lists of "radiazioni",
that is, members whose membership is canceled due to lack of payment. For
obvious confidentiality reasons, the lists include only the member's number and
not the name. Most of these "expelled members" were German, because of the
tremendous inflation in Germany after the war.[43] Despite all difficulties, new
members were still joining the Circolo: 28 joined in 1919, 47 in 1920, 16 in
1921, 18 in 1922, 13 in 1923, 32 in 1924, 35 in 1925, 22 in 1926 and 31
in 1927. Despite all the difficulties, in June of 1928 the society still had 584
members.[44]

Regarding the Steering Council, on Sunday January 15, 1928, elections to the
Steering Council for the 1928–1930 term took place. It was the first time since 1912
that elections in the Circolo were organized.

The forty-five members elected were:

Albeggiani	Fejér	Pascal
Angelitti	Fubini	Picard
Appell	Gebbia	Picone
Bertini	Hadamard	Pincherle
Bianchi	Hardy	Plans
Birkhoff	Hilbert	Scorza
Bohr	Landau	Severi
Borel	Lefschetz	Somigliana
Carathéodory	Levi-Civita	Tonelli
Castelnuovo	Lichtenstein	Valiron
Cipolla	Littlewood	Vallée Poussin, de la
Courant	Love	Vivanti
De Franchis	Marcolongo	Volterra
Dickson	Moore, E.H.	Weyl
Enriques	Nörlund	Wirtinger

---

[41] Suppl. Rend. Circ. Mat. Palermo, 15 (1926–1928), p. 5.

[42] Annuario Circ. Mat. Palermo 1928, pp. 128–129.

[43] Suppl. Rend. Circ. Mat. Palermo, 15 (1926–1928), p. 5.

[44] Annuario Circ. Mat. Palermo 1928, pp. 115–127.

There were 587 members of the Circolo at that time, of which 277 voted, that is to say, 47 percent, which still was a high figure for elections within a scientific society. As had been common in the past, there was little dispersion of the voting. The list of voters reveals that the society still had the support of the international mathematical community.[45]

The elections for the Steering Council for the 1931–1933 term took place at the beginning of 1931, and the committee remained very much the same. Four members had passed away during the previous period (Angelitti, Appell, Bianchi and Gebbia); they were replaced by G. Giorgi and F. Cantelli. This time, the committee had only forty-three members. We have no information on the details of the voting. In fact, we just know about the members of the committee from the inside cover of the volumes of the Rendiconti, as the Supplemento was no longer being published at the time.

Unfortunately, there were no new elections for the Steering Council of the Circolo once the three-year 1931–1933 term ended. The reason was that Mussolini's Fascist regime issued a law in 1933 that subjected all academies, institutes and associations devoted to sciences, letters and arts to the tutelage and surveillance of the State. The last statute of the Circolo had been legally approved in 1915 by the Italian State. The statute was the last step in the legal process of converting the Circolo into a charitable institution, which was a necessary condition for receiving the annual subsidy from the bequest of Guccia.[46] In accordance with Mussolini's 1933 law, the statute of the Circolo was repealed in 1934 and a new one was imposed on the society.

The new statute was absolutely contrary to the spirit that had inspired the Circolo since Guccia created it. It ended the international character by drastically limiting the number and the role of foreign members. The new statute gave control of the society to the Minister of National Education, who was now capable of controlling the election of the President of the society; requiring the President to swear to the Prefect of Palermo fidelity to the Fascist regime; entrusting the scientific guidance of the society to a committee appointed by the President; requiring annual activity reports be sent to the Minister; and giving the Minister the capacity to revoke the membership of "unworthy" members. One year later, Michele de Francis was appointed President of the Circolo. The Circolo Matematico di Palermo as created by Giovanni Battista Guccia in 1884 ceased to exist.

The Rendiconti continued its publication for a few years, almost half way into World War II. The last volume published was number 63, corresponding to the years 1940–1941. It contained ten articles occupying 176 pages, and it was printed in the Tipografia Matematica. The printing date of the last article was January 20, 1942, and the volume appeared on May 30, 1942. Some months later, the bombardments of Palermo destroyed the Tipografia Matematica.

---

[45] Suppl. Rend. Circ. Mat. Palermo, 15 (1926–1928), pp. 21–23.

[46] Letter from Bagnera to Volterra, November 19, 1914. Vito Volterra Collection, folder Guccia, Archive of the Accademia Nazionale dei Lincei.

After World War II, the Circolo Matematico di Palermo was reconstructed and the Rendiconti commenced the publication of its second series. The efforts of the Sicilian mathematician Eduardo Gugino were instrumental in these two processes. But, that is another story.[47]

7.3 Final Curtain

To end this story, we return to our starting point: the Palermo bombardments of 1943. The Palazzo Lampedusa of the Tomasi family was near the port facilities. For this reason, it was struck by bombs from the beginning of the bombardments, and was severely damaged. The letters of Giuseppe Tomasi di Lampedusa, the author of *The Leopard*, to his wife narrate the day-by-day destruction of the palace. On January 7, doors and windows were shattered; on March 23, a bomb hit an ammunition storehouse in the port causing a devastating explosion that collapsed part of the roof; on April 5, bombs fell in the central part of the palace, and the mountain of rubble inside the palace was two stories high.[48] The palace remained in ruins for seventy years. Today, it is being renovated to be sold as individual apartments.

The Palazzo Guccia, though not far away from the Palazzo Lampedusa, was not in the port area. In any case, bombs were falling in its vicinity since the beginning of 1943. Some nearby buildings were severely damaged, and some collapsed. It is commonly accepted that "the bombs demolished the Palazzo Guccia, irreparably destroying a sizeable part of the library and completely the typography."[49] This information is consistent with the recollection by Luigi Cardamone, a Sicilian mathematician born in 1919, who was studying at the University of Palermo during the bombardments. In an article published in 1962 in a local newspaper of Palermo, Cardamone explained that

> In the catastrophe of the last war, the seat of the Circolo and the Tipografia Matematica di Palermo were both totally destroyed. On the initiative of Prof. Eduardo Gugino, an assiduous and laborious search effort was made to recover the remains of the library of the Circolo that were scattered in the rubble.[50]

However, the data available in the State Archive in Palermo corresponding to the records of "War Damages" provide no evidence that bombs hit the Palazzo Guccia. In fact, there is plenty of evidence to the contrary. First, it is notable that no claim for compensation of damages by the owners of the palace was found. This, however, could be due to the loss of the documents. The State Archive does contain several compensation claims filed by the owners of the businesses located on the ground

---

[47] A. Brigaglia, G. Masotto, *op. cit.*, pp. 382–386.

[48] A. Vitello, *op. cit.*, pp. 208–216.

[49] A. Brigaglia, G. Masotto, *op. cit.*, p. 382.

[50] L. Cardamone, *Il Circolo Matematico di Palermo*, L'Ora, October 4–5, 1962, p. 4.

floor of the palace. From the very beginning of the bombardments, these claims
were regularly filed, requesting the payment for windows, showcases and small
furniture damaged by the shrapnel from the bombs falling in the streets. Some of
these claims were filed as much as three years after the bombardment.[51] This would
not be the case if the whole palace had been destroyed. Additionally, in one of the
dossiers there is a map of the area around the palace, drawn by the official in charge
of the dossier, marking the places where bombs had fallen (Fig. 7.6). None of the
markings appear over the palace.[52] Furthermore, the death certificate of Giovanni
Battista Guccia's sister in law Adriana Fatta records that she died six months after
the bombardments on December 28, 1943, in the Guccia Palace. It is also notable
that the photographs of the Palazzo Guccia in the 1950s and 1960s exhibit a full
building with no signs of any major reconstruction process.

The origin of this confusion probably lies in the misinterpretation of the words
of Cardamone when he referred to the total destruction of "the seat of the Circolo
and the Tipografia Matematica di Palermo." We have already explained that the
Circolo moved its premises from the Palazzo Guccia to Guccia's building on Via
Villareale and the Tipografia moved to Via Guccia, both of which are at the rear of
the palace. In the "War Damages" records in the State Archive of Palermo, there is a
dossier concerning the total destruction of the building adjacent to the one where the
Circolo had its seat, and situated precisely in front of the premises of the Tipografia
Matematica.[53] It is highly probable that the collapse of this building caused the
damages to the Circolo and the Tipografia described by Cardamone.

There is a whole collection of stories by mathematicians from Palermo concerning the recovery of the books of the library of the Circolo from the ruins left by
the bombardment. Several thousands of books and the boxes containing the Archive
of the Circolo were rescued. The library materials were certainly rescued from the
premises of the Circolo, but those premises were not in the palace. The books and
the boxes recovered were moved to the new university area where the Schools of
Sciences were located, on Via Archirafi.

The stories of the Guccia family and of the Palazzo Guccia run parallel to
each other. Let us recall that almost the full fortune of the old marquis Giovanni
Battista Guccia e Bonomolo went to the parents of Giovanni Battista Guccia e
Guccia, thanks to strategic marriages within the family, well-considered wills and
wise agreements with other heirs. In 1899, the company "Acque di Casa Guccia"
was sold for a sizeable amount to a firm from Turin.[54] Giovanni Battista's father
died in 1900; in his will, he followed the old-style tradition and left almost all his

---

[51]Folder 14, n. 1510; folder 17, n. 2677; folder 35, n. 5532; folder 38, n. 7872. Section "Danni di
Guerra", Archivio di Stato di Palermo, section "La Gancia".

[52]*Ibid.,* folder 35, n. 5532.

[53]*Ibid.,* claim by Agatha Spadafora, folder 27, n. 7094.

[54]Notar Francesco Cammarata, August 31, 1899. Section "Notai". Archivio di Stato di Palermo,
section "La Gancia".

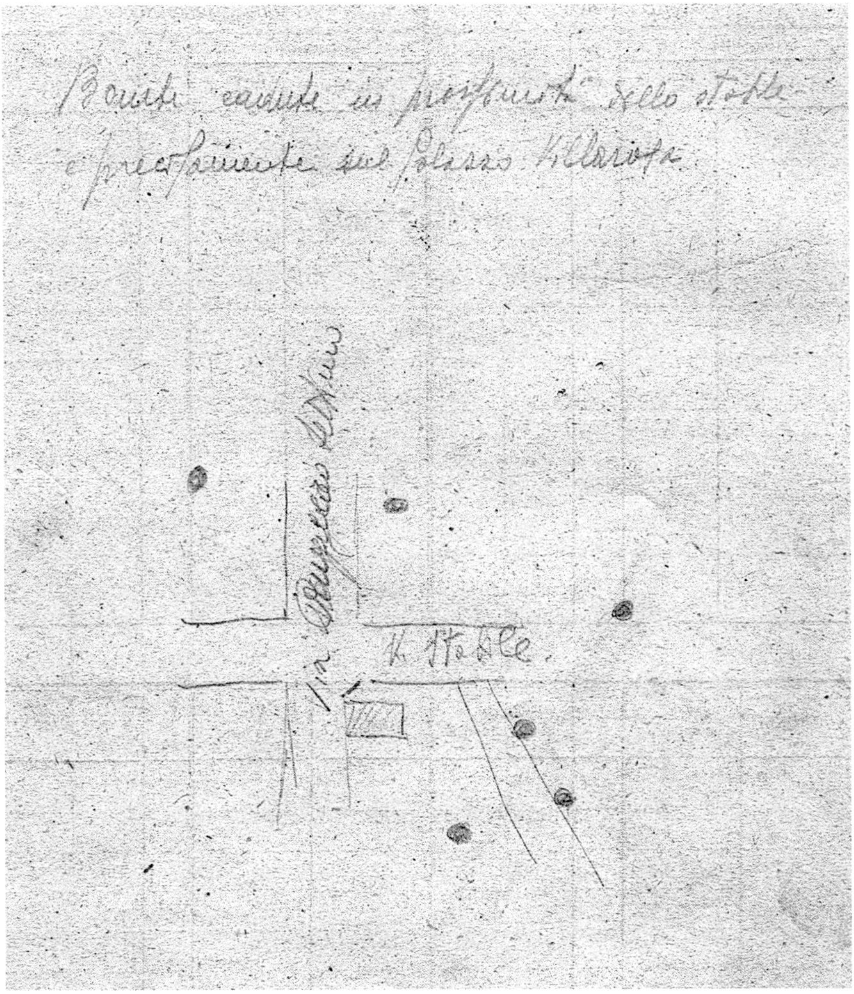

Fig. 7.6 Bombs over Via Ruggiero Settimo in 1943. Courtesy of the Archivio di Stato di Palermo

fortune to his oldest son, Giovanni Battista Guccia e Guccia, the mathematician.[55] When Giovanni Battista died in 1914, he left the bulk of his fortune to his bother Francesco,[56] who had not studied at university and had devoted his life to managing the family businesses. In 1924, Francesco died and left his fortune to his youngest

[55]Notar Salvatore Spinoso, March 1, 1900. Section "Notai". Archivio di Stato di Palermo, section "La Gancia".

[56]Notar Ferdinando Lionti, October 31, 1914. Section "Notai". Archivio di Stato di Palermo, section "La Gancia".

daughter, twenty-eight-year-old Carmela, with smaller bequests to his wife Adriana
Fatta and his older daughter Chiara.[57] The three women lived in the huge palace;
neither Carmela nor Chiara married. Chiara died in 1939, and her mother died in
December 1943, six months after the bombardments of Palermo. Carmela Guccia
was forty-seventy years old when she was left alone, living in the second floor of
the Palazzo Guccia.

The Guccia family had been deeply religious since its origins. We have already
commented that Gaspare Guccia, the grandfather of the family patriarch Giovanni
Battista Guccia e Bonomolo, Marquis of Ganzaria, had three daughters who became
nuns, and that he paid for having a daily and perpetual mass for his soul at the
Monastery of Saint Rosalia. The patriarch had a sister and three daughters who were
nuns, and in his will he left several bequests to religious institutions. Indeed, inside
the "Chiesa dei Cappuccini" in Palermo, in the "cappella della Madonna", there is
a family tomb under a magnificent marble gravestone where seven members of the
Guccia family are buried: the marquis, his wife, three sons and two grandsons. The
marquis had two granddaughters who became nuns and a grandson who became
a priest. Giovanni Battista, the mathematician, had a sister who was also a nun.
The exception to this general family atmosphere was the mathematician's father,
Giuseppe Maria Guccia e Vetrano. He studied in a religious seminary, and later
married his relative Chiara Guccia e Cipponeri, following the family's interest. His
experiences with the Catholic Church certainly must have been traumatic, as during
his life he published several books and articles of strong anticlerical content. One
of these books was a detailed description of the maneuvers of his family to prevent
one of the nun granddaughters of the marquis (Agata, daughter of Luigi, see the
genealogical tree of the Guccia family on p.21) from receiving her inheritance. The
role played by several priests in the episode—which caused a court case that lasted
several years—was quite dubious.[58] The other book and the article exhibit a strong
criticism of the Catholic Church and a proposal for a major reform. The book is an
essay with 500 pages.[59]

Guccia's niece Carmela was very devout; perhaps in the same immoderate way in
which the three daughters of the Prince of Salina, Carolina, Caterina and Concetta,
were depicted at the end of the novel *The Leopard*. It is not surprising, then, that
when Carmela Guccia privately wrote her personally handwritten will in 1952
she thought about favoring a religious institution. The situation regarding housing
in Palermo was absolutely disastrous after the war. The number of hungry and
homeless children in the city was very high; they lived in the streets and slept

[57] *Ibid.,* June 7, 1924.

[58] G. Guccia, *Osservazioni sulla vertenza Guccia e Demanio,* Tipografia Amenta, 1888, Palermo.
Biblioteca Centrale della Regione Siciliana, Palermo, MISC. C 15.6, 15.7.

[59] G. Guccia, *La quistione religiosa in Italia risoluta a Campo dei Fiori o il vero cristianesimo,*
Tipografia Diretta da Sani Andó, 1891, Palermo. Biblioteca Centrale della Regione Siciliana,
Palermo, 5.8.A.137, and G. Guccia, *Ancora una parola sulla convenienza e possibilità di una
Riforma Religiosa,* Pensiero Italiano, 16 (1894), Biblioteca Centrale della Regione Siciliana,
Palermo, Fondo Tumminelli, IX.E.51.5.

in the doorways. This made a deep impression on Carmela, who was by then living by herself in the Palazzo Guccia. The religious institute Congregazione Don Orione was at the time building a home for these homeless children in Palermo, the "Villaggio del fanciullo". They were under a number of financial and legal difficulties, especially regarding the plot of land for the home. On December 8, 1954, all the debts concerning the buying of the plot were unexpectedly paid. The principal and main officers of the Congregazione Don Orione were called up to the Palazzo Guccia in early January 1955. There, Carmela Guccia explained her interest in their work and that she had paid the debt.[60] She died one week later, on January 14, 1955.

The opening of the will of Carmela Guccia drew a lot of attention from the large Guccia family; all of them, however, were distant relatives of Carmela. The amount to be divided up was enormous. After some minor bequests to various religious institutions and to a limited number of her relatives, she declared the Congregazione Don Orione as "universal heir" of all her estate. It was a shocking surprise for everybody. The author of *The Leopard*, Giuseppe Tomasi di Lampedusa, who was still alive (though he would die two years later) commented to his adoptive son, Gioacchino Lanza Tomasi, that it was a mockery of destiny that the legacy of a declared atheist—he was referring to the mathematician—would end up in the hands of the Catholic Church.[61] The Circolo Matematico di Palermo, which was being reconstructed at that time, received as legacy from Carmela Guccia the wooden shelves of the personal library of her uncle Giovanni Battista Guccia as well as his portrait (Fig. 7.7) painted by the renowned Italian painter of the Liberty style Ettore De Maria Bergler.[62]

The Congregazione Don Orione worships its benefactor Carmela Guccia. Indeed, she is buried in a personal chapel in the Parrocchia Madonna della Provvidenza in Palermo, side-by-side to the "Villaggio del fanciullo" (Fig. 7.8). They referred to her as the "marchesina", a term of endearment for a young and unmarried marquise. This is another irony of destiny. Despite the fact that the mathematician's grandfather from his father's side was the first Marquis of Ganzaria and his grandfather from his mother's side was the third Marquis of Ganzaria, Giovanni Battista Guccia did not inherit the noble title. A cousin of his was the fifth marquis. He was never too keen on titles, but this issue must have been uncomfortable for him in a certain moment, because in 1901 he requested and obtained the noble mention of *Nobile dei Marchesi di Ganzaria* (in English, Noble of Marquis of Ganzaria).[63]

---

[60]Congregazione Don Orione, *La storia del "Santuario S. Rosalia" e "Villaggio del fanciullo"*.

[61]Personal communication from Gioacchino Lanza Tomasi.

[62]Notar F.P. Lionti, January 14, 1955. Section "Notai". Archivio di Stato di Palermo, section "La Gancia".

[63]From G. Capaccio and E. Tomarchio, *op. cit.*, section 5.5: On February 23, 1901, Giovanni Battista Guccia requested the title of noble. On March 2, 1901, the Sicilian heraldry commission considered the request favorably; it also allowed Guccia the use of the coat of arms of his grandfather, Giovanni Battista Guccia e Bonomolo, Marquis of Ganzaria. On September 18, 1901, a King's decree granted Guccia of the title of "Noble of Marquis of Ganzaria", transmisable via

Fig. 7.7 Giovanni Battista Guccia, portrait by Ettore De Maria Bergler. Courtesy of the Circolo Matematico di Palermo

Thus, calling his niece Carmela Guccia the "marchesina" is more irony than it is poetic.

We now come to the reason for the lack of personal and family records, and the scarcity of images, of Giovanni Battista Guccia (apart from the documents in the Archive of the Circolo Matematico di Palermo and the information in public and legal archives). In 1957, two years after receiving the legacy of Carmela Guccia, the Congregazione Don Orione sold the Palazzo Guccia to a real estate company. The palace was demolished in 1968 (Fig. 7.9). An office building for the Banco di Sicilia was built on the plot.[64] Nothing was kept, and all family archives and personal objects from the Guccia family were lost. This explains why so little is known about Guccia's childhood, youth, and personal life.

---

direct masculine line to the legitimate and natural descendants of both sexes. On October 9, 1901, the King's commissioner informed the Sicilian heraldry commission. On October 7, 1901, the Prefect of the province of Palermo delivered the nobility's degree to Giovanni Battista Guccia. He also received, after paying 30 lire, a miniature of the coat of arms.

[64] A. Chirco, M. Di Liberto, *op. cit.,* p. 84.

Fig. 7.8 Tomb of Carmela Guccia in the Parrocchia Madonna della Provvidenza in Palermo

As explained in the Introduction, via the family photo album of a relative of the Prince of Lampedusa, recently, some images of Giovanni Battista Guccia and members of his family were found.

The gold medal given to Guccia in the 1914 celebration of the anniversary of the Circolo, and paid for by an international collection, remains lost.

7.4 An Evaluation of the Work of Giovanni Battista Guccia

Chance was the driving force behind Giovanni Battista Guccia's introduction into mathematics. It was by chance, and most likely by the influence of his uncle the Prince of Lampedusa, that he attended in 1875 the Palermo conference of the Società Italiana per il Progresso delle Scienze where he met Luigi Cremona. This encounter caused Guccia to change from studying engineering to mathematics, and prompted him to move to Rome. It was by chance that in 1880, after the completion of his thesis, Guccia followed Cremona's suggestion to visit Paris and London. The visit to Paris and his attendance to the meeting of the Association Française pour l'Avancement des Sciences in Reims marked his view of mathematics, and were instrumental in developing his strong links with French mathematicians. And it was

Fig. 7.9 Palazzo Guccia around 1968 before demolition. Courtesy of Mario Di Liberto

also by chance that Guccia was in Paris when the auction for the library of Chasles took place in 1881, where he represented Cremona to buy books for the Vittorio Emanuele Library. There he discovered the insights of scientific publications.

These influences gave Guccia a modern view of what was needed—and what had to be avoided—for the development of mathematical research. This view he transformed into a project, and from the project followed a plan. At first, he implemented his plan in his closest environment, Palermo. What resulted from this plan was the creation in 1884 of the Circolo Matematico di Palermo. The scientific sessions, the existence of a library and the search for international contacts were all part of his plan. The success of the first plan inspired Guccia to formulate a second plan, which was more advanced, more complicated but more ambitious than before. The 1888 statute of the Circolo and the creation of the Rendiconti were part of this new plan. The management of a scientific journal was a big milestone for him. It placed him in a key position within the mathematical research community, and he became aware of the new results in all areas of research, and made multiple new contacts. Following his drive for perfection, he established in 1893 a printing house for the Rendiconti, the Tipografia Matematica di Palermo. This allowed him to have complete control over the printing of the journal and to adjust its publication according to its interests, and ultimately to the interests of the society.

By that time, the developments that eventually led to the first International Congress of Mathematicians were starting to yield results. The Circolo was sufficiently well known by then, so that Guccia was asked to collaborate in the preparations for the Congress. The Congress was held in Zurich in 1897. The discussions of the Congress implicitly outlined a program for the structuring of mathematics at an international level in the new era that was beginning. There was a need for international cooperation and agreements to address issues such as the creation of a review journal that would report promptly on the mathematical literature; devising a general classification of mathematics aimed at aiding mathematicians dealing with the bibliographic growth; creating an international directory of mathematicians, specifying addresses and field of specialty; and completing the mathematical literature by publishing the works of important mathematicians from the past. Guccia was not able to attend the Zurich 1897 Congress, but was aware of those discussions, which were recorded in the proceedings of the Congress. The International Congresses themselves were not able to address those duties. This was clearly seen in the Paris Congress of 1900, when the lack of progress on the issues discussed in Zurich (creating an international review journal for mathematics, a general classification of mathematics, an international directory of mathematicians, among others) was evidenced. Guccia attended the Paris Congress and witnessed the lack of improvements in those issues.

All these experiences and influences made Guccia become a true internationalist of mathematics. He was aware of the need to set the scope of action at the international level in order to address many important issues. Mathematical societies and journals were limited if they restricted themselves to the national level. The success of the Circolo and the Rendiconti at the international level were for him signs that his efforts were aimed in the right direction. This confidence in his mission motivated him to embark on and accomplish many projects, such as the creation of an award associated to the International Congresses, the Medaglia Guccia; the expansion of the Rendiconti by publishing the Annual directory (with name, address, and brief curriculum of the society members), the "Supplemento" (which included the table of contents of all journals received in the Circolo, as well as mathematical announcements), and the "Indici" (which listed all articles published in the Rendiconti); the publication of the works of Ruffini; and the preliminary discussions for creating a journal devoted to high quality Applied Mathematics.

Convinced that he was working in the right direction, Guccia placed all his expectations on the 1908 International Congress in Rome. He thought that then the Circolo Matematico di Palermo could be turned into the institution that mathematics needed: the International Association of Mathematicians. The failure of that attempt was a sad experience for him. Nevertheless he persevered, and in 1909 launched a change in the statute of the Circolo aimed at enhancing the international character of the Rendiconti. The definitive end to what he called "his dream" occurred in the Cambridge 1912 International Congress, when the project of creating and international association of mathematicians was buried.

As a mathematician, Giovanni Battista Guccia started his scientific career in the right place and in the right direction: as a student of Cremona, he was properly trained and advised, well oriented and received the best possible influences. He lived during a time when Geometry was undergoing important changes, mainly through the introduction of algebraic tools. Michele de Franchis in his obituary of Guccia commented that he had a "too purist conception of Geometry, which kept him far from the treasure of the theory of linear series and from the important results obtained by Brill and Noether."[65] At a certain moment in his life, Guccia decided to divide his energies between research, building a mathematical society and managing a research journal. The latter activities turned into an absorbing and exciting endeavor, which eventually required all his effort; moreover, they turned into a lifetime project. The indirect consequence of this was that Guccia abandoned active research.

In order to evaluate Guccia's project, it is important to note that many of the components of his project were realized many years after his death. An international society uniting mathematicians from all countries was created: the International Mathematical Union was established in 1920, although its activity was tainted by political aims during the interwar period, which eventually led to its dissolution. It was reestablished in the 1950s. Under its umbrella, an international award was created, the Fields Medal, and an international listing of mathematicians was published in the World Directory of Mathematicians, which existed from 1958 to 2002. The definitive solution to the problem of timely reviews of publications came via Mathematical Reviews and Zentralblatt für Mathematik.

Guccia's goals were the right ones, as proved by the fact that they were eventually realized. He failed to achieve them because of a variety of reasons; two of them were external and stand out. The first reason lies in the history and the geography of Italy. The main Italian academic institutions and scientists were in the North of Italy. It was thus mortifying for the Northern establishment to accept that there was a strong mathematical society with high international recognition in Palermo, which was so distant, both geographically and culturally, from the historical, cultural and political core of Italy. This attitude was apparent in the library crisis of the Circolo that occurred in 1904; in the disdain with which the Circolo was treated in the Rome 1908 Congress; and in the opposition within Italy to Guccia's plans of creating an international association based on the Circolo (as proved, in particular, by the failure of Guccia's strategy to obtain the support for his project from the Italian Faculties of Science and hence, the support of the Italian mathematical community). The existence and success of the Circolo also prevented the creation of any other Italian mathematical society. It was not until 1922, eight years after the death of Guccia and four years after the end of World War I, when there was a decline in the Circolo, that Salvatore Pincherle founded in Bologna the Unione Matematica Italiana. Even on that occasion, many Italian mathematicians opposed Pincherle's

---

[65]M. De Franchis, *op. cit.,* p. iii.

initiative, since they argued that the new society could only be a cheap copy of the already existing one, the Circolo Matematico di Palermo.

In particular, Tullio Levi-Civita on March 16, 1922, wrote to Pincherle

> I cannot avoid the impression that the real and desirable analogue of the "Société Math. de France", "American Math. Society", "Deutsche Math. Ver." and others, will always remain the Circolo that has truly honored Italy when, living Guccia, it was in full efficiency. Why kill it with a new society? Would not be much better to vivify it, continuing and exploiting its good traditions and indisputable merits?[66]

And on April 17, 1922, Guido Castelnuovo wrote to Pincherle

> We have [in Italy] a real international journal of mathematics, the Rend. del Circolo Matematico di Palermo. It may have decayed in recent years, possibly due to the current director not having the skills for the post. But the journal has great tradition and can be raised once more. Why don't we attempt to do this, rather than the other, which will more or less damage the Circolo, removing part of its members and attracting them to the Unione [Matematica Italiana] and its Bolletino? (In Italy few are the people who are members of two similar societies or subscribe to two similar journals.) Our two really good mathematical journals (the Rendiconti and the Annali) today are struggling for life. It is necessary to create a third journal, the Bolletino [della Unione Matematica Italiana], which will not be able to survive without damaging the other two?[67]

Despite the divergent opinions, Pincherle, with the decisive support of Volterra, gradually managed to convince the Italian mathematical community and the new society became the unique and true scientific association for Italian mathematicians.[68]

The second main external cause for the failure of Guccia's project was the international atmosphere before 1914. It was not the most conducive for a project like Guccia's that aimed to overcome national barriers, while strong and narrow nationalism was the dominant attitude at the time. It was quite surprising that after the war mathematicians from many different countries agreed to share membership on the Editorial Board of the Rendiconti. The survival of the Circolo and the Rendiconti, even after Guccia's death and under the difficult circumstances of the aftermath of World War I, reveals that Guccia's project of internationalization was well conceived.

However, some of the reasons for the failure of Guccia's project were not external. The Circolo Matematico di Palermo as a scientific society had a narrow social basis since, due to the remote geographic location of Sicily within Italy, the number of members that could actively participate in the daily running of the society was confined to those living in Palermo. This placed a burden on Guccia as

---

[66]Letter from Levi-Civita to Pincherle, March 16, 1922, Archive of the Unione Matematica Italiana, Dipartimento di Matematica, Università di Bologna.

[67]Letter from Castelnuovo to Pincherle, April 17, 1922, Archive of the Unione Matematica Italiana, Dipartimento di Matematica, Università di Bologna.

[68]P. Nastasi, R. Tazzioli, *I matematici italiani e l'internazionalismo scientifico (1914–1924)*, La Matematica nella Società e nella Cultura, Rivista dell'Unione Matematica Italiana, 6 (2013), pp. 355–405, p. 390.

he had to deal with most of the work himself. Guccia's own personal investment in the management of the society also made it difficult to involve other colleagues in managing the Circolo and the Rendiconti. These two factors constituted a danger for the continuity of the society and the journal after Guccia's death. It is remarkable how well both the society and the journal survived in the absence of their founder and through the effects of World War I.

There were other initiatives of Guccia which were not well conceived. One of them was the Medaglia Guccia. The idea of establishing an international award above all national academies and mathematical societies was laudable, as the later success of the Fields Medal showed. However, there were two critical problems with the medal: it was restricted to a specific problem in one area of mathematics, and it had Guccia's name attached to it. It is true that it was Guccia's personal money which paid for the award, but naming the award after Guccia himself was criticized for the excessive glorification of its creator. It should be recalled that in the memorandum written by John Charles Fields in 1932 where he proposed creating "International Medals for Outstanding Discoveries in Mathematics" it was explicitly stated that

> The medals should be of a character as purely international and impersonal as possible. There should not be attached to them in any way the name of any country, institution or person.[69]

The medal was named after Fields only four years after his death, at the Oslo 1936 International Congress of Mathematicians with the awarding of the first medals.[70]

To conclude, we recall a phrase from Hardy's obituary notice for Mittag-Leffler. After reviewing the scientific achievements of Mittag-Leffler, his personality and the honors that he received, Hardy wrote, "But to the outside world he was, above everything, the editor of the 'Acta Mathematica', the famous journal which he founded, and, with the co-operation of a committee of the four Scandinavian countries, edited for forty-five years."[71] Giovanni Battista Guccia did not have the mathematical stature of his Swedish friend, but following Hardy's phrase, we can say that

> To the outside world of the time Giovanni Battista Guccia was, above everything, the founder of the Circolo Matematico di Palermo and editor of its Rendiconti, the famous journal which he founded, and, with the co-operation of an international committee, edited for almost thirty years.

---

[69] G.P. Curbera, *op. cit.*, p. 115.

[70] *Comptes rendus du congrés international des mathématiciens*, Oslo 1936, A.W. Brogers Boktrykkeri A/S, 1937, Oslo. See pp. 7, 45, 48 and 308.

[71] G.H. Hardy, *Obituary Notice: Gösta Mittag-Leffler, 1846–1927*, Proc. Royal Society of London A, 119 (1928), pp. v–viii, p. v.

Chapter 8
Appendixes

8.1 Mathematical Works of Giovanni Battista Guccia

We include the list of works of Giovanni Battista Guccia. It is based on the list that Michele de Franchis included in his article *G.B. Guccia: Cenni biographici*, Rend. Circ. Mat. Palermo, 39 (1915), i–x, but it follows the numbering given in the paper *L'attività scientifica di Giovan Battista Guccia,* by A. Barlotti, F. Bartolozzi, G. Zappa, published in the Suppl. Rend. Circ. Mat. Palermo, 67 (2001), pp. xi–xxvi, where the scientific activity of Giovanni Battista Guccia is analyzed in detail. On pages 64–68, 93–94, 110–114, 138, 139 and 149 we have discussed the mathematical work of Guccia.

We use the abbreviations for the names of journals used by the Jahrbuch über die Fortschritte der Mathematik. For the articles that were reviewed in the Jahrbuch, we include the reference to the review.

1. *Sur une classe de surfaces représentables, point par point, sur un plan.*
   Franc. Ass. (1880) 191–200. JFM 13.0652.03 [(1881)].
2. *Formole analitiche per la trasformazione Cremoniana.*
   Palermo Rend. I. (1884–87) 17–19. JFM 16.0065.03 [(1884)].
3. *Formole analitiche per la trasformazione Cremoniana.*
   Palermo Rend. I. (1884–87) 20–23 JFM 16.0065.03 [(1884)].
4. *Formole analitiche per la trasformazione Cremoniana.*
   Palermo Rend. I. (1884–87) 24–25 JFM 16.0065.03 [(1884)].
5. *Estensione di un teorema sul tactinvariante di due curve algebriche.*
   Palermo Rend. I. (1884–87) 23, 26.
6. *Sopra alcune formole del Sig. Picquet.*
   Palermo Rend. I. (1884–87) 26–27.
7. *Sur les transformations géométriques planes birationnelles.*
   C. R. CI. (1885) 808–809. JFM 17.0794.04 [1886].

© Springer International Publishing AG, part of Springer Nature 2018
B. Bongiorno, G. P. Curbera, *Giovanni Battista Guccia,*
https://doi.org/10.1007/978-3-319-78667-4_8

 8. *Sur les transformations Cremona dans le plan.*
      C. R. CI. (1885) 866–869. JFM 17.0794.02 [1886].
 9. *Formole analitiche per la trasformazione Cremoniana.*
      Palermo Rend. I. (1884–87) 50–53. JFM 18.0787.02 [1887].
10. *Teoremi sulle transformazioni Cremoniane nel piano.*
      Palermo Rend. I. (1884–87) 27, 56–59, 66. JFM 18.0787.01 [1887].
11. *Restituzione di prorità.*
      Palermo Rend. I. (1884–87) 68.
12. *Teoremi sulle transformazioni Cremoniane nel piano.*
      Palermo Rend. I. (1884–87) 119–132. JFM 18.0787.01 [1887].
13. *Sur une question concernant les points singuliers des courbes algébriques planes.*
      C. R. CIII. (1886) 594–596. JFM 18.0672.01 [1887].
14. *Generalizzazione di un teorema di Noether.*
      Palermo Rend. I. (1884–87) 139–156. JFM 18.0671.02 [1887].
15. *Sulle superficie algebriche le cui sezioni piane sono unicursali.*
      Palermo Rend. I. (1884–87) 165–168. JFM 19.0787.01 [1887]
16. *Sulla riduzione dei sistemi lineari di curve ellittiche e sopra un teorema generale delle curve algebriche di genere p.*
      Palermo Rend. I. (1884–87) 169–189. JFM 19.0704.02 [1887].
17. *Sui sistemi lineari di superficie algebriche dotati di singolarità base qualunque.*
      Palermo Rend. I. (1884–87) 338–349. JFM 19.0612.01 [1887].
18. *Osservazioni sopra una comunicazione del Dott. Segre.*
      Palermo Rend. I. (1884–87) 386–387.
19. *Due sistemi lineari d'ordine minimo di genere $p = 2$.*
      Palermo Rend. I. (1884–87) 388–389.
20. *Théorème sur les points singuliers des surfaces algébriques.*
      C. R. CV. (1887) 741–743. JFM 19.0686.03 [1888].
21. *Un teorema sulle curve singolari delle superficie algebriche.*
      Palermo Rend. II. (1888) 79–80.
22. *Sur l'intersection de deux courbes algébriques en un point singulier.*
      C. R. CVII. (1888) 656–658. JFM 20.0705.01 [1889].
23. *Théorème général concernant les courbes algébriques planes.*
      C. R. CVII. (1888) 903–904. JFM 20.0705.03 [1889].
24. *Sulla classe e sul numero dei flessi di una curva algebrica dotata di singolarità qualunque.*
      Rom. Acc. L. Rend. V1. (1889) 18–25. JFM 21.0700.01 [(1889)]
25. *Su una proprietà delle superficie algebriche dotate di singolarità qualunque.*
      Rom. Acc. L. Rend. (4) V1. (1889) 349–353. JFM 21.0773.01 [(1889)].

26. *Sulla intersezione di tre superficie algebriche in un punto singolare e su una questione relativa alle transformazioni razionali nello spazio.* (Italian)
    Rom. Acc. L. Rend. (4) V1. (1889) 456–461. JFM 21.0773.02 [(1889)].
27. *Nuovi teoremi sulle superficie algebriche dotate di singolarità qualunque.*
    Rom. Acc. L. Rend. (4) V1. (1889) 490–497. JFM 21.0774.01 [(1889)].
28. *Sopra un recente lavoro concernente la riduzione dei sistemi lineari di curve algebriche.*
    Palermo Rend. III. (1889) 233–234. JFM 21.0620.01 [1889].
29. *Sulle singolarità composte delle curve algebriche piane I.* (Italian)
    Palermo Rend. III. (1889) 241–259. JFM 21.0698.02 [1889].
30. *Su un esempio addotto del Sig. Castelnuovo.*
    Palermo Rend. IV. (1890), 71.
31. *Due proposizioni relative alle involuzioni di specie qualunque, dotate di singolarità ordinarie.*
    Palermo Rend. VII. (1893) 49–60. JFM 25.0968.01 [1893].
32. *Ricerche sui sistemi lineari di curve algebriche piane, dotati di singolarità ordinarie.*
    Palermo Rend. VII. (1893) 193–255. JFM 25.1097.01 [1893]
33. *Una definizione sintetica delle curve polari.*
    Palermo Rend. VII. (1893) 263–272. JFM 25.0979.02 [1893].
34. *Sulle involuzioni di specie qualunque, dotate di singolarità ordinarie.* (Italian)
    Palermo Rend. VIII. (1894) 227–247. JFM 25.0969.01 [1884].
35. *Ricerche sui sistemi lineari di curve algebriche piane, dotati di singolarità ordinarie.*
    Palermo Rend. IX. (1895) 1–64. JFM 25.1097.01 [1895].
36. *Sur une question concernant les points singuliers des courbes gauches algébriques.*
    C. R. CXX. (1895) 816–819. JFM 26.0727.05 [1894].
37. *Sur les points doubles d'un faisceau de surfaces algébriques.*
    C. R. CXX. (1895) 896–899. JFM 26.0727.04 [1894].
38. *Sur une expression du genre des courbes gauches algébriques douées de singularités quelconques.*
    S. M. F. Bull. XXIII. (1895) 101–102. JFM 26.0728.01 [(1895)].
39. *Sui vincoli esistenti fra i punti di contatto delle tangenti condotte da un punto a k curve algebriche piane.*
    Palermo Rend. IX. (1895) 268–269. JFM 26.0658.01 [1895].
40. *Sulle curve algebriche piane. Sulle superficie algebriche.*
    Palermo Rend. 16, (1902) 204–208.
41. *Sulle curve algebriche piane. Sulle superficie algebriche.*
    Palermo Rend. 16, (1902) 286–293. JFM 33.0604.01 [(1902)].
42. *Un théorème sur les courbes algébriques planes d'ordre n.*
    C. R. 142, (1906) 1256–1259. JFM 37.0597.03 [1906].

43. *Un théorème sur les surfaces algébriques d'ordre n.*
    C. R. 142, (1906) 1494–1497. JFM 37.0597.04 [1906].
44. *Sopra una nuova espressione dell'ordine e della classe di una curva gobba algebrica.*
    Palermo Rend. 21, (1906) 389–390. JFM 37.0597.05 [1906].
45. *Teoria generale delle curve e delle superficie algebriche. Lezioni di Geometria superiore date da G.B. Guccia nella R. Università di Palermo nell'anno scolastico 1889–90.* (1890) A. Longo e C. Palermo. JFM 22.0689.01 [1890].

8.2   The 1884 Statute of the Circolo Matematico di Palermo

Approved on March 2, 1884

Article 1
The aim of the Circolo is the increase of mathematical studies in Palermo. It will contribute to that aim with:

a) Original communications by its members on topics on *Pure Analysis and Geometry,* and moreover on *Rational Mechanics, Mathematical-Physics, Geodesy* and *Theoretical Astronomy.*
b) The formulation and discussion of mathematical questions, excluding its elementary parts.
c) Bibliographical discussions on recent mathematical publications.

Article 2
Communication or discussions on matters not pertaining to the aim of the Circolo are not allowed.
Article 3
To be admitted to the Circolo it is necessary:

1. To be proposed by two members, who should present to the President a written request specifying the titles of the candidate.
2. To obtain in the next session the majority of votes of the members participating in the session.

Article 4
The number of members is unlimited.
Article 5
Each member is obliged to pay in cash in the Treasurer's house:

1. *Ten* lire as entry rights, to be paid at the time of admission.
2. An annual contribution of *fifteen* lire, to be paid at the beginning of each four-month period: on January 1st, on May 1st, and September 1st.

New members start paying from the current four-month period.

Article 6

Those who shall not pay at the established time, will be, after a warning from the Treasurer, considered as resigning.

Article 7

The Presidency of the Circolo is composed by the *President,* the *Vice-President,* the *Treasurer* and two *Secretaries,* in office for a one-year term.

Article 8

The voting for the election of the Presidency will be done by secret ballot among the participating members. In the case of equal number of votes, the eldest candidate will be elected.

Article 9

The sessions of the Circolo will be held every two weeks, from November to June inclusive. The President, whenever it considers it necessary, may convene the members to an extraordinary session.

Article 10

At the end of October the Presidency will communicate to the members the date of the first session in November.

Article 11

The first session in January will be devoted to the election of the Presidency, to audit the past year and to the approval of the budget.

The funds of Circolo are applicable to:

1. The office expenses.
2. The subscriptions of mathematical journals.

Article 13

The Presidency, whenever it considers it appropriate, may propose the purchase of mathematical books.

Article 14

The books and journals, purchased or received in donation, remain of exclusive property of the Circolo. Donations from members will be returned in the event of dissolution of the Circolo, as long as decade had not passed from the time of the donation.

Article 15

The members may use the books and journals in the premises of the Circolo.

Under special regulations, approved by the members, it would be possible to establish rules for granting the home loan.

Article 16

The order of precedence for the communications of each session will be established in the previous session.

Article 17

The proceedings of the sessions will give a comprehensive account of the communications by members and will be drafted, from one session to the next one, by the Secretaries.

Those communicating original work they will care that the proceedings do not contain errors and will assume full scientific responsibility. The approval of the proceedings by the Circolo will be given once is verified that the communications and the discussions of the members have been faithfully reported.

Article 18

The proceedings once approved will be recorded in a specific registry and will be signed by the President and the Secretaries. The registry will be kept in the archive of the Circolo. The members can consult it in the premises of the Circolo and obtain a copy, given the permission of the Presidency.

Article 19

In case of absence of the President and the Vice-President, the senior among the members present at the session will serve as President. In case of absence of both Secretaries, the President of the session will invite one of the members present in the session to act as Secretary.

Article 20

Those who have been members of the Circolo for at least two years and change residence, will retain, after consideration by the Circolo, all rights of the other members, despite being exempted from the annual contribution during their absence.

Article 21

Non-members of the Circolo who wish to attend its sessions must, each time, be introduced by a member.

Article 22

Requests for amending the statute require the backing of at least one third of the members, and will be directed to the President, who will convene with that aim the members for an extraordinary session. In this case the resolutions will be valid only provided the presence of at least two-thirds of the members residing in Palermo.

Article 23

This statute is temporary.

Provisionally the seat of the Circolo, on via Ruggiero Settimo n. 28, left side in the ground floor, is offered by the member Dr. Giovanni Guccia.

8.3   The 1888 Statute of the Circolo Matematico di Palermo

Discussed and approved in the general assembly of members on February 26, 1888.

Purpose of the Society—Seat

Article 1

The aim of the scientific society Circolo Matematico di Palermo is the increase and the dissemination of mathematical sciences in Italy.

Article 2
To that aim the Circolo:

a) Holds sessions in its premises.
b) Prints a mathematical journal called Rendiconti del Circolo Matematico di Palermo.

Moreover, it may establish competitions for awards and promote scientific congresses in the cities of the Kingdom.

Article 3
The seat of the Society is in Palermo, and it is immovable.

Members—Admission—Fees

Article 4
The Circolo has two types of members: resident and non-resident. The first type comprises those having usual residence in Palermo.
Foreigners can belong to the society.

Article 5
The number of members, resident and non-resident, is not limited.

Article 6
To be admitted to the Circolo it is necessary: 1st To be proposed, during a session of the Society, by two (resident or non-resident) members who present a written request to the President; 2nd To obtain, in the next session, the majority of votes of the members attending the session.

Article 7
Each resident member is required to pay: 1st an entrance fee of ten lire to be paid at the time of admission; 2nd an annual contribution of fifteen lire, to be paid at the beginning of each four-month period: on January 1st, on May 1st, on September 1st. Newly admitted members start paying from the beginning of the current year.

Article 8
Each non-resident member is required to pay only the annual contribution of fifteen lire, which will be paid at the beginning of each year. Newly admitted members start paying from the beginning of the current year.

Article 9
The resignation of member of the Circolo is not valid until the annual contribution has not been fully paid. The resignation must be addressed to the President, who will communicate it to the society in the next session.
Those who resign, through a written request, may reenter the society after a voting in accordance to Article 6.

Article 10
The defaulting member, six months after the payment period and after an early warning by the Treasurer, will be expelled from the group of members.

Article 11
The payment in one installment of 300 lire confers the title of perpetual member, exempting from the payment of the annual contribution. In case of dissolution of the society there is no entitlement for refund.

Article 12

A resident member who changes his usual residence is considered as a non-resident member, without being entitled to a refund of the entrance fee. The entrance fee is due to non-resident members who, for the first time, become resident members.

Article 13

All members, resident and non-resident, will receive for free the Rendiconti del Circolo Matematico di Palermo. Every new member is entitled to receive the volume in press at the time of admission.

On the Presidency

Article 14

The representation and the administrative direction of the society is done by the Presidency, consisting of the officials of the society:

1 president
1 vice-president
2 secretaries
2 vice-secretaries
1 treasurer
2 librarians

who are elected, among them, by the resident members in secret ballot.

The Presidency remains in office for two years. All members can be re-elected.

Article 15

The election of the Presidency will be held in an extraordinary session in the first Sunday of January. Only members attending the session can vote.

In the event that during the two-year period there is a vacancy in the Presidency, the resident members will be convened to an extraordinary session for election to the post.

On the Steering Council

Article 16

The scientific direction of the Society is entrusted to the Steering Council that will act as Editorial Board for the journal Rendiconti del Circolo Matematico di Palermo, according to the rules of its internal regulations.

Article 17

The Steering Council is composed of 20 members, 5 resident members and 15 non-resident, elected by the society as a whole in secret ballot. In each group will be elected the members having the largest number of votes. In case of an equal number of votes between two or more candidates, there will be a tiebreaker voting in which only the members attending the session will participate.

The Steering Council holds office for three years. All members can be re-elected.

Article 18

The election of the Steering Council will be held during an extraordinary session in the third Sunday of January.

Every non-resident member, and every resident member that cannot participate in the session, will send, inside a signed letter addressed to the President, a sealed envelope indicating 20 names of members: 5 resident and 15 non-resident.

The letters that do not meet all the above conditions or that are received by the Presidency after the 3 pm of the aforementioned day will be considered as null votes.

The President assisted by the Secretaries will do the counting of votes.

Article 19

There is no incompatibility between members of the Presidency and resident members of the Steering Council.

Article 20

After taking office the Steering Council will delegate in one of its resident members the management of the publication of the Rendiconti.

On the Sessions

Article 21

The ordinary sessions of the Circolo will be held the second and the fourth week of the month. The society takes two months of vacations: September and October. The President, if he considers it necessary, may convene the resident members in extraordinary session.

Article 22

In case of absence of the President and the Vice-President, the senior among the members present at the session will serve as President. In case of absence of both Secretaries, the President of the session will invite one of the members present in the session to act as Secretary.

Article 23

By the end of January each year, the President will convene the resident members in an extraordinary session to audit the past year and for the approval of the budget.

Article 24

Communications or discussions on subjects unrelated to the scientific nature and the aim of the society are not allowed in the sessions of the Circolo.

Article 25

Matters relates to the administration of the Circolo will be treated exclusively in the extraordinary sessions, for which the President will prepare and communicate in advance to the resident members the session's agenda.

Article 26

Resignation as a member of the Circolo is not subject to discussion or voting.

Article 27

Ordinary sessions of the Circolo are legally constituted whatever the number of present members attending.

In the extraordinary sessions, except for those referred to in Articles 15 and 18, a second call is required in the event that the first call did not include at least half plus one of the resident members.

Article 28
Communications by resident members in ordinary sessions will follow the order of their entrance in the society. This will be preceded by the reading by the Secretary of the titles of the communications by non-resident members received by the Presidency since the previous session.
Article 29
All matters related to the regulations of ordinary and extraordinary sessions fall under the special powers of the President.
Article 30
Non-resident members present temporarily in Palermo enjoy all rights of resident members and can participate in all votings, except those mentioned in Articles 15, 23 and 45.
Article 31
Non-members who wish to attend the ordinary sessions of the Circolo must, each time, be introduced by a member.

On the Rendiconti
Article 32
The Rendiconti del Circolo will compulsorily publish:

1. Summaries of the minutes of the sessions (written by the Secretaries) containing the title and, if necessary, a brief abstract of the communications by members.
2. The original notes and memoirs communicated by members and accepted by the Editorial Board.
3. A bibliographic bulletin of the national and international mathematical production containing:

   a. The abstract of all mathematical articles published in the periodicals (proceedings of academies, journals, newspapers etc.) that the Circolo receives in exchange with its Rendiconti.
   b. The list of all non-periodic mathematical publications (books, memoirs, notes,...) received as donations to the Library of the Circolo.

Article 33
For all matters concerning the journal Rendiconti del Circolo Matematico di Palermo, the Steering Council can implement whatever reforms and extensions considered to be appropriate in order to enhance its scientific importance and to better meet the needs of the practitioners of mathematical sciences.
Article 34
The summaries of the minutes do not reproduce the scientific discussions that occurred during the sessions of the Circolo; however, if the members involved wish it to be mentioned, they are required to transmit to the Secretary, during the same session, a written note that, in any case, cannot exceed the size of one page of the Rendiconti.

Article 35

The notes, the memoirs and the bibliographic reviews for the Rendiconti must be unpublished and written in one of the following languages: Italian, Latin, Spanish, French, German, English; they cannot be published separately or in other scientific journals once they are published by the Circolo. Authors assume, solely, the scientific responsibility.

On the Library

Article 36

The Circolo will constitute a Library and exchanges its Rendiconti with national and foreign scientific collections.

Article 37

The books, booklets and the periodic collections purchased or received in donation, or in exchange with the Rendiconti, remain the exclusive property of the Circolo; in case of dissolution of the society they will lawfully belong to the City Library of Palermo.

Article 38

A special regulation for the Library, approved by the resident members, could establish the rules for granting to resident and non-resident members home temporary use of scientific works.

Article 39

Funds from the annual contributions of members cannot be allocated to the Library.

On Income and Expenses

Article 40

The incomes of the Circolo are:

1. The fees of members.
2. The profit obtained from the sales of the Rendiconti.
3. The subsidies and donations receive by the Circolo from private and non-profit organizations.

Article 41

The expenses of the Circolo are divided into ordinary and extraordinary. The ordinary expenses are:

1. The office expenses, comprising the annual rent for the premises of the Circolo.
2. The expenses for printing and sending the Rendiconti.

Each extraordinary expense needs a discussion by the assembly of resident members.

Article 42

The Treasurer has the management of social affairs, both active and passive; uses the available funds; takes care of the ordinary expenses and the extraordinary ones voted by the assembly of resident members.

Article 43

All articles of the provisional statute of March 2, 1884, are revoked.

Article 44

Proposals for additions or amendments to this statute must be presented by at least 30 members and approved by the society with the majority of at least 2/3 of the voters.

Article 45

In case of dissolution of the society, the general assembly of resident members, with the participation of at least 2/3 of them, will decide on the allocation of the remaining social funds.

Transitional Article

The first elections for the Presidency and the Steering Council will be done by the assembly of resident members, convened into an extraordinary session.

8.4   The Reform of the Statute in 1908

All articles of the 1888 statute remain unchanged except for articles 17 and 18, which become:

Article 17

The Steering Council is composed of 45 members, 5 resident members and 40 non-resident, elected by the society as a whole in secret ballot. In each group will be elected the members having the largest number of votes. In case of an equal number of votes between two or more candidates, there will be a tiebreaker voting in which only the members attending the session will participate.

The Steering Council holds office for three years. All members are eligible for reelection.

Article 18

The election of the Steering Council will be held during an extraordinary session in the third Sunday of January.

Every non-resident member, and every resident member that cannot participate in the session, will send, inside a signed letter addressed to the President, a sealed envelope indicating 45 names of members: 5 resident and 40 non-resident.

The letters that do not meet all the above conditions or that are received by the Presidency after the 3 pm of the aforementioned day will be considered as null votes.

The President assisted by the Secretaries will do the counting of votes.

8.5   Members Joining the Circolo Matematico di Palermo Until 1914

ORDINE D'AMMISSIONE

(2 marzo 1884 — 24 maggio 1914).

I nomi in carattere *corsivo* sono di coloro che non appartengono più alla Società
(per decessi, dimissioni, cessazioni di pagamento o radiazioni).

1884

1. Alagna (R.).
2. *Albeggiani (G.).*
3. Albeggiani (M.L.).
4. *Arioti (A.).*
5. Bontade (G.).
6. *Cacciatore (G.).*
7. Caldarera (F.).
8. *Capelli (A.).*
9. *Capitò (M.).*
10. *Damiani (G.).*
11. *Fileti (E.).*
12. Gambera (P.).
13. Gebbia (M.).
14. *Giardina (A.).*
15. Guccia (G.B.) *).
16. *Guidotti (G.).*
17. *Maggiacomo (F.).*
18. Maisano (G.).
19. Paternò (F.P.).
20. Pepoli (A.).
21. Pintacuda (C.).
22. *Pittaluga (G.).*
23. Politi (G.).
24. *Porcelli (S.).*
25. *Righi (A.).*
26. *Salemi-Pace (G.).*
27. *Scichilone (S.).*

1885

28. *Di Simone (G.).*
29. *Cavallaro (F.).*
30. *Conti (I.).*
31. *Albeggiani (C.).*
32. *Armò (E.).*

Soscrittori del primo Statuto della Società (2 marzo 1884), per ordine alfabetico.

33. *Albanese (V.).*
34. *Basile (Ed.).*
35. *Cantone (A.).*

1886

36. Martinetti (V.).
37. Giudice (F.).
38. D'Arone (G.D.).
39. *Zona (T.).*
40. *Rotigliano (S.).*
41. *Cantone (M.).*
42. Taschetta (G.).
43. Macaluso (D.).
44. La Manna (A.).
45. *Spina (R.).*
46. Catalan (E.).
47. *Battaglini (G.).*
48. *Cerruti (V.).*
49. Del Pezzo (P.).

1887

50. *Hirst (T.A.).*
51. Cesàro (E.).
52. *Vaněček (J.S.).*
53. *Vaněček (M.N.).*
54. *Amato-Pojero (G.).*
55. La Mensa (G.).
56. *Dainelli (U.).*
57. *La Manna (D.).*
58. *Pertica (E.).*
59. *Masoni (U.).*
60. Del Re (A.).
61. Rindi (S.).
62. *Tirelli (F.).*
63. Retali (V.).
64. Amodeo (F.).

65. Jung (G.).
66. *Mastricchi (F.).*
67. *La Farina (E.).*
68. Segre (C.).
69. Gerbaldi (F.).
70. *Brioschi (F.).*
71. *Cremona (L.).*
72. D'Ovidio (E.).
73. *Siragusa (A.).*
74. Berzolari (L.).
75. *Brambilla (A.).*
76. *Ciollaro (G.).*
77. Costa (G.).
78. Fouret (G.).
79. Humbert (G.).
80. Loria (G.).
81. *Mollame (V.).*
82. Morera (G.).
83. Peano (G.).
84. *Piuma (C.M.).*
85. Salvatore-Dino (N.).
86. Tonelli (A.).
87. Volterra (V.).
88. *Betti (E.).*
89. *Platania (G.).*
90. Vivanti (G.).

1888

91. *Murer (V.).*
92. Grimaldi (G.P.).
93. *D'Angelo (F.P.).*
94. Della Rocca (G.).
95. *Gianforme (A.).*
96. Venturi (A.).
97. Breglia (E.).
98. *Ruggiero (P.).*

*) Fondatore della Società.

26 febbrajo 1888:

*Entra in vigore il vigente
Statuto della Società.*

99. *Beltrami* (E.).
100. *Casorati* (F.).
101. Bertini (E.).
102. *De Paolis* (R.).
103. *Sannia* (A.).
104. *Arzelà* (C.).
105. *Bassani* (A.).
106. *Chizzoni* (F.).
107. Lazzeri (G.).
108. Pincherle (S.).
109. *Ruffini* (F.P.).
110. Veronese (G.).
111. Paternò (E.).
112. *Tripiciano* (G.).
113. *Starkoff* (A.).
114. Mittag-Leffler (G.).
115. *Le Paige* (C.).
116. Marcolongo (R.).
117. Montesano (D.).
118. *Panizza* (F.).
119. *Schlegel* (V.).
120. *Reggio* (G.Z.).
121. *Merlani* (A.).
122. Torelli (G.).
123. Sforza (G.).
124. *Gambardella* (F.).
125. *Cónigliaro* (S.).
126. *Morisani* (E.).
127. Amaturo (E.).
128. Basile (Ern.).
129. *Fatta* (G.).
130. *Russo* (G.).
131. *Tomasini* (F.).
132. Bordiga (G.).
133. Jadanza (N.).
134. *Bresslau* (L.).
135. Certo (L.).
136. *Previtera* (C.).

1889

137. *Padelletti* (D.).
138. Castelnuovo (G.).
139. Moore (E.H.).
140. *Visalli* (P.).

141. Pittarelli (G.).
142. *Pieri* (M.).
143. Burali-Forti (C.).
144. Bottino (F.).
145. Lebon (E.).
146. Sadun (E.).
147. *Frattini* (G.).
148. Jordan (C.).
149. Bortolotti (E.).
150. Krause (M.).
151. Porro (F.).
152. *Del Vecchio* (G.).
153. Bagnera (G.).

1890

154. *Floridia* (G.).
155. Reina (V.).
156. *Messina* (A.).
157. Giuliani (G.).
158. *Monroy* (G.).
159. Macrì (V.).
160. André (D.).
161. *Poincaré* (H.).
162. Picard (É.).
163. Mukhopâdhyây (A.).
164. *Delitala* (G.).
165. Lanza di Mazzarino (G.).
166. *Bettazzi* (R.).
167. Pennacchietti (G.).

1891

168. *Marraffa* (S.)
169. Mendizábal Tamborrel (J. de).
170. *Fiorentino* (G.).
171. *Paci* (P.).
172. *Pagliani* (S.).
173. *Martone* (A.).
174. Appell (P.).
175. *Rozzolino* (G.).
176. Laisant (C.-A.).
177. Galdeano (Z.G. de).
178. Vries (J. de).

1892

179. Soler Balsano (E.).
180. Kerbedz (E. de).

181. Lori (F.).
182. *Mattina* (C.).
183. *Miceli* (L.).
184. *Legnazzi* (E.N.).
185. *Weyr* (Em.).

1893

186. *Garibaldi* (C.).
187. *Bigiavi* (C.).
188. Fano (G.).
189. *Agnello* (F.).
190. *Lauricella* (G.).
191. *Zanotti-Bianco* (O.).
192. *Fiorentino* (A.).
193. *Barone* (E.).
194. Valeri (D.).
195. Bianchi (L.).

1894

196. Mancini (E.).
197. Viola (A.).
198. *Romeo* (A.).
199. Somigliana (C.).
200. *Castellano* (F.).
201. Amici (N.).
202. *Studnička* (F.J.).
203. *Vailati* (G.).
204. *Amanzio* (D.).
205. Halsted (G.B.).
206. *Porcelli* (O.).
207. *Massimi* (P.).
208. Burgatti (P.).
209. Enriques (F.).
210. *Bucca* (F.).
211. *Macfarlane* (A.).
212. Ricci-Curbastro (G.).
213. *Crescini* (E.).

1895

214. *Terzi* (G.).
215. De Franchis (M.).
216. *Menabrea* (L.F.).
217. Levi-Civita (T.).
218. Di Pirro (G.).
219. *Neppi Modona* (A.).
220. Cordone (G.).
221. Painlevé (P.).
222. *Guimarães* (R.).

223. Massarini (I.).
224. Spelta (C.).

1896

225. Blaserna (P.).
226. Cosserat (E.).
227. Autonne (L.).
228. *Durán Loriga* (*J.J.*).
229. *Conoscente* (*E.*).
230. Lo Monaco Aprile (L.).
231. *Saladino* (*D.*).
232. Verde (F.).
233. Viterbi (A.).

1897

234. *Spanò* (*D.*).
235. Carslaw (H.S.).
236. *Castelli* (*E.*).
237. Petrovitch (M.).
238. *Traverso* (*N.*).
239. *Gugliuzzo* (*A.*).
240. *Correnti* (*V.*).
241. *Cassarà Cabasino* (*G.*).
242. Martin (A.).

1898

243. Scott (C.A.).
244. *Cosserat* (*F.*).
245. *La Motta di S. Silve-*
      stro (*G.*).
246. Severini (C.).
247. Ronco (N.E.).
248. Conti (A.).
249. Weber (E. von).
250. Medolaghi (P.).
251. Rudio (F.).
252. Levi (B.).
253. Perna (A.).
254. Puglisi (M.).
255. Ciani (E.).
256. Bourget (H.).
257. *Cassani* (*P.*).
258. Daniele (E.).
259. Angelitti (F.).

1899

260. Fazzari (G.).
261. Guerra (E.P.).

262. Del Giudice (M.).
263. Lovett (E.O.).
264. Vacca (G.).
265. Vassilief (A.).
266. *Morale* (*M.*).
267. *Barack* (*K.A.*).
268. *Hardcastle* (*F.*).
269. *Bourlet* (*C.*).
270. Gyldén (O.).
271. Almansi (E.).
272. Tagliarini (R.).
273. Fisher (G.E.).
274. *Schwatt* (*I.J.*).
275. *Martone* (*M.*).
276. Calapso (P.).
277. Pascal (E.).

1900

278. *Lugaro* (*E.*).
279. *Guggino* (*F.*).
280. Vitali (G.).
281. Toja (G.).
282. Dini (U.).
283. Alasia de Quesada (C.).

1901

284. Montessus de Ballore
        (R. de).
285. *Gallucci* (*G.*).
286. Nobile (V.).
287. Severi (F.).
288. *Arista* (*A.*).
289. *Bonola* (*R.*).
290. De Donder (T.).
291. *Perazzo* (*U.*).
292. Cattaneo (P.).
293. Barresi (F.P.).
294. *Barbieri* (*U.*).
295. Amaldi (U.).
296. Strazzeri (V.).
297. *Veneroni* (*E.*).
298. *Silberstein* (*L.*).
299. *Newson* (*H.B.*).
300. Tedone (O.).
301. Marletta (G.).

1902

302. Mineo (C.).

303. Cipolla (M.).
304. Mignosi (G.).
305. *Sansone* (*G.*).
306. Sinigallia (L.).
307. Scheffers (G.W.).
308. Zaremba (S.).
309. *Ferretti* (*G.*).
310. Gigli (D.).
311. *Carrone* (*C.*).
312. Giulotto (V.).

1903

313. Gerrans (H.T.).
314. *Toffoletti* (*C.*).
315. Composto (S.).
316. Fubini (G.).
317. Giorgi (G.).
318. Ackermann-
                    Teubner (A.).
319. Insolera (F.).
320. *Pexider* (*J.V.*).
321. Dickstein (S.).
322. *Caldarera* (*G.M.*).
323. *Orlando* (*L.*).
324. Amato (V.).
325. Aguglia (G.).
326. *Bozal y Obejero* (*A.*).

1904

327. *Trafelli* (*L.*).

2 marzo 1904:

20° anniversario
della fondazione della Società.

328. *Ripamonti* (*M.*).
329. Rius y Casas (J.).
330. Zappetta (A.).
331. Licata di Bauci-
                    na (G.B.).
332. *Sofio* (*L.*).
333. *Monroy* (*A.*).
334. Menabrea (C.).
335. *Barbieri* (*A.*).
336. Noether (M.).
337. Stéphanos (C.).
338. Fouët (É.).
339. Brill (A. von).
340. Boutroux (P.).

341. Hermann (A.).
342. Maggi (G.A.).
343. *Guldberg (A.)*.
344. Brusotti (L.).
345. Sannia (G.).
346. *Sbrana (U.)*.
347. *Schoute (P.H.)*.

1905

348. Boggio (T.).
349. *Giampaglia (N.)*.
350. Schur (F.).
351. *Siacci (F.)*.
352. *Gordan (P.)*.
353. Klein (F.).
354. Krazer (A.).
355. Mange (F.).
356. Nielsen (N.).
357. Zeuthen (H.G.).
358. Adhémar (R. d').
359. Alvarez Ude (J.G.).
360. Caminati (P.).
361. *Chini (M)*.
362. Dall'Acqua (F.A.).
363. D'Amico (F.).
364. Fehr (H.).
365. Ferrari (A.).
366. Guareschi (G.).
367. *König (G.)*.
368. Kœnigsberger (L.).
369. Laura (E.).
370. Pannelli (M.).
371. Pensa (A.).
372. Pompeiu (D.).
373. Prym (F.).
374. Repetto (G.).
375. Settimo di Fitalia (G.).
376. Wirtinger (W.).
377. Young (W.H.).
378. Forsyth (A.R.).
379. Kohn (G.).
380. *Pisati (L.)*.
381. Torelli (R.).
382. Osgood (W.F.).
383. Pintacuda (M.).
384. Eiesland (J.).
385. *Pietra (G.)*.
386. Cantor (G.).

387. Berry (A.).
388. *Lüroth (J.)*.
389. Rosanes (J.).
390. Emch (A.).
391. Haskell (M.W.).
392. Finkel (B.F.).
393. Gauthier-Villars (A.).
394. Veblen (O.).
395. Cantelli (F.P.).
396. Papè-Lanza (S.).
397. Silla (L.).
398. Kneser (A.).
399. Darboux (G.).
400. Pringsheim (A.).
401. Stäckel (P.).
402. Study (E.).
403. Schumacher (R.).
404. Hočevar (F.).
405. Graefe (F.).
406. Landau (E.).
407. *Muth (P.)*.
408. Graf (J.H.).
409. Sturm (A.).
410. Ovazza (E.).
411. Jensen (L.).
412. Schotten (H.).
413. Ludwig (W.).
414. Störmer (C.).
415. Keyser (C.J.).
416. Landis (W.W.).
417. *Polignac (C. de)*.
418. Petrovitch (S.).
419. Bullard (W.G.).
420. Moore (C.L.E.).
421. Wilson (E.B.).
422. Sommerville (D.).
423. Wiman (A.).
424. Falk (M.).
425. Sharpe (J.W.).
426. Frizell (A.B.).
427. Joffe (S.A.).
428. Graber (M.E.).
429. *Barton (S.M.)*.
430. *Miller (E.)*.
431. Shaw (J.B.).
432. White (H.S.).
433. Moulton (F.R.).
434. Greenwood (G.W.).

435. Gutzmer (A.).
436. Hancock (H.).
437. Lehmer (D.N.).
438. Saltykow (N.).
439. Snyder (V.).
440. *Doyle (P.)*.
441. Bauer (G.N.).
442. Oseen (C.W.).
443. Mason (M.).
444. Ena (S.).
445. Farkas (G.).
446. Lanza (G.).
447. Rothrock (D.A.).
448. Prasad (G.).
449. Leonardi (G.).
450. Vassilas-Vitalis (J.).
451. Gavrilovitch (B.).
452. Stoïanovich (C.).
453. Di Dia (G.).
454. Eisenhart (L.P.).
455. See (T.J.J.).
456. Westlund (J.).
457. Padoa (A.).

1906

458. *Tweedie (C.)*.
459. Mancinelli (F.).
460. Maroni (A.).
461. Brand (E.).
462. Denizot (A.).
463. Metzler (W.H.).
464. Cajori (F.).
465. Roe (E.D.).
466. Clariana y Ricart (L.).
467. Stasi (F.).
468. Nicoletti (O.).
469. Bocchetta (G.).
470. Michel (P.).
471. Teixeira (F. Gomes).
472. *Vassura (G.)*.
473. Teofilato (P.).
474. Basset (A.B.).
475. *Ruggeri (C.)*.
476. De Zolt (A.).
477. Kiseljak (M.).
478. Hagemann (G.A.).
479. Deimel (R.F.).
480. *Gambardella (G.)*.

481. Juel (C.).
482. Paranjpye (R.P.).
483. Cowley (E.B.).
484. Manning (H.P.).
485. Dingeldey (F.).
486. Fredholm (I.).
487. Garcia (C.V.).
488. Hayashi (T.).
489. Lebeuf (A.V.).
490. *Pettegrew (D.L.)*.
491. Young (J.W.).
492. *Molk (J.)*.
493. *Webb (J.B.)*.
494. *Helguero (F. de)*.
495. Stuyvaert (M.).
496. Pirondini (G.).
497. Giordano (D.).
498. Pagliero (G.).
499. Varvaro (E.).
500. Whitaker (J.I.S.).
501. Sestilli (G.).
502. Alagona (G.).
503. Bliss (G.A.).
504. Lampe (E.).
505. *Bosi (L.)*.
506. *Mayer (A.)*.
507. Pizzetti (P.).
508. French (J.S.).
509. Akers (O.P.).
510. Stone (O.).
511. Lowber (J.W.).
512. Travis (J.F.).
513. Borel (É.).
514. Escherich (G. von).
515. Fréchet (M.).
516. *Stringham (I.)*.
517. Clifford (H.E.).
518. Barthels (K.L.).
519. Carathéodory (C.).
520. Brooke (W.E.).
521. Ashton (C.H.).
522. Hedrick (E.R.).
523. *Quintili (P.)*.
524. Rados (G.).
525. Kürschák (J.).
526. Hilbert (D.).
527. Quanjel (J.).
528. Da Rios (L.S.).

529. *Weber (H.)*.
530. Halphen (C.).
531. Fileti (M.).
532. Reye (Th.).
533. *Ricci (F.)*.
534. Hallett (G.H.).
535. Hadamard (J.).
536. Maclagan-Wedderburn
               (J.H.).
537. Mansion (P.).
538. Jacona Guccia (N.).
539. Lucchesi Palli (P.).
540. Mari (T.).
541. Blumenthal (O.).
542. *Tafelmacher (A.)*.
543. Wangerin (A.).
544. Rodenberg (C.).
545. Koch (H. von).
546. Candido (G.).
547. Corey (S.A.).
548. *Manning (W.A.)*.
549. *Fiorentini (P.)*.
550. *Mannheim (A.)*.
551. *Champreux (A.J.)*.
552. Beman (W.W.).
553. *Rockwood (C.G.)*.
554. Costanzi (C.).
555. Varićak (V.).
556. Kœnigs (G.).
557. Suták (J.).
558. Léauté (H.).
559. Frege (G.).
560. Smith (O.A.).
561. Pellet (A.).
562. Rémoundos (G.).
563. Zervos (P.).
564. Lefèvre (É.).

1907

565. Roberts (W.R.W.).
566. Stewart (A.B.).
567. Sylow (L.).
568. Haberland (M.).
569. Kasner (E.).
570. Graham (W.J.).
571. *Friesendorff (T.)*.
572. Levi (E.E.).
573. Smith (A.W.).

574. Yanney (B.F.).
575. *Hagen (G.)*.
576. Coddington (E.).
577. Mollerup (J.).
578. Jeans (J.H.).
579. *Ford (W.B.)*.
580. Wilczynski (E.J.).
581. Ette (C.R.).
582. Gummere (H.V.).
583. Wieleitner (H.).
584. Walker (B.M.).
585. Kragh (O.).
586. Giambelli (G.Z.).
587. Sellerio (A.).
588. Tavani (M.).
589. Carlheim-
             Gyllensköld (V.).
590. Séguier (J. de).
591. Greco (M.).
592. Greenstreet (W.J.).
593. Zahradník (K.).
594. *Maltese (N.)*.
595. *Ruiz-Castizo
             y Ariza (J.)*.
596. Pucciano (G.).
597. Frankland (F.W.).
598. Wilding
             v. Königsbrück (E.).
599. Scarborough (J.H.).
600. Carmichael (R.D.).
601. Brenke (W.Ch.).
602. Lo Giudice (G.).
603. Hawkesworth (A.S.).
604. Obear (G.B.).
605. Gunther (C.O.).
606. Slichter (C.S.).
607. Cantor (M.).
608. Pick (G.).
609. Grossi (P.).
610. Rossi (C.).
611. Tzitzéica (G.).
612. Mattson (R.).
613. Drach (J.).
614. Santangelo (G.B.).
615. Constantius (B.).
616. Giraud (G.).
617. Cappello (C.).
618. Gleijeses (M.).

619. Neumann (E.R.).
620. *Trier (B.)*.
621. Godeaux (L.).
622. *Urzi (G.)*.
623. Mainardi (M.).
624. Zanca (A.).
625. *Olivier (L.)*.
626. Duhem (P.).
627. Lerch (M.).
628. *Baire (R.)*.
629. Wölffing (E.).
630. Fulco (P.).
631. Vallée Poussin
      (Ch.-J. de la).
632. Mehmke (R.).
633. Bucca (R.).
634. Slaught (H.E.).
635. Hensel (K.).
636. Żorawski (K. von).
637. Cisotti (U.).
638. Cecioni (F.).
639. Knopp (K.).
640. Chella (T.).
641. *Peirce (B.O.)*.
642. Jolles (S.).
643. Schmidt (E.).
644. *Reber (A)*.
645. Lebesgue (H.).
646. Holmgren (E.).
647. Hobson (E.W.).
648. Pitrè (G.).
649. Lindelöf (E.).
650. *Wigert (C.S.)*.
651. Stekloff (W.).
652. Liapounoff (A.).
653. Tummarello (A.).
654. Scorza (G.).
655. Medici (S.).
656. *Schiaparelli (G.)*.
657. *Höchberg (K.)*.
658. Bernstein (F.).
659. Korn (A.).
660. Haton de la
      Goupillière (J.).
661. London (F.).
662. Majcen (G.).
663. *Ball (R.S.)*.
664. Conway (A.W.).

665. Crespi Baccin (I.).
666. Meyer (Fr.).
667. Lasker (E.).
668. Hanna (U.S.).
669. Jung (H.W.E.).
670. Hahn (H.).
671. Webster (A.G.).

1908

672. Haussner (R.).
673. Riesz (F.).
674. Cellérier (G.).
675. *Colpitts (E.C.)*.
676. Nannei (E.).
677. Müller (F.).
678. Rosati (C.).
679. Picone (M.).
680. *Picciati (G.)*.
681. Palatini (F.).
682. Beloch (M.).
683. Lorey (W.).
684. Gmeiner (J.A.).
685. *Dulac (H.)*.
686. Smith (W.B.).
687. Demoulin (A.).
688. Gérardin (A.).
689. Padova (E.).
690. Torka (J.).
691. Appelrot (G.).
692. Caronia (S.).
693. Olds (G.D.).
694. Baker (R.P.).
695. Koebe (P.).
696. *Barchi (A.)*.
697. Bolza (O.).
698. Mitchell (H.B.).
699. König (R.).
700. *Ajello (C.)*.
701. *Licopoli (G.)*.
702. Dodds (J.M.).
703. Toeplitz (O.).
704. Müller (J.O.).
705. Colonna di Cesarò (G.).
706. Fueter (R.).
707. Dehn (M.).
708. Smith (D.E.).
709. König (D.).
710. Hartwell (G.W.).

711. Kellogg (O.D.).
712. *Cobb (H.E.)*.
713. Schnee (W.).
714. Abraham (M.).
715. Grove (Ch.C.).
716. Sobotka (J.).
717. Salkowski (E.).
718. Epstein (P.).
719. Oriani (A.).
720. *Macry Correale (F.)*.
721. Rovetti (C.).
722. Mertens (F.).
723. Riesz (M.).
724. Jacobsthal (E.).
725. Ocagne (M. d').
726. Phragmén (E.).
727. *Minkowski (H.)*.
728. Suppantschitsch (R.).
729. Goldziher (Ch.).
730. Haseman (Ch.).
731. Sintsov (D.).
732. Zermelo (E.).
733. Dyck (W. von).
734. Fejér (L.).
735. Morley (F.).
736. *Brückner (M.)*.
737. Lipka (J.).
738. Jahnke (E.).
739. Wellstein (J.).
740. Bonaparte (Prince R.).
741. Weyl (H.).
742. Kyrillopoulos (D.).
743. Rogel (F.).
744. Finazzi (G.).
745. Sansone (G.).
746. Buttafarri (G.).
747. Paci (E.).
748. Occhipinti (R.).
749. Love (A.E.H.).
750. Noether (E.).
751. Westfall (W.D.A.).
752. Pierpont (J.).
753. Hartogs (F.).
754. Block (H.G.)
755. *Orlandi (C.)*.
756. Kriloff (A.).
757. Coialowitsch (B.).
758. Hamel (G.).

759. Sonin (N.).
760. *Gundelfinger* (*S.*).
761. Wallner (C.R.).
762. Roth (P.).
763. *Von der Mühll* (*K.*).
764. Zöllich (H.).
765. Blaschke (W.).
766. Kagan (B.).
767. *Ptaszycki* (*J.*).
768. Neuendorff (R.).
769. Roever (W.H.).
770. Mantegna di Gangi (G.).
771. Schoenflies (A.).
772. Tonolo (A.).
773. Müller (C.H.).
774. Schur (I.).
775. Lanza di Trabia (P.).
776. Broggi (U.).

1909

777. Herglotz (G.).
778. Chillemi (G.).
779. Myers (G.W.)
780. Fiske (T.S.).
781. *Sittignani* (*M.*).
782. Dowling (L.W.).
783. *Bates* (*W.H.*).
784. Merlin (É.).
785. Cassinis (G.).
786. Amoroso (L.).
787. *Lewent* (*L.*).
788. Hurwitz (W.A.).
789. Haentzschel (E.).
790. Loewy (A.).
791. *Fock* (*G.*).
792. Got (Th.).
793. Faber (G.).
794. *Sayre* (*H.A.*).
795. Lunn (A.C.).
796. *Dohmen* (*F.J.*).
797. Tonelli (L.).
798. Griffin (F.L.).
799. Dunkel (O.).
800. *Born* (*M.*).
801. Sós (E.).
802. Voghera (G.).
803. Richardson (R.G.D.).
804. Richmond (H.W.).

805. Comessatti (A.).
806. Degel (O.).
807. Durell (F.).
808. Pagano (F.).
809. Vries (H. de).
810. Finsterbusch (J.).

2 marzo 1909 :

25° anniversario
della fondazione della Società.

811. Bäcklund (A.V.).
812. Hilb (E.).
813. Vogt (W.).
814. Lichtenstein (L.).
815. Allen (F.E.).
816. *Sadow-Pittard*
                  (*H.A.P. de*).
817. *Tessari* (*D.*).
818. Petrén (L.).
819. Rietz (H.L.).
820. Puccini (A.).
821. Weitzenböck (R.).
822. Bohr (H.).
823. *Mauthner* (*E.*).
824. Perron (O.).
825. Lewickyj (W.).
826. Bauer (M.).
827. Sommer (J.).
828. Goursat (É.).
829. Dávid (L. de).
830. Gram (J.P.).
831. Mohrmann (H.).
832. Maillet (E.).
833. *Janzen* (*O.*).
834. Liebmann (H.).
835. *Brocard* (*H.*).
836. Garnier (R.).
837. Teege (H.).
838. *Markoff* (*A.*).
839. Sierpiński (W.).
840. Fujiwara (M.).
841. Sterneck (R. von).
842. Ivanoff (I.I.).
843. Gernett (N. von).
844. Pezzinga (A.).
845. *Pirozzi* (*A.*).
846. *Romanazzi* (*D.*).
847. Marotte (F.).

848. Denjoy (A.).
849. Fleck (A.).
850. Stożek (W.).
851. Grousinzeff (A.).
852. Carosi (E.).
853. Wieferich (A.).
854. *D'Amico* (*G.*).
855. Pál (J.).
856. Comstock (E.H.).
857. Pavanini (G.).
858. Steffensen (J.F.).
859. Mulder (P.).
860. Takagi (T.).
861. Dumas (G.).
862. Brouwer (L.E.J.).
863. Hellinger (E.).
864. Zecca (G.B.).
865. Plancherel (M.).
866. Svensson (B.).
867. Rosenblatt (A.).
868. Gast (P.).
869. Banachiewicz (T.).
870. Crudeli (U.).
871. Tortorici (P.).
872. Dintzl (E.).
873. Cunha (P.J. da).
874. Hoborski (A.).
875. Kryloff (N.).
876. Terradas é Ylla (E.).
877. Pöschl (T.).

1910

878. Pattillo (N.A.).
879. Bobylew (D.).
880. Jackson (F.H.).
881. *Ewing* (*W.F.*).
882. Butts (W.H.).
883. Moore (Ch.N.).
884. *Risack* (*M.*).
885. Bieberbach (L.).
886. Nalli (P.).
887. Brillouin (M.).
888. Colonnetti (G.).
889. Haar (A.).
890. Nörlund (N.E.).
891. Richard (J.).
892. Aprile (G.).
893. Dell'Agnola (C.A.).

894. *Landsberg* (G.).
895. Zignago (I.).
896. Bellesini (L.).
897. Russocki (K.).
898. Argento (L.).
899. Parisi di S. Bartolomeo (V.).
900. Sibirani (F.).
901. Bernardi (G.).
902. Simon (H.).
903. Sire (J.).
904. Ott (W. d').
905. Degenhart (H.).
906. Phillips (H.B.).
907. Bernstein (S.).
908. Slobin (H.L.).
909. Barton (R.M.).
910. Krygowski (Z.).
911. Bucht (G.).
912. Pangrati (E.-A.).
913. Andreoli (G.).
914. Caspar (M.).
915. Signorini (A.).
916. Bianchedi (R.).
917. Jolovicz (L.).
918. Courant (R.).
919. Meidell (B.).
920. Aguilar (F.).
921. Beccadelli di Bologna (P.).
922. Albanese (G.).
923. Bálint (E.).
924. Vergerio (A.).
925. Hatzidakis (N.).
926. Richarz (F.).
927. Servant (M.).
928. Hausdorff (F.).
929. Bompiani (E.).
930. Szily (K. de).
931. Hardy (G.H.).
932. Hatton (J.L.S.).
933. Burnside (W.S.).
934. Tietze (H.).
935. Guadet (J.).
936. Hulburt (L.S.).
937. Ondracek (J.).
938. Mirimanoff (D.).
939. Boulad (F.).

940. Cherubino (S.).
941. Ravaioli (C.).
942. Stott (W.).
943. Bassani (V.).
944. Lévy (P.).

1911

945. Plemelj (J.).
946. Precchia (L.).
947. Abbìa (R.).
948. Calonghi (M.).
949. Rothe (H.).
950. Schrutka (L.).
951. Genese (R.W.).
952. Truhelka (B.).
953. Frank (Ph.).
954. Gaba (M.).
955. Mises (R. von).
956. Uven (M.J. van).
957. Crelier (L.).
958. Cancelliere (G.).
959. Trudo (C.).
960. Conner (J.R.).
961. Villat (H.).
962. Terracini (A.).
963. Myller-Lebedeff (V.).
964. Puzyna (J.).
965. Axer (A.).
966. Jackson (D.).
967. Galvani (L.).
968. Tamarkine (J.).
969. Malmquist (J.).
970. Sparrow (C.M.).
971. Oxiński (T. von).
972. Tancredi (F.).
973. Scuto (G.).
974. Friedmann (A.).
975. Kirchner (W.H.).
976. Watson (G.N.).
977. Schlesinger (L.).
978. Yoshiye (T.).
979. Coolidge (J.L.).
980. Luckhaub (J.).
981. *Haret* (S.C.).
982. Ouspensky (J.).
983. Chohatt (J.).
984. Bouliguine (B.).
985. Fischer (E.).

986. Grober (M.).
987. Hudson (H. Ph.).
988. Isnardi (T.).
989. Loyarte (R.).
990. Collo (J.).
991. Weichardt (G.).
992. Killing (W.).
993. *Restivo* (G.).
994. Rey Pastor (J.).
995. *Cerrella* (F.).
996. Lusin (N.).
997. Magliano (H.).
998. Spadafora di Bissana (M.).
999. Pellizzari (N.).
1000. Niemeyer (H.).
1001. Silverman (L.L.).
1002. Partridge (E.A.).
1003. Engel (F.).
1004. Wiener (F.).
1005. Remak (R.).
1006. Naraniengar (M.T.).
1007. Bouton (Ch.L.).
1008. Noether (F.).
1009. Beck (H.).
1010. Smirnoff (W.).

1912

1011. Ranum (A.).
1012. Usai (G.).
1013. Minetola (S.).
1014. Craig (J.I.).
1015. Kowalewski (G.).
1016. Baldus (R.).
1017. Curtiss (D.R.).
1018. Bertrand (G.).
1019. Speiser (A.).
1020. Evans (G.C.)
1021. Allen (E.S.).
1022. Schiff (W.).
1023. Palomby (A.).
1024. Russell (B.).
1025. Githens (C.E.).
1026. Rothe (R.).
1027. Bottasso (M.).
1028. Wedekind (C.).
1029. Fekete (M.).
1030. Pólya (G.).

1031. Pétéline (M.).
1032. Errera (A.).
1033. Braude (L.).
1034. Monard (J.).
1035. Galbrun (H.).
1036. Levi (F.).
1037. Paine (G.P.).
1038. *Rothschild* (G.).
1039. Larmor (Sir J.).
1040. Wiarda (G.).
1041. Libický (V.).
1042. Lussan (É.).
1043. Rayleigh
    (J.W. Strutt, Lord).
1044. Szász (O.).
1045. Baker (H.F.).
1046. Stecker (H.F.).
1047. Gronwall (T.H.).
1048. Jordan (Ch.).
1049. Roncagli (W.).
1050. Maccaferri (E.).
1051. Rosenthal (A.).
1052. Melikoff (C.).
1053. Podetti (F.).
1054. Heywood (H.B.).
1055. Van Vleck (E.B.).
1056. Lecat (M.).
1057. Pérès (J.).
1058. Nencini (D.).

1913

1059. Steinhaus (H.).
1060. Berwald (L.).
1061. Sicca (T.).
1062. Cartan (É.).
1063. Buchanan (D.).
1064. Delaunay (N.).
1065. Geöcze (Z. de).
1066. Turrière (É.).
1067. Müller (E.).
1068. Szücs (A.).
1069. Browne (P.J.).
1070. Rudnicki (J.).
1071. Bratu (G.).
1072. Regis (B.F.).
1073. Palmer (E.S.).
1074. Della Casa (L.).

1075. Kottler (F.).
1076. Poli (C.).
1077. Martinotti (P.).
1078. Spitzer (G.).
1079. Spunar (V.M.).
1080. Birkhoff (G.D.).
1081. Rabinovitch (G.).
1082. Kruppa (E.).
1083. Cutrera (I.).
1084. Feemster (H.C.).
1085. Montel (P.).
1086. Tocchi (L.).
1087. McMahon (J.).
1088. Nagy (G. von Sz.).
1089. Sárközy (P.).
1090. Mazurkiewicz (É.).
1091. Vasconcellos (F. de).
1092. Ragusa (S.).
1093. Stouffer (E.B.).
1094. Schweitzer (A.R.).
1095. Galle (A.).
1096. Voigt (A.).
1097. Happel (H.).
1098. Arany (D.).
1099. Fichtenholz (G.).
1100. Finzel (A.).
1101. *Brehm* (E.).
1102. Lüders (O.).
1103. Coppedè (C.).
1104. Contini (A.).
1105. Prange (G.).
1106. Jacobsthal (W.).
1107. Ruiz-Feduchy (F.).
1108. Panoff (A.).
1109. Berger (A.).
1110. Cebrián (F.).
1111. Careddu (C.).
1112. Kiveliovitch (M.).
1113. Jiménez (E.).
1114. Ascoli (G.).
1115. Capra (A.).
1116. Mariares (F.).
1117. Bilimovitch (A.).
1118. Hepites (S.C.).
1119. Tedesco (G.).
1120. Cirera (R.).
1121. Griffith (J.H.).

1122. Uribe (D.M.).
1123. Pasquino (E.).
1124. Torner
    de la Fuente (J.).
1125. Kryjanovsky (D.).
1126. Lindwart (E.).
1127. Pidoll (M. von).
1128. Dines (L.L.).
1129. Green (G.M.).
1130. Santos Lucas (A.).
1131. Tafani (G.).
1132. Riabouchinsky (D.).
1133. Leonelli (S.).
1134. Toscano (S.A.).
1135. Fraser (P.).
1136. Rauber (A.).
1137. Chittenden (E.W.).
1138. Glause (H.).

1914

(11 gennaio-24 maggio)

1139. Williams (K.P.).
1140. Marseguerra (V.).
1141. Lefschetz (S.).
1142. Littlewood (J.E.).
1143. Schöll (E.).
1144. Zambler (P.).
1145. Behrens (W.).
1146. Sciolette (E.).
1147. Hecke (E.).
1148. Rice (L.H.).
1149. Carlbaum (T.).
1150. Chisini (O.).
1151. Galachine (S.A.).
1152. Cauer (D.).
1153. Agostini (A.).
1154. Colonna (A.).
1155. Sofio (F.).
1156. Cecconi (A.).
1157. Adinolfi (V.).
1158. Mayr (K.).
1159. Seiberl (R.).
1160. Cavallaro (V.G.).
1161. Barrow (D.F.).
1162. Palatini (A.).
1163. Caldonazzo (B.).

172   ANNUARIO BIOGRAFICO DEL CIRCOLO MATEMATICO DI PALERMO: 1914.

2 marzo 1914:	1170. Schouten (J.A.).	1180. Morgutti (A.).
30° anniversario della fondazione della Società.	1171. Vâlcovici (V.).	1181. Galajikian (H.).
	1172. Dickson (L.E.).	1182. Chiaramonte Bordonaro (G.).
	1173. Fraenkel (A.).	
1164. Brodėn (T.).	1174. Togliatti (E.).	1183. Michnik (H.).
1165. Pitcher (A.D.)	1175. Barraco (V.R.).	1184. Lindström (S.).
1166. Baer (W.S.).	1176. Gateaux (R.).	1185. La Rosa (M.).
1167. Cavaccini (A.).	1177. Gawriloff (A.).	1186. Vannucci di Petrulla (G.).
1168. Boehm (K.).	1178. Torroja y Miret (A.).	
1169. Fôppl (L.).	1179. Burnengo (G.).	

RIASSUNTO.

Soci ammessi dal 2 marzo 1884 al 24 maggio 1914 1186
Decessi, dimissioni, cessazioni di pagamento o radiazioni, di soci ammessi negli anni:

			109		160
1884	16	1894	10	1904	7
1885	8	1895	4	1905	13
1886	7	1896	3	1906	20
1887	21	1897	6	1907	15
1888	27	1898	3	1908	15
1889	6	1899	6	1909	16
1890	7	1900	2	1910	3
1891	6	1901	9	1911	3
1892	4	1902	3	1912	1
1893	7	1903	5	1913	1
	109		160		254

254

Soci al 24 maggio 1914 932

SOCI DEFUNTI

Cfr. *Annuario 1912.*

* **BOURLET** (C.) [269]. † Annecy (Haute-Savoie) (France): 12.8.1913.
BREHM (E.) [1101]. † Neuruppin (Preussen) (Deutschland): 15.7.1913.
COSSERAT (F.) [244]. † Paris (France): ... 3.1914.
DOYLE (P.) [440]. † Bombay (India): 27.3.1907.
FRIESENDORFF (T.) [571]. † St.-Pétersbourg (Russie): 6.3.1913.
GORDAN (P.) [352]. † Erlangen (Bayern) (Deutschland): 21.12.1912.
HARET (S.C.) [981]. † Bucuresti (România): 17/30.12.1912.
KÖNIG (G.) [367]. † Budapest (Magyarország): 8.4.1913.
LANDSBERG (G.) [894]. † ... 14.9.1912.
LAURICELLA (G.) [190]. † Catania (Italia): 9.1.1913.
LICOPOLI (G.) [701]. † ... 8.1913.
MACFARLANE (A.) [211]. † Chatham (Ontario) (Canada): 28.8.1913.
MOLK (J.) [492]. † Nancy (Meurthe-et-Moselle) (France): 7.5.1914 *).
PEIRCE (B.O.) [641]. † Cambridge (Mass.) (U.S.A.): 14.1.1914.
PETTEGREW (D.L.) [490]. † Worcester (Mass.) (U.S.A.) 19.1.1914.
PIERI (M.) [142]. † S. Andrea di Compito (Lucca) (Italia): 1.3.1913.
POINCARÉ (H.) [161]. † Paris (France): 17.7.1912.
POLIGNAC (C. de) [417]. † Paris (France): 15.11.1913.
PTASZYCKI (J.) [767]. † St.-Pétersbourg (Russie): ... 5.1912.
RESTIVO (G.) [993]. † S. Flavia (Palermo) (Italia): 10.5.1912.
SADOW-PITTARD (H.A.P. de) [816]. † Calcutta (India): 27.2.1913.
SCHOUTE (P.H.) [347]. † Groningen (Nederland): 18.4.1913.
SOFIO (L.) [332]. † Messina (Italia): 1.10.1913.
WEBER (H.) [529]. † Strassburg (Elsass) (Deutschland): 17.5.1913.

*) Veggasi la Nota della Redazione a pag. 161.

SOCI PERPETUI

ART. 11 DELLO STATUTO: Il versamento, in unica volta, di L. 300 conferisce il titolo di *socio perpetuo,* ed esonera dal pagamento della contribuzione annua. In caso di scioglimento della Società non si ha alcun diritto a rimborso.

BLUMENTHAL. — BOURLET †. — GERRANS. — GUCCIA. — HADAMARD. — HAGE-MANN. — HALSTED. — HUMBERT. — JORDAN. — KÜRSCHÁK. — PAINLEVÉ. — PARISI di SAN BARTOLOMEO. — WILDING von KÖNIGSBRÜCK.

8.6   Statistical Data of the Circolo Matematico di Palermo Until 1914

NOTE STATISTICHE

REDATTE DAL TESORIERE DELLA SOCIETÀ [1]).

I. – Stato della Società al 24 maggio 1914.

604

Argentina	soci 8	Misr	soci 3
Australia.	» 1	Nederland	» 6
Belgique.	» 12	New Zealand	» 1
Canada	» 1	Nippon	» 4
Danmark.	» 12	Norge	» 3
Deutschland.	» 141	Österreich-Ungarn	» 78
Ἑλλάς	» 6	Portugal	» 5
España	» 12	România.	» 6
France	» 67	Rossiya.	» 44
Great Britain and Ireland.	» 29	Schweiz.	» 12
India	» 4	Srbija.	» 3
Italia *(Residenti 75 ; Non residenti 234).*	» 309	Sverige	» 22
Mexico	» 2	United States of America	» 141
	604	Totale	932

[1]) Le Note **IV, V** e **VI** sono estratte dai « Conti consuntivi » della Società. Cfr., per gli Esercizi 1884-1903, l'opuscolo : « *Prospetti dei Conti consuntivi del Circolo Matematico di Palermo dei 20 Esercizi 1884-1903, etc. »*, compilati dai sigg. Ing. S. PORCELLI, *tesoriere della Società, e* Prof. G. B. GUCCIA, *fondatore della Società* (Tipografia Matematica di Palermo, 1904), con le relative « Note » esplicative. Cfr. inoltre : *Rendiconti*, t. XIX (1905), p. 118; *Supplemento*, vol. II (1907), p. 40 ; vol. III (1908), p. 44 ; vol. V (1910), pp. 3, 6-7; vol. VI (1911), pp. 67-68 ; vol. VII (1912), p. 3 ; vol. IX (1914), pp. 91-92, 97 ; *Annuari* : 1907, p. 109 ; 1908, pp. 130-132 ; 1910, p. 167 ; 1911, p. 40 ; 1912, pp. 190-191.

II. — Movimento nel numero dei Soci dal 2 marzo 1884 (fondazione della Società) al 24 maggio 1914.

Soci ammessi nelle sedute degli anni:

	178		277		671
1884 ... 27 [2])	1892 ... 7	1900 ... 6	1908 ... 105
1885 ... 8	1893 ... 10	1901 ... 18	1909 ... 101
1886 ... 14	1894 ... 18	1902 ... 11	1910 ... 67
1887 ... 41	1895 ... 11	1903 ... 14	1911 ... 66
1888 ... 46	1896 ... 9	1904 ... 21 [3])	1912 ... 48
1889 ... 17	1897 ... 9	1905 ... 110	1913 ... 80
1890 ... 14	1898 ... 17	1906 ... 107	
1891 ... 11	1899 ... 18	1907 ... 107	1138
178	277	671	

Nuovi soci ammessi nelle sedute dall'11 gennajo al 24 maggio 1914. . . .   48

Totale dei soci ammessi a tutto il 24 maggio 1914.   1186

Decessi, dimissioni, etc., dal 2 marzo 1884 al 24 maggio 1914 . . . . .   254 [4])

Soci al 24 maggio 1914.   932

III. — Progressi della Società dalla sua fondazione [5]).

2 marzo 1884: soci 27	10 luglio 1898: soci 174	25 agosto 1907: soci 510
31 luglio 1887: » 63	5 aprile 1900: » 180	9 agosto 1908: » 605
11 marzo 1888: » 102	17 aprile 1903: » 188	22 agosto 1909: » 709
9 febbrajo 1890: » 143	2 marzo 1904: » 195	28 agosto 1910: » 745
31 marzo 1892: » 158	26 marzo 1905: » 255	9 giugno 1912: » 826
26 gennajo 1896: » 171	8 luglio 1906: » 403	24 maggio 1914: » 932

8.7   Members of the Steering Council from 1888 to 1914

MEMBRI DEL CONSIGLIO DIRETTIVO DAL 1888.

ALBEGGIANI (G.): 1888-1892 († 16.9.1892).
ALBEGGIANI (M.L.): 1888-.
BAGNERA (G.): 1909-.
BATTAGLINI (G.): 1888-1893 († 29.4.1894).
BELTRAMI (E.): 1888-1900 († 18.2.1900).
BERTINI (E.): 1888-1893, 1909-.
BETTI (E.): 1888-1892 († 11.8.1892).
BIANCHI (L.): 1894-.
BOREL (É.): 1909-.
BRIOSCHI (F.): 1888-1897 († 13.12.1897).
CALDARERA (F.): 1888-1893.
CAPELLI (A.): 1894-1908 († 28.1.1910).
CARATHÉODORY (C.): 1909-.
CASORATI (F.): 1888-1890 († 11.9.1890).
CASTELNUOVO (G.): 1912-.
CERRUTI (V.): 1888-1908 († 20.8.1909).
CREMONA (L.): 1888-1903 († 10.6.1903).
DE FRANCHIS (M.): 1909-.
DEL PEZZO (P.): 1888-1908.
DEL RE (A.): 1900-1908.
DE PAOLIS (R.): 1888-1892 († 24.6.1892).
DINI (U.): 1903-.
D'OVIDIO (E.): 1888-1893, 1906-1908.
ÉNRIQUES (F.): 1909-.
FEJÉR (L.): 1909-.
FORSYTH (A. R.): 1909-.
FREDHOLM (I.): 1909-.
GEBBIA (M.): 1888-.
GERBALDI (F.): 1894-1904.
GUCCIA (G. B.): 1888-.
HADAMARD (J.): 1909-.
HILBERT (D.): 1909-.
HUMBERT (G.): 1909-.
JUNG (G.): 1888-1899.

KLEIN (F.): 1909-.
LANDAU (E.): 1909-.
LEVI CIVITA (T.): 1909-.
LIAPOUNOFF (A.): 1909-.
LORIA (G.): 1894-.
LOVE (A. E. H): 1909-.
MAISANO (G.): 1894-1899.
MARCOLONGO (R.): 1909-.
MERTENS (F.): 1909-.
MITTAG-LEFFLER (G.): 1894-.
MOORE (E.H.): 1909-.
NOETHER (M.): 1909-.
OSGOOD (W.F.): 1909-.
OVAZZA (E.): 1906-1911.
PASCAL (E.): 1900-.
PEANO (G.): 1894-1908.
PICARD (É.): 1909-.
PINCHERLE (S.): 1888-.
POINCARÉ (H.): 1891-1912 († 17.7.1912).
SCORZA (G.): 1909-.
SEGRE (C.): 1888-1893, 1909-.
SEVERI (F.): 1909-.
SOMIGLIANA (C.): 1909-.
STÄCKEL (P.): 1909-.
STEKLOFF (W.): 1909-.
STÉPHANOS (C.): 1909-.
TONELLI (A.): 1900-1908.
TORELLI (G.): 1894-1908.
VALLÉE POUSSIN (Ch.-J. de la): 1909-.
VIVANTI (G.): 1909-.
VOLTERRA (V.): 1888-1908.
WIRTINGER (W.): 1909-.
ZEUTHEN (H.G.): 1909-.

8.8   Original Text of Excerpts of Documents

Introduction: Note 2 *Discorso del Prof. Dr. Edmund Landau*, XXX anniversario della fondazione del Circolo Matematico di Palermo, Adunanza solenne del 14 aprile 1914, Suppl. Rend. Circ. Mat. Palermo, 9 (1914), pp. 16–17.

Mesdames et Messieurs, Chers Confrairès du Circolo Matematico, Cher M. Guccia, je m'excuse de ne pas pouvoir me servir de votre belle langue italienne; mon allemand n'étant pas non plus suffisamment connu ici, malgré la grande amitié qui unit nos deux pays, ma seule ressource est de m'exprimer en français. Néanmoins, ce que j'ai à dire me vient bien profondément du cœur et c'est pour le dire que je suis venue à Palerme.

Nous célébrons le jubilé d'une société n'ayant que la grande minorité de ses membres dans la ville où elle siège, mais qui a réuni presque un millier de mathématiciens dans tous les pays du monde et, parmi ceux-ci, les plus grands et les plus illustres savants de l'Italie, de l'Allemagne, de l'Angleterre, de la France, des États-Unis, de la Hongrie, et de toutes les nations où l'on cultive notre science. C'est la seule organisation internationale permanente que nous ayons; aussi considérons-nous Palerme comme le centre du monde mathématique. Ce n'est pas non plus pour le plaisir et l'honneur que nous éprouvons à etre en relation avec le mathématicien bien connu et l'homme charmant qu'est M. Guccia. La raison en est principalement dans le journal, les Rendiconti, que le Circolo Matematico publie sous la direction de son fondateur, M. Guccia, qui a consacré á cette direction le travail entier de ses trente dernières années. Étant lui-même au courant de tous les chapitres des sciences mathématiques, nul ne pouvait, mieux que lui, se vouer á une telle tâche. Ses *Rendiconti* sont maintenent le meilleur journal mathématique du monde. Monsieur Guccia a réussi á conquérir comme amis de son journal les mathématiciens sérieux de tous les pays. Il faudrait entrer dans des détails qui n'intéressent pas cette assemblée, pour pouvoir expliquer pourquoi on aime mieux publier les meilleures de ses recherches dans les *Rendiconti* de Palerme plûtot que dans tous les autres périodiques du monde. Je ne veux donc mentioner que deux détails, en soulignant d'abord la façon si aimable et si encourageante dont les autheures sont traités par vous, et en mentionnant surtout la grande rapidité avec laquelle les mémoires se composent et se publient ici. Il faut trois jours pour venir de ma ville à Palerme, et il m'est arrivé d'avoir les épreuves huit jours après l'expédition du manuscrit; et grâce à la perfection de l'imprimerie, il n'y avait pas grand'chose à corriger.

Le grand monde s'interéresse peu aux mathématiques; cependant je suis convençu qu'à Palerme on savait déjà, avant cette fête, ce qu'est le Circolo Matematico; mais vous ignorez sans doute qu'il y a en Allemagne une ville appelée Göttingen (plus petite que votre grande et illustre cité), ville dans laquelle on fait beacoup de mathématiques, et qui est en relations intimes avec le Circolo. Il me faut bien parler de cet endroit, puisque je suis chargé de vous apporter deux félicitations.

La première, de notre Société Mathématique de Göttingen, le centre régulier de notre vie scientifique, que vous avez bien voulu inviter à se faire représenter à cette fête, et qui m'a fait l'honneur de m'en charger.

J'apporte égualment les félicitations des quatre membres de votre Comité de Rédaction qui abitent Göttingen: Klein, Hilbert, Carathéodory et moi-même, qui sommes, en même temps, les quatre seuls professeurs de mathématiques pures de notre Université; il n'y a certainement, en dehors de Palerme, aucune autre ville réunissant un nombre supérieur de vos collègues de rèdaction.

Mais ne croyez pas que ce soit à dessein que M. Guccia a choisi comme collaborateurs tant de professeurs d'une même université, université jouissant, depuis les temps de Gauss, Dirichlet, Riemann, Minkowski et autres, d'une grande réputation mathématique. Au contraire, deux d'entre nous avaient été invité par lui à faire partie de son Comité de Rédaction, alors que nous demeurions dans des endroits bien différents, et avant même nous n'ayons eu aucune position officielle dans l'enseignement de notre pays. Peut-être les liens par lesquelles la collaboration pour le Circolo nous a unis, ont-ils contribué à nous réunir pour l'enseignement de nos élèves, et pour continuer ensemble la direction de l'École mathématique de Göttingen. En tout cas -et c'est parmi les grands et immortels mérites de M. Guccia un des plus remarquables- il a toujours jugé les mathématiciens uniquement d'aprés leurs traveaux, sans s'inquiéter de leur âge ni de leur rang officiel, et il a secouru bien des commençants -comme j'en étais un, il y a une dizaine d'années- à publier leurs recherches dans son important journal, et à avoir confiance en eux-même. Je suis venu ici pour remercier le Circolo Matematico, c'est-à-dire M. Guccia, de ce qu'il à créé, à bien de mathématiciens, leur situation dans la science. Puisse-t-il voir longtemps, et en parfait santé, les fruit du travail de sa vie laborieuse, la gratitude de nous tous, qui sommes fiers d'être ses collaborateurs, et surtout les progrès de notre chère science mathématique, à laquelle il a consacré sa vie.

Chapter 2: Note 32 V. Volterra, *Betti, Brioschi, Casorati, trois analystes italiens et trois manières d'envisager les questions d'analyse.* In: E. Duporcq (ed.), *Compte rendu du Deuxième Congrès Internationale des Mathématiciens*, Gauthier-Villar, 1902, Paris, pp. 43–57.

Dans l'automne de l'année 1858, trois jeunes géomètres italiens partaient ensemble pour un voyage scientifique. Leur but était de visiter les Universités de France et d'Allemagne, d'entrer en rapport avec les savants les plus remarquables, d'en connaître les idées et les aspirations scientifiques et, en même temps, de répandre leurs traveaux.

Ce voyage entrepis par Betti, Brioschi et Casorati marque une date qu'il est bon de rappeler. L'Italie allait devenir une nation. Elle entre à partir de cette époque dans le courant des grands travaux scientifiques et, par un nombre de travailleurs toujours croissant, apporte une contribution à l'hœuvre commune.

Dans ce jour où tant de mathématiciens se réunissent en inaugurant un échange fécond d'idées, j'aime à rappeler ce souvenir. Il serait impossible de comprendre et de suivre les progrès de l'Analyse en Italie, dans la seconde moitié du XIXe siècle, sans connaître à fond l'œuvre poursuivie avec patience et avec énergie pendant un grand nombre d'années par les trois géomètres dont je viens de rappeler les noms, secondés par les efforts de leurs meilleurs élèves.

C'est à leur enseignement, à leur traveaux, au dévouement infatigable avec lequel ils poussaient les élèves et les jeunes savants vers les recherches scientifiques, à l'influence qu'ils ont établis entre notre patrie et l'étranger que nous devons d'avoir vu naitre en Italie une jeune école d'analystes.

Chapter 4: Note 11 Letter from Cremona to Guccia, August 3, 1880. Folder Cremona, Archive of the Circolo Matematico di Palermo.

Solo vi consiglio di essere breve, conciso, quanto più è possibile [...] Io so per esperienza che in simili occasioni pochi degli astanti si interessano a un dato soggetto. Cercate dunque di esporre la sostanza della cosa senza entrare in dimostrazioni o in particolari minuti.

Chapter 4: Note 23 Letter from Cremona to Guccia, June 25, 1881. Folder Cremona, Archive of the Circolo Matematico di Palermo.

Vorreste rappresentarmi dal 4 al 18 luglio asta libri Chasles acquistando conto Biblioteca Vittorio Emanuele? Domani scrivo.

Chapter 4: Note 25 Letter from Cremona to Guccia, June 26, 1881, bis. Folder Cremona, Archive of the Circolo Matematico di Palermo.

Sento che la Biblioteca Nazionale di Parigi vuol acquistare quel che le manca. Spero che ciò ci recherà poco danno. Primo perché la Nazionale di Parigi dovrebbe essere già ben provveduta, mentre noi siamo mancanti di tutto; e secondo perché noi non possiamo darci il lusso di andare a caccia di rarità. Dobbiamo contentarci del necessario e di quello che può esserci utile.

Chapter 4: Note 26 Letter from Cremona to Guccia, July 10, 1881. Folder Cremona, Archive of the Circolo Matematico di Palermo.

Voi vi rendete altamente bene merito de' nostri studi nel nostro Paese. Mi ha fatto sommo piacere la vittoria che avete riportata sulla Faculté des Sciences [...] Bravo davvero!

Chapter 4: Note 27 Letter from Cremona to Guccia, July 16, 1881. Folder Cremona, Archive of the Circolo Matematico di Palermo.

Carissimo Guccia, Vi ringrazio della costante premura colla quale mi mandate le notizie delle successive vacations. Voi fate le cose eccellentemente; e godo assai che vi troviate soddisfatto del sig. Lorenz: al quale spedisco oggi stesso direttamente un acconto di lire cinquemila. Vi ripeto che sono soddisfattissimo dell'opera vostra e che vene sono riconoscentissimo.

Chapter 4: Note 28 Letter from Cremona to Guccia, July 31, 1881. Folder Cremona, Archive of the Circolo Matematico di Palermo.

Il servizio che voi, per amor mio, avete reso alla Biblioteca Vittorio Emanuele è grandissimo: ed io non so trovare parole sufficienti per attestarvene la mia profonda gratitudine. E questo mio sentimento non è scompagnato da rimorso, quando penso che, solo per cagion mia, siete rimasto tanti giorni a Parigi a godervi un calore insopportabile! [...] Le casse contenenti le cose acquistate a Parigi cominciano ad arrivare. Ne arrivarono già due, ne debbono arrivare altre quattro. In ogni cosa ho seguito i vostri consigli: compresi quelli riguardanti le ordinazioni Cauchy allo Hermann. Voi avete fatto bene in ogni cosa: ed io vene rendo ampia testimonianza.

Chapter 4: Note 31 Letter from Guccia to Cremona, November 12, 1882. Folder Cremona, Archive of the Mazzini Institute of Genoa.

[Ci ho una zia (sorella del prof. Cannizzaro)] che pare creata per il flagello mio ! Quest'anno è sortita fuori con una mezza pretesa, mercè la quale si prometteva di ottenere dalla mia famiglia qualcosa come 600.000 lire. Meno male che i magistrati finiscono poi sempre per rigettare le sue strane domande e per condannarla alle spese, ma con ciò non mi viene restituito il tempo sciupato a correre su e giù per i tribunali.

Chapter 4: Note 37 Letter from Guccia to Cremona, April 13, 1884. Folder Cremona, Archive of the Mazzini Institute of Genoa.

Il concorso dell'anno passato per la cattedra di Matematica al Liceo Umberto I mi aprì gli occhi sullo stato miserando in cui trovansi abbandonati gli studi matematici in Palermo. Quando lei volesse fermare la sua attenzione su questo stato di cose, potrei fornirle documenti e raccontarle fatti, da non sembrare credibili; ma mi crederà sulla parola! Giovani e Professori pubblicano e pubblicano continuamente, senza avere alcun indirizzo scientifico e solo animati dalla fretta di acquistare titoli e notorietà. E dire che in Italia non si trova nessuno che si dia la pena di aprire loro gli occhi e disilluderli!

Non lo fanno nemmeno quei Geometri che sono stati citati a sproposito. Ciò che caratterizza principalmente la natura di queste produzioni, sono le ingenuità, le quali, a parer mio, in scienza sono più condannabili degli stessi spropositi. Frattanto è fuori di dubbio che l'Università dá vita ed applaude a questa sconsigliata attività scientifica. Né (bisogna pure avere la franchezza di dirlo) i bravi professori del continente che si sono succeduti in breve spazio di tempo, quali il Padelletti, il Tonelli, il Capelli, sono riusciti a modificare questo stato di cose; chi per mancanza di energia e di abilità nel lottare coi vecchioni della Facoltà, chi per naturale egoismo che non lo fa sortire dai limiti dello stretto dovere, il fatto si è che tutti hanno lasciato le cose per come si trovano. [. . .]

Come spiegare infatti che quel tale libro dove, tra le mille corbellerie, si assume "essere lo spazio di retta a 5 dimensioni" con quel che segue, fu l'oggetto di un rapporto favorevole firmato dal Tonelli e dal Padelletti in occasione di una domanda di libera docenza? La frase con cui il lavoro è lodato suona presso a poco così: "l'autore mostra di avere un'esatta (sic) cognizione dei metodi moderni".

Come spiegare ancora che il Paternò (ing. F.co Paolo), dichiarato ineleggibile nel concorso pel Liceo Umberto I, è stato, immediatamente dopo, nominato professore incaricato della Geometria Projettiva nell'Università?

A queste considerazioni si aggiunge nella mente mia l'impressione avuta dalla lettura della "Relazione della Commissione Universitaria" che lei mi diede l'occasione di avere tra le mani, nella quale questi signori hanno la spudoratezza di domandare due insegnamenti complementari, onde essere autorizzati a conferire la Laurea in Matematica.

Chapter 4: Note 38 Letter from Guccia to Cremona, April 13, 1884. Folder Cremona, Archive of the Mazzini Institute of Genoa.

Fu allora che mi venne l'idea di un Circolo Matematico che riunisse in un locale comodo e fornito di un buon gabinetto di lettura e di studio (non meno di 16 pubblicazioni periodiche di Matematica) tutte le persone, vecchi e giovani, che da vicino o da lontano avessero avuto commercio colle matematiche. Il mio caldo appello fu ascoltato, tutti accorsero al mio programma il quale conteneva due punti essenzialissimi su cui non intendevo recedere:

1° esclusione assoluta di tutti gli argomenti che hanno riguardo alle matematiche applicate;
2° esclusione a priori di qualunque lontana idea di pubblicazione per parte del Circolo, il cui primo scopo doveva essere quello (nella mente mia) di evitare o meglio di scongiurare le pubblicazioni.

(Non starò a dirle quanto ho dovuto lottare e lotto tuttavia per persuadere questa brava gente che non è il caso di andare in giro per l'Europa con un *Bollettino del Circolo Matematico di Palermo*). È così che è sorto fuori questo statuto la cui natura ibrida ha la sua ragion di essere per le circostanze locali di cui dovevo necessariamente tener conto.

Chapter 4: Note 40 Letter from Guccia to Cremona, April 13, 1884. Folder Cremona, Archive of the Mazzini Institute of Genoa.

Lei non può immaginare quante difficoltà ho dovuto sormontare ed a qual tirocinio di simulazione ho dovuto sobbarcarmi per portare la nave in porto. Fin dal principio era nel mio programma segreto di non urtare la suscettibilità di nessuno, lo scopo che io mi sono prefisso colla fondazione del Circolo essendo quello di migliorare le cose di Palermo per via di persuazione anziché colla violenza. Ond'è che composi nella mia mente il Consiglio direttivo tal quale restò eletto nella 1ª seduta del 20 marzo, dove si trovano riguardati i titoli di anzianità.

Chapter 4: Note 41 Letter from Guccia to Cremona, April 13, 1884. Folder Cremona, Archive of the Mazzini Institute of Genoa.

Veniamo ora alla parte finanziaria. Se il Circolo doveva provvedere all'affitto del locale e alle spese di fondazione dei suoi mobili, è certo che i suoi fondi sarebbero stati assorbiti. È così che ho pensato che bisognava far qualche sacrificio per venire in ajuto della istituzione. Ho offerto gratis il locale (due buone stanze a pianterreno) l'ho mobigliato a spese mie e fornito di tutto il necessario per l'uso a cui serve e l'ho messo a completa disposizione dei soci dalle 11 a.m. alle 5 p.m. Così i fondi del Circolo i quali provengono dalle contribuzioni annuali dei soci e dal diritto di ammissione servono esclusivamente a provvedere agli abbonamenti ed all'acquisto di qualche opera. È dunque la formazione di una buona biblioteca, uno dei principali obbiettivi della istituzione da me creata. L'art. 14 dello statuto mi serve a stimolare i soci a far doni. Ma i doni dei soci non bastano ! Ecco il punto su cui mi permetto di rivolgerle il più caldo appello.

Chapter 4: Note 42 Letter from Guccia to Cremona, April 13, 1884. Folder Cremona, Archive of the Mazzini Institute of Genoa.

Se Lei approva l'istituzione e la reputa veramente utile ad un paese dove (per la parte matematica) l'Università non esiste, e le Biblioteche sono sfornite di tutto, sono certo che non saprà negarmi il suo valevole appoggio, spiegando a pró della stessa tutta la sua energica operosità. Sarebbe forse indiscreto di rammentarle in questa occasione che anch'io (nell'estate del 1881) prestai i miei servizi alla Biblioteca V.E.?

Trattandosi di far qualcosa per la Scienza, non deve essere permesso di fare appello ai sentimenti di riconoscenza? Non le pare che questa nota toccata con garbo possa riuscire a commuovere i caporioni del Ministero?

Ma in questo capitolo se c'è da fare qualche cosa è Lei che mi darà la direzione. La questione è impregiudicata non avendo io scritto a nessuno, appunto perché attendo da Lei i consigli opportuni, dettati dal suo amore per la Scienza e dalla sua esperienza. Così mi sono fin'anco astenuto dal rivolgermi alle Accademie italiane onde tentare di ottenere gratis i loro atti e le loro pubblicazioni.

Non ho che una pratica in corso (segreta) con Ministère d'I.P. di Francia per estorcergli le opere di Lagrange o di Laplace.

È utile che io la prevenga che in quanto ad ottenere incoraggiamenti dal Comune o dalla Provincia per ora è opera vana o per lo meno pericolosa, giacché questi enti locali mettono solitamente alle loro largizioni delle condizioni che potrebbero far degenerare l'indole e lo scopo del Circolo. Ed il Circolo (è buono a sapersi) è nelle mani mie e seguirà le mie vedute (che credo Lei non avrà disapprovate) finché è povero e finché non sarà adescato da incoraggiamenti che ne falsino l'indirizzo.

Dunque io conto sopra di Lei, esclusivamente sopra di Lei! E per cominciare mi pare che la sua escursione nel Nord potrebbe procurarmi dei regali di memorie, per il Circolo, dai principali Geometri che Lei avrà occasione di vedere. Questa comincerebbe per essere un'opera utilissima. Al suo ritorno in Roma poi Lei mi dirà cosa si potrà fare in Italia col Ministero, colle Accademie, colle Biblioteche, coi privati.

Frattanto sarebbero utilissime due paroline sue officiali dirette al Consiglio Direttivo del Circolo per mostrare che Lei approva l'istituzione, nei limiti che si è imposto il nostro programma. Potrebbe prendere occasione dall'invio dello "Statuto" ovvero da una lettera mia privata in cui si suppone che io le ho partecipato la fondazione del Circolo. Giacché è bene inteso che tutto ciò che le ho scritto oggi resta in linea confidenzialissima.

Chapter 4: Note 44 Letter from Guccia to Cremona, July 12, 1884. Folder Cremona, Archive of the Mazzini Institute of Genoa.

In tal caso le proporrei la pensione del Rigi-Staffel, la di cui posizione è sorprendente, la cucina ottima ed i prezzi modestissimi. Sono certo che Lei colla famiglia vi si troverebbero ottimamente. Era questo il posto dove volevo condurre i Manneheim. È inutile che insista sulle escursioni che Ella potrebbe fare a piedi partendo dallo Staffel, perché forse in parte le avrà già fatte; ciò che posso aggiungere per mia propria esperienza è che una residenza al Rigi, per un mese, è di gran giovamento alla salute e non credo che a questo riguardo si possa trovare meglio in Svizzera. Aggiunga a ciò la facilità che si ha coi mezzi ferroviari di potersi recare in qualunque posto della Svizzera. Per parte mia, dato che Ella seguisse i miei consigli, le sarei fedele compagno sulle escursioni e metterei a sua disposizione una piccola biblioteca (da viaggio) dove Lei troverebbe la Collezione completa delle note e delle memorie del Cremona.

Chapter 4: Note 46 Letter from Guccia to Cremona, October 22, 1884. Folder Cremona, Archive of the Mazzini Institute of Genoa.

A proposito del Circolo credo di avere lavorato bene. Tutti, nessuno eccetto, si sono interessati alla Biblioteca della società matematica di Palermo ed i doni piovono da tutte le parti. Vediamo ora cosa sa fare Lei a Roma!! Sappia bene che le mie più grandi speranze sono riposte su di Lei che dispone di tanta influenza presso il Ministero. In varie circostanze Ella ha mostrato che quando vuole riesce ad ottenere quattrini a profitto della Scienza.

Or a Palermo il caso è grave e va ponderato. Mai denari saranno stati meglio spesi che a favore di un'istituzione che si propone principalmente tre scopi di somma importanza.

1° Impedire la pubblicazione di libri ed opuscoli che disonorano il paese!
2° Fornire agli studiosi della nuova generazione una biblioteca che li metta al corrente della Scienza;
3° Stabilire, colla risoluzione e discussione di quesiti, quella gara necessaria ai giovani per far loro acquistare il vero Amore della Scienza.

Il 1° scopo che era il più urgente ad ottenere, ha dato, fin dalle prime sedute, le migliori speranze! Procurerò di spiegarmi meglio a Roma quando avrò il piacere di vederla, ma per ora mi creda salla parola. È urgente tanto più, in quanto che il Governo, a proposito dell'Università, non lascia sfuggire un'occasione per commettere corbellerie. Mi scrivono infatti che il Ministro Coppino ha richiesto alla Facoltà Matematica di Palermo una lista di nomi per istituire colà quest'anno stesso, una Scuola Superiore di Matematica. Evidentemente, e senza tema di andare errati, si può presumere, fin da ora, che gli individui proposti ad insegnare matematica, saranno quelli stessi che uscirono malconci (mercè le lotte disperate sostenute da alcuni membri della Commissione) dal Concorso pel Liceo Umberto I.

Chapter 4: Note 50 Letter from Guccia to Cremona, March 19, 1885. Folder Cremona, Archive of the Mazzini Institute of Genoa.

Allo stato attuale delle cose vi è molto, ma molto, da fare in Geometria e non è verosimile che la scienza debba rassegnarsi a rinunciare al soccorso del suo potente ingegno. Non si tratta di sviluppare ed applicare metodi già noti (a questo servizio vi è tutto un popolo di geometri che vi accudisce con prudenza e coscienza) ma bensì di crearne di nuovi per riparare di urgenza a delle lacune, le quali trascurate potrebbero forse un giorno portare al discredito e la sfiducia su tutto ciò che è stato sin'ora. La geometria del piano è tutta da rifare. la curva generale d'ordine n è una forma ternaria completa, è dominio dell'algebra moderna, la quale incontestabilmente su questo terreno perviene agli stessi risultati della geometria con maggiore speditezza e spesso con più rigore. Bisogna dunque, pare a me, rifare tutte le teoria del piano avendo di mira la curva d'ordine n affetta da singolarità qualunque.

È dunque la definizione geometrica della singolarità qualunque di una curva piana che s'impone come problema della massima importanza, su cui dovrebbero tendere gli sforzi delle menti creative come la sua! I risultati ottenuti dall'Halphen hanno per punto di partenza i metodi di Cramer e Poiseux. Proseguendo su quella via è l'algebra che incidentalmente, vi guadagnerebbe, ma non certo la geometria.

Chapter 4: Note 52 Letter from Guccia to Cremona, July 6, 1886. Folder Cremona, Archive of the Mazzini Institute of Genoa.

È uno spettacolo indecente! Una Facoltà che si rispetta avrebbe domandato i concorsi. Punto. Domanda invece che si distribuiscano incarichi a 1200 lire in famiglia. L'appetito viene mangiando ed è così che gli incarichi del 1885-86 saranno probabilmente (per alcuni la richiesta è fatta) straordinari pel 1886-87. Di guisa a perpetuare, a rendere stabile, la sconcezza.

Chapter 4: Note 56 Letter from Guccia to Cremona, July 6, 1886. Folder Cremona, Archive of the Mazzini Institute of Genoa.

[Mi diceva che] l'intelligenza lo abbandonava, non era più capace di creare nulla, non ritrovava la solita energia mentale per condurre a fine qualsiasi ricerca, si vedeva raggiunto e sorpassato da altri, curiosava intorno a diversi argomenti ma non si sentiva la forza di approfondirne alcuno; vedeva estinguersi in lui il fuoco sacro della Scienza, era scoraggiato.

Chapter 4: Note 64 Letter from Guccia to Cremona, August 8, 1886. Folder Cremona, Archive of the Mazzini Institute of Genoa.

Se non vi fosse altro da rimproverare alla dinastia dei Borboni in Sicilia, per conto mio non esiterei a condannare il Re Ferdinando I delle due Sicilie per avere regalato (in parte venduto) il cespite delle Acque ad un mio antenato, il Marchese Guccia! Una simile macchina amministrativa non si può definire a parole: Bisogna vedere cosa è! Basti il dirle che io preferirei mille volte di amministrare da me solo, in luogo delle Acque, il Comune, la Provincia e tutte le opere Pie create e da crearsi! quante volte non ho meditato il progetto di abbandonare tutto e correre in un angolo oscuro d'Italia per dare lezioni di Matematica e dedicarmi esclusivamente alla Scienza.

Mi ha trattenuto la speranza, che tutt'ora nutro, di cedere onorevolmente (non regalarla al Comune ben inteso!) la mia proprietà. Tutti i miei sforzi tendono a questo scopo [...] Non voglio specularci sopra, non chiedo una lira di più di ciò che prendo annualmente, ma voglio sortirne a qualunque costo!

Chapter 4: Note 77 Letter from Guccia to Cremona, September 6, 1886. Folder Cremona, Archive of the Mazzini Institute of Genoa.

Gent.mo Professore, non resisto alla tentazione di comunicare a Lei per primo un teorema sui punti singolari delle curve algebriche piane. Non starò a ripeterle quali e quanti tentativi siano stati fatti dai più distinti geometri allo scopo di fare un pó di luce su questo intricatissimo argomento. Quasi tutti ne son sempre sortiti stanchi, nauseati e spesso disillusi. Nei limiti modesti delle mie forze sarò tornato alla carica almeno una decina di volte; e se da 5 anni a questa parte mi sono spesso rivolto a delle ricerche in altro campo, ciò ha avuto per iscopo di farmi riposare, per qualche tempo, dai punti singolari. Ma posso ben dire di non averli abbandonati mai. Comprenderà dunque come io sia contento di vedere ricompensate largamente le fatiche ed il tempo spesi in questi difficili problemi, con un risultato di grande importanza per il futuro svolgimento della teoria.

Chapter 4: Note 92 Letter from Guccia to Cremona, May 6, 1888. Folder Cremona, Archive of the Mazzini Institute of Genoa.

Ho potuto convincermi che di 5 o 6 membri del Consiglio posso fare grande assegnamento, di alcuni altri non mi fido troppo, perché da noi si è deboli per natura, difficilmente si resiste alla insistente sollecitazione. [...] I manoscritti finora pervenuti (dopo la costituzione definitiva della Società) superano ciò che potrei pubblicare nel tomo II, cioè a tutto dicemdre 1888. Siamo dunque sullo stretto dovere di fare una buona scelta! A tutt'oggi, dietro revisione, ho respinto 10 manoscritti!

Chapter 4: Note 100 E. de Amicis, *Ricordi di un viaggio in Sicilia*, Giannotta Editore, 1908, Catania. Newly edited: Il Palindromo, 2014, Palermo.

Palermo [...] è una grande città. I nuovi quartieri eleganti, le nuove vaste piazze alberate, i nuovi magnifici passeggi pubblici, veri luoghi di delizie, degni di Parigi e Londra [...] Ma [Palermo] è tutta una città di violenti contrasti [...] Dal grande viale marino del Foro Italico, [...] dove corrono centinaia di carrozze aristocratiche, si riesce in pochi passi lungo la [...] dove [...] vi rappresentano tutte le miserie e le calamità [...] dei secoli passati. Usciti da quell'enorme labirinto di viuzze oscure e sudice, [...] dove brulica una popolazione poverissima in migliaia di fetidi covi, che sono ancora quei medesimi in cui si pigiavano gli Arabi di nove secoli or sono, e vi trovate dinanzi a un "Teatro Massimo", il più grande e il più splendido teatro d'Italia [...] e di cui fu decretata la costruzione quando Palermo non aveva un ospedale che rispondesse ai suoi più stretti bisogni.

Chapter 5: Note 1 *Discorso del Senatore Prof. Dr. Vito Volterra*, XXX anniversario della fondazione del Circolo Matematico di Palermo, Adunanza solenne del 14 aprile 1914, Suppl. Rend. Circ. Mat. Palermo, 9 (1914), pp. 17–19.

Ricordo. Ventinove anni or sono io facevo il mio primo viaggio in quest'isola incantevole e, venuto a Palermo, dopo avere ammirato le superbe bellezze naturali, le splendide opere d'arte di questa meravigliosa città, un mio amico, Ignazio Conti, che purtroppo oggi non è più fra noi, mi invitava a visitare una istituzione scientifica palermitana sorta da poco tempo. Venni in via Ruggiero Settimo e fui presentato a G.B. Guccia. Pochi erano allora, sebbene zelanti e volenterosi, i soci del Circolo Matematico, non ancora incominciato il primo fascicolo di pubblicazioni, modesta la Biblioteca. Ma grande era l'entusiasmo, che animava Guccia. Pieno di speranza e di ardore, egli aveva fin da allora la visione che qualcosa di vitale e di utile doveva uscire da quei modesti principi. La sua fede sincera e profonda rivelava una di quelle energie che creano le grandi istituzioni e danno origine alle opere durature e feconde. Ne rimasi colpito fin da quella prima visita.

Chapter 5. Note 38 Letter from Segre to Castelnuovo, November 5, 1893. Guido Castelnuovo Collection, Archive of the Accademia Nazionale dei Lincei.

Sono molto disgustato del modo come Cremona ha presa la sconfitta del suo "protetto".

Chapter 5: Note 42 Letter from Segre to Guccia, November 28, 1887. Folder Segre, Archive of the Circolo Matematico di Palermo.

Mi dolse molto che Ella non si sia fermata qui qualche giorno. Avrei udito volentieri le notizie da Lei portateci dalla Francia, ma spero che Ella mene darà per lettera. Ha udito leggere qualche lavoro importante di Geometria? Ha visto l'Hirst e gli ha detto che, venendo in Italia, si fermi a Torino?

Chapter 5: Note 48 Letter from Guccia to Dini, January 9, 1906. Folder Dini, Archive of the Circolo Matematico di Palermo.

...ho potuto ... dare attuazione ad un programma che da tanto tempo accarezzavo e che tende ad imprimere alla nostra Società un progresso tale da fargli occupare, fra pochi anni, il primo posto fra le 5 grandi associazioni di matematici del mondo (Londra, Parigi, New York, Lipsia, Palermo)!

Chapter 5: Note 50 Letter from Guccia to the President of the Circolo Matematico di Palermo, July 23, 1904. Folder Guccia, Archive of the Circolo Matematico di Palermo.

Fó noto alla S.V. Ch.ma di avere stabilito un premio per un Concorso Internazionale rivolto ad incoraggiare qualcuna delle alte e moderne teorie della Geometria. Siffatto premio, consistente in una mediaglietta ed in una somma di lire 3000 (tremila) in oro, sarebbe conferito dal Circolo Matematico di Palermo nell'anno 1908, giusto il giudizio di una Commissione Internazionale composta di tre membri, nominati dal Presidente della nostra Società, alla quale Commissione si affiderebbe anzitutto l'incarico di formulare, in tutti i suoi particolari, il corrispondente Programma del Concorso.

Chapter 5: Note 61 Adunanza August 27, 1905. Rend. Circ. Mat. Palermo, 20 (1905), p. 377.

Le memorie destinate al concorso dovranno essere: inedite, redatte in lingua italiana, o francese, o tedesca, o inglese, e scritte (tranne che per le formole) con una macchina da scrivere. Munite di una epigrafe, esse dovranno pervenire, in tre esemplari al Presidente del Circolo Matematico di Palermo prima del 1º Luglio 1907, accompagnate da un plico suggellato portante sulla busta l'epigrafe adottata e nell'interno il nome e l'indirizzo dell'autore. La memoria premiata sarà inserita nei Rendiconti o altra pubblicazione del Circolo Matematico di Palermo. L'autore riceverà 200 estratti.

Qualora nessuna delle memorie presentate al concorso fosse ritenuta degna di premio, questo potrà essere aggiudicato ad una memoria sulle teorie anzidette, che venisse pubblicata dopo la pubblicazione del presente programma e prima del 1º luglio 1907.

Il premio verrà conferito dal Circolo Matematico di Palermo in conformità al giudizio di una commissione internazionale di tre membri, composta dai signori: Prof. Max Noether, dell'Università di Erlangen; Prof. Henri Poincaré, dell'Università di Parigi; Prof. Corrado Segre, dell'Università di Torino.

La lettura del rapporto della Commissione, nonché la proclamazione del nome dell'autore premiato ed il conferimento del premio si faranno a Roma, nel 1908, durante una seduta del IV Congresso Internazionale dei Matematici.

Chapter 5: Note 69 Letter from Guccia to Mittag-Leffler, September 11, 1905. Folder Mittag-Leffler, Archive of the Circolo Matematico di Palermo.

Je désire avant tout vous voir, et conférer tranquillement avec vous sur quelqu'un de mes projects, ayant pour but de bien internationaliser, diffonde et répandre la production mathématique du monde entier, en profitant des progrès accomplis par la civilisation moderne dans les rapports internationaux. Je compte beaucoup sur les conseils que vous pouvez me donner là-dessus.

Chapter 5: Note 71 Letter from Castelnuovo to Volterra, July 29, 1905. Vito Volterra Collection, Archive of the Accademia Nazionale dei Lincei.

Trovo opportunissima l'idea di Guccia di riunirci a Roma in Novembre per accordarci sul lavoro di organizzazione. Noi intanto ci vedremo anche prima della venuta del Guccia, e concorderemo ciò che in quella seduta dovrà decidersi.

Chapter 5: Note 73 Letter from Guccia to Volterra, June 30, 1906. Folder Volterra, Archive of the Circolo Matematico di Palermo.

La politica: microbo che in Italia entra da per tutto e … uccide tutto! in particolar modo la scienza! La politica! Ecco il gran nemico della scienza in Italia! Ecco perché istituzioni che prosperano e fioriscono in altri paesi non possono attecchire in Italia!

Chapter 5: Note 74 Letter from Guccia to Volterra, November 6, 1907. Vito Volterra Collection, Archive of the Accademia Nazionale dei Lincei.

La nostra Società non riceve alcun sussidio dallo Stato Italiano! La nostra Società, sorta e sostenuta dall'iniziativa privata, non ha stato civile a Roma. Di ciò, il pubblico matematico internazionale (che guarda ai fatti e non alle chiacchiere) comincia a rendersi conto! Io sono il primo ad addolorarmene, come italiano, per quanto abbia la coscienza di aver fatto a suo tempo, in proposito, tutto ciò che potevo fare per risparmiare al mio paese questa cattiva figura. Non sono riuscito, perché a Roma non avevo, come non ho e come non desidero di avere, nessuna di quelle investiture ufficiali che ci conferiscono il grado di persona autorevole! A Roma mi considerano come uno stampatore!!! un tipografo! e nulla più! Ciò le spieghi, una volta per tutte, il fatto che io a Roma mi fermo raramente e solo per casi di assoluta necessità! ovvero. tra un treno e l'altro!

Chapter 5: Note 92 Letter from Guccia to Mittag-Leffler, July 12, 1908. Folder Mittag-Leffler, Archive of the Circolo Matematico di Palermo.

Quant au Congrès de Rome, ainsi que vous avez pu le remarquer sur place, notre Société s'est effacée dès que dans la séance d'inauguration, l'Académie des Lincei a fait annoncer par son Président qu'elle prenait la direction du Congrès. Cette circostance, aussi imprévue que bizarre, étant en désaccord avec les circulaires précédentes, a un peu étonné le public. Il fallait donc, sans faire du bruit, sauvegarder la dignité de notre grande société international (600 membres), qui jouit, à tous les points de vue, d'une grande indépendence, et à laquelle appartiennent les plus illustres et célèbres mathématiciens du monde entier. Je ne crois donc pas avoir mal agi, dans ma qualité de délégué du Circolo au Congrès, en résiliant, sous un prétexte avouable et vraisemblable (et, par conséquent, sans potins) notre accord avec le Comité d'organisation. [...] J'ai reçu de Rome, il y a un mois et demis, sur la question, une longue lettre d'excuses. Mais comme cette lettre ne pouvait pas changer les faits, ni l'impression reçu a Rome par le public mathématique, j'ai répondu, par dépêche ... "que je ne répondais pas"! Voila tout.

La seule chose qui m'est vraiment pénible dans cette affaire, c'est que, parmi les membres du Congrès "organisateur" (le seul résponsable vis-à-vis du public) il y avait trois membres du Conseil de Direction du Circolo, qui sont de mes bons et excellent amis, mais qui, dans la circostance, ont oublié tout-à-fait notre Société! Entre l'Académie des Lincei, illustre et célèbre, qui s'occupe de tout, même des maladies du fromage "Gorgonzola" (Rendiconti dei Lincei, 26 avril 1908, page 568), et notre modeste société, qui ne s'occupe que de mathématiques, le chois n'etait pas douteux pour eux: ils se sont tournés vers l'Académie, pour accroître sa gloire, à l'occasion de ce congrés. Ils ont peu-être eu raison, à leur point de vue. Mais je trouve que dans ce cas, à leur place, j'aurais préalablement donné ma démission de membre du Conseil de Direction du Circolo!

Chapter 6: Note 52 Letter from Guccia to Volterra, January 7, 1914. Vito Volterra Collection, Archive of the Accademia Nazionale dei Lincei.

1º   Che la cerimonia rimanga limitata e circoscritta alla famiglia matematica. [...] accetto tutto dai matematici, nulla dagli altri.

2º   Che nel caso in cui venisse in mente a qualcuno di chiedere delle decorazioni per me, Ella impedirà che ciò avvenga.

Chapter 6: Note 54 *Discorso del Senatore Prof. Dr. Vito Volterra*, XXX anniversario della fondazione del Circolo Matematico di Palermo, Adunanza solenne del 14 aprile 1914, Suppl. Rend. Circ. Mat. Palermo, 9 (1914), pp. 18–19.

... il Guccia, dando al suo periodico una schietta impronta internazionale, lo ha posto sopra basi sicure ed incrollabili. Egli ha fatto convergere verso i Rendiconti, mercè la stima e la simpatia acquistate, le memorie di matematici di tutte le parti del mondo. Con opera assidua, perseverante e geniale, con fine discernimento, ha saputo scegliere fra gli autori, fra le scuole, fra gli indirizzi, tanto che il Circolo ha pubblicato dei lavori, che hanno fatto epoca nel mondo scientifico e sono destinati a rimanere classici e fondamentali. Oggi il Circolo di Palermo è la prima società matematica del mondo pel numero di suoi membri fra i quali si contano i nomi più belli della scienza; e i trentasette volumi che ha pubblicato contengono memorie di alta analisi, di geometria, di meccanica e di fisica matematica, attraverso le quali si può seguire la storia dei mirabili progressi compiuti dalle metematiche nell'ultimo trentennio. I matematici più insigni si tengono onorati di pubblicarvi i loro scritti; i Rendiconti penetrano ovunque esiste il culto della Geometria e vi diffondono rapidamente le più recenti scoperte.

Chapter 6: Note 62 Letter from Guccia to Cerruti, June 15, 1905. Folder Cerruti, Archive of the Circolo Matematico di Palermo.

Ma molto, molto ancora, ci è da fare! Le dirò francamente ciò che mi passa per la testa, incoraggiato come sono del successo che, ogni giorno di più, ottiene all'estero la nostra Società. Dunque, appena ci saremo bene costituiti, con 400 soci, di tutti i paesi, e che ci sentiamo abbastanza forti, ... un gran passo, molto ardito, ci sarebbe da fare secondo me. Ma è un passo talmente difficile e ardito, che prima di darlo bisogna pensarci molto!

Chapter 6: Note 63 Letter from Guccia to Cerruti, June 15, 1905. Folder Cerruti, Archive of the Circolo Matematico di Palermo.

...non può lasciarci indifferenti! Sorge dunque naturale l'idea che anche in Italia qualche cosa di serio si faccia per la matematica applicata, [...] tale compito dovrebbe spettare al Circolo Matematico di Palermo, per ovvie ragioni:

$1°$ perché (quando avremo 400 soci) la nostra stessa clientela all'estero ci ajuterà, ne sono sicuro, a svolgere con successo questo $2°$ programma;

$2°$ perché i vecchi nostri Rendiconti (organo per le matematiche pure), già tanto accreditati, servirebbero ad accreditare in poco tempo i nuovi Rendiconti, che si pubblicherebbero parallelamente col titolo: "Organo per le matematiche applicate", o altro equivalente.

S'imporrebbe allora una riforma di Statuto in questo senso:

$1°$ che il Circolo pubblicherebbe due "Rendiconti": i Rendiconti per le matematiche pure e i Rendiconti per le matematiche applicate;

$2°$ che ogni socio con le quindici lirette di contribuzione annua avrebbe diritto (a sua scelta) agli uni o agli altri, ovvero ad entrambi mediante un supplemento di spesa. [...]

Le organizzazioni e i metodi sarebbero gli stessi, ma ciò non basta per assicurare il successo dell'impresa. Bisogna che sorta fuori un uomo che abbia speciale ed estesa competenza, grande facoltà di lavoro, numerose relazioni personali all'estero, etc.. [...] Quest'uomo poi dovrebbe avere la voglia e il tempo di dedicarsi interamente a questa missione, senza ambire altro guiderdone se non quello di avere reso un gran servizio al nostro paese.

Chapter 6: Note 66 Letter from Guccia to Cerruti, June 15, 1905. Folder Cerruti, Archive of the Circolo Matematico di Palermo.

...debbo portare in porto quel disgraziatissimo "Repertorio Bibliografico delle Scienze Matematiche in Italia", che vorrei far sortire nel 1906 (il $1°$ vol. ben'inteso) in volume a parte. Ma lei non potrà certo immaginare (glielo dico confidenzial-mente) quante corbellerie mi hanno fatto gli amici che vi collaborarono. In proposito mi trovo in imbarazzi serissimi.

Chapter 6: Note 67 Letter from Guccia to Dini, August 4, 1907. Folder Dini, Archive of the Circolo Matematico di Palermo.

La lettura dell'importante Prefazione, nonché dell'Introduzione e di qualche capitolo del suo libro, mi ha fatto pensare alla mancanza di un servizio di recensioni nelle pubblicazioni periodiche della nostra società. Le confesso che mi sono sempre spaventato di affrontare questa difficoltà! [...] Non vediamo forse tutti i giorni opere piene di errori, ingenuità e plagi lodate oltre misura da compiacenti recensori, mossi a far ciò da interessi che non sono sempre quelli della scienza e dell'insegnamento? Ed allora, dico io, quali e quante sarebbero le difficoltà contro le quali si urterebbe un direttore di periodico se si ponesse in mente di disciplinare e moralizzare siffatto servizio?

Chapter 6: Note 68 Letter from Guccia to Poincaré, December 19, 1905. Folder Poincaré, Archive of the Circolo Matematico di Palermo.

Je n'ai pas abandonné l'idée dont je vous parlais dernièrment à l'hôtel Continental, savoir de publier chaque année dans l'Annuaire (qui se répand gratuitment à un très grand nombre d'exemplaires) un article d'un maître de la Science sur un argument qui puisse intéresser même ceux qui n'ont qu'une modeste éducation scientifique. Cela devrait, suivant moi, donner une idée aux grand publique des grandes conquètes de la science dans ces dernieres temps, et augmenter le nombre des amis des mathématiques. Le lien entre la science pure et les sciences appliquées échappe au grand public et, parfois, même au savants!

Il serait dons utile, suivant moi, de bien faire ressortir (en s'adressant au grand public) tout ce qui revient de plain droit aux hautes mathématiques dans les grandes découvertes moderne d'ordre pratique. Mais un argument aussi délicat et profond en même temps ne peut être traité que par un savant comme vous.

Chapter 6: Note 75 Letter from Guccia to Bortolotti, May 12, 1907. Folder Bortolotti, Archive of the Circolo Matematico di Palermo.

Disgraziatamente il nostro bilancio non offre alcun margine per onorari ed indennizzi ai nostri collaboratori: si tratterebbe, quindi, di lavorare per l'onore di far cessare l'ingiusto oblio nel quale, per un secolo, gli italiani hanno lasciato il sommo geometra di Modena. [...]

Tenga presente che la diffusione dei nostri Rendiconti supera quella di ogni altro dei più reputati periodici di matematica del mondo! Bisognerebbe quindi prevedere il caso di lettori cui il nome del Ruffini riesce assolutamente nuovo. Dunque [...] bisognerebbe informare brevemente ma chiaramente i lettori dei Rendiconti delle vicende della famosa questione della "impossibilità" etc.. e della parte che vi prese il Ruffini, del recente movimento di rivendicazione (Burkhardt, Bortolotti) a favore del Ruffini e della opportunità di mettere sotto gli occhi dei lettori un'opera che nessuno ha potuto leggere, perché non è più in commercio, e che fu una gran luce su un punto importantissimo della storia dell'Algebra! Ma tutto ciò senza aver l'aria di volere iniziare una campagna contro Cauchy o altri.

Chapter 6: Note 77 Letter from Cauchy to Ruffini, September 20, 1821. In: E. Bortolotti, *Opere matematiche di Paolo Ruffini*, Vol. III (mathematical correspondence), Unione Matematica Italiana, Ed. Cremonese, 1954, Rome, pp. 88–89.

Votre mémoire sur la résolution générale des équations est un travail qui m'a toujours paru digne de fixer l'attention des géomètres, et qui, à mon avis démontre complètement l'insolubilité algébrique des équations générales d'un dégré supérieur au quatrième. [...] Votre travaille sur l'insolubilité algébrique des équations est précisément le titre que j'ai fait valoir dernièrement en Votre faveur auprés de quelque membres de l'Académie, lorsqu'il s'est agi de nommer un corréspondant pour la section de géométrie.

Chapter 6: Note 78 Adunanza, August 31, 1907. Suppl. Rend. Circ. Mat. Palermo, 2 (1907), p. 41.

1° Di pubblicare in volumi a parte, per cura del Circolo Matematico di Palermo e sotto la direzione del Prof. Guccia, Direttore dei Rendiconti, le *Opere matematiche* di Paolo Ruffini ed il suo "Carteggio" con gli scienziati del suo tempo;

2° Di accettare, con plauso, l'offerta del Prof. Guccia di apprestare egli i fondi necessari per "detta pubblicazione";

3° Di versare al Prof. Guccia quanto sarà ricavato dalla vendita di dette *Opere*; e ciò sino alla "concorrenza della somma erogata per detta pubblicazione", dovendo, in conformità ed analoga proposta del Prof. Guccia, l'eventuale profitto rimanere a beneficio del Circolo e l'eventuale perdita ad esclusivo carico del Prof. Guccia.

Chapter 7: Note 5 M. De Franchis, *G.B. Guccia, Cenni biografici,* Rend. Circ. Mat. Palermo, 39 (1915), pp. i–x, pp. ix–x.

Ricordo ancora quei giorni. Il Guccia era stato duramente colpito, fin dall'autunno 1913, dalla malattia che doveva rapidamente spegnerne l'esistenza. In quei giorni dell'aprile, la malattia aveva fatto progressi giganteschi. Pure, il vedere universalmente riconosciuta l'importanza per la Scienza del Circolo Matematico, da parte dei più antichi ed illustri consessi scientifici e da parte degli uomini più insigni per dottrina, gli infuse un nuovo vigore. Effimero vigore! Il giorno della consegna della medaglia, egli volle parlare per ringraziare i numerosi intervenuti: un fremito di commozione percorse l'adunanza, ché la sua voce sembrava venire d'oltre tomba e sul suo viso scarno già aleggiava la morte inesorabile.

Tutto fu inutilmente tentato per risanarlo, consulti di illustri medici e dimore in sanatori. Fu appunto mentre si trovava nel sanatorio di Valmont, che scoppiò l'orrenda guerra che ancora sconvolge l'Europa. Essa sconvolse anche il suo ben fatto animo, per le stragi e per gli odi che essa scatenava tra popoli da lui stimati e tra i quali contava diletti amici, e per la visione delle immancabili ripercussioni che ne avrebbe risentito una Società scientifica internazionale qual'è la nostra.

Chapter 7: Note 22 Suppl. Rend. Circ. Mat. Palermo, 11 (1919–1920), pp. 5–6.

Malgrado che le difficoltà di ogni genere, specialmente di ordine economico, rendano il nostro lavoro sempre più penoso e che gli effetti del grande cataclisma per il quale l'Europa si è da se stessa quasi annientata, lungi dall'essere leniti dal tempo, crescano d'intensità, ripercuotendosi in maniera allarmante sulla nostra società scientifica internazionale, noi vogliamo dare prova di buona volontà, riprendendo la pubblicazione, interrotta dal 1914, del "Supplemanto ai Rendiconti del Circolo Matematico di Palermo". Speriamo con ciò di fare cosa gradita ai soci. [...]

Non possiamo tacere l'augurio che il Circolo Matematico di Palermo riesca a superare il disagiato periodo attuale, tornando alle floride condizioni che esso aveva durante la vita del suo grande fondatore G.B. Guccia e possa così sempre costituire uno degli organi della collaborazione tra gl'intellettuali di ogni paese, condizione prima e necessaria per la pacifica fraterna collaborazione dei popoli, cioè per il bene della Umanità intera.

Chapter 7: Note 25 Suppl. Rend. Circ. Mat. Palermo, 11 (1919–1920), pp. 5–6.

La Redazione sente di dovere pubblicamente tributare un caldo ringraziamento al tesoriere, ing. E.P. Guerra, il quale ha accuratamente fatto gran parte del lavoro occorrente a questa pubblicazione, e ciò malgrado le grandi cure che occorrono per l'amministrazione della società, che pesa tutta su di lui. Una lode va pure tributata al direttore proprietario della Tipografia Matematica di Palermo, sig. Gaetano Senatore, il quale ha saputo superare le difficoltà nelle quali si dibatte l'industria tipografica, mantenedo alle nostre pubblicazioni, fino a quanto è possibile, la tradizionale precisione tipografica.

Chapter 7: Note 35 Letter from Bianchi to De Franchis, June 9, 1919. Folder Bianchi, Archive of the Circolo Matematico di Palermo.

I sottoscritti mentre soddisfano con l'accluso vaglia alla richiesta ad essi rivolta dal cassiere di codesto Circolo Matematico in ordine al pagamento delle quote 1918 e 1919, si permettono di osservare che il mancato invio di dette quote sembrava giustificato dall'avvenuta sospensione di ogni qualsiasi attività sociale, e che avrebbero gradito che la richiesta in discorso fosse stata accompagnata dalla esplicita promessa di una prossima pubblicazione dei volumi arretrati dei Rendiconti.

Chapter 7: Note 36 Letter from De Franchis to Bianchi, June 15, 1919. Folder Bianchi, Archive of the Circolo Matematico di Palermo.

Ella avrà visto che quando era vivente il compianto prof. Guccia, egli rimetteva annualmente, con i suoi averi personali e per cifre non indifferenti, al deficit di cassa. Si era in tempi normali ed il costo della mano d'opera e della carta ... non era quello attuale. Il mio periodo di direzione ebbe invece inizio col disastro che tutti sappiamo. Noi adesso dobbiamo anche pagare l'affitto dei locali e le spese per la stampa sono triplicate. Dallo scoppio della guerra gran numero di soci non paga.

Chapter 7: Note 66 Letter from Levi-Civita to Pincherle, March 16, 1922, Archive of the Unione Matematica Italiana, Dipartimento di Matematica, Università di Bologna.

Non mi posso sottrarre all'impressione che il vero e desiderabile analogo della "Société Math. de France", della "American Math. Society", della "Deutsche Math. Ver." ecc. resta sempre il Circolo, che ha fatto veramente onore all'Italia quando, vivente il Guccia, era in piena efficienza. Perché ammazzarlo o intisichirlo con un nuovo sodalizio?

Chapter 7: Note 67 Letter from Castelnuovo to Pincherle, April 17, 1922, Archive of the Unione Matematica Italiana, Dipartimento di Matematica, Università di Bologna.

Noi abbiamo [in Italia] un vero organo internazionale di matematica, i Rend.[iconti] del Circolo di Palermo. Può esser decaduto in questi ultimi anni, può darsi che l'attuale direttore non abbia le attitudini per tener quel posto. Ma il periodico ha grandi tradizioni e può esser rialzato. Perché non tentiamo quest'opera, piuttosto che l'altra che porterà un danno più o meno remoto al Circolo, togliendo ad esso dei soci attratti dall'Unione e dal suo Bollettino? (In Italia pochi sono soci di due Società consimili o abbonati a due giornali affini). I due nostri giornali veramente buoni di matematica, i detti Rendiconti e gli Annali stentano oggi la vita. è proprio necessario di creare un terzo organo, il Bollettino, che non potrà vivere senza danneggiare quei due?

Bibliography

150 anni di attività tra passato e futuro 1862–2012, Istituto Tecnico Statale per Geometri "Filippo Parlatore", Annuario 2012–2013, Palermo.

E.S. Allen, *Periodicals for mathematicians,* Science, N.S. 70, n. 1825 (1929), pp. 592–594.

E.S. Allen, *The Scientific Work of Vito Volterra,* The American Mathematical Monthly, 48 (1941), pp. 516–519.

R.C. Archibald, *Georg Hermann Valentin (1848–1926).* In: *Atti del Congresso Internazionale dei Matematici: Bologna, 3–10 Settembre 1928 (VI),* Nicola Zanichelli, 1929–1932, Bologna, pp. 465–470.

Atti del duodecimo congresso degli scienziati italiani tenuto in Palermo nel settembre del 1875, Rome, 1879.

E. Ausejo, M. Hormigón (eds.), *Messengers of mathematics: European mathematical journals (1800–1946),* Siglo XXI España Ed., 1993, Madrid.

R.G. Ayoub, *Paolo Ruffini's Contributions to the Quintic,* Arch. Hist. Exact Sci., 23 (1980), pp. 253–277.

A. Barlotti, F. Bartolozzi, G. Zappa, *L'attività scientifica di Giovan Battista Guccia,* Suppl. Rend. Circ. Mat. Palermo, 67 (2001), pp. xi–xxvi.

J.E. Barrow-Green, *Gösta Mittag-Leffler and the foundation and administration of Acta Mathematica.* In: K. Hunger Parshall, A.C. Rice, Adrian (eds.), *Mathematics unbound: the evolution of an International Mathematical Research Community, 1800–1945,* American Mathematical Society/London Mathematical Society, 2002, Providence, pp. 138–164.

E. Bertini, *Della vita e delle opere di Luigi Cremona,* In: *Opere matematiche di Luigi Cremona,* Hoepli, 1917, Milano, t. III, pp. V–XXII.

E. Bortolotti, *Opere matematiche di Paolo Ruffini,* Vol. I, Circolo Matematico di Palermo, 1915, Palermo.

E. Bortolotti, *Opere matematiche di Paolo Ruffini,* Vol. II, Unione Matematica Italiana, Ed. Cremonese, 1943 (facsimile edition, 1953), Rome.

E. Bortolotti, *Opere matematiche di Paolo Ruffini,* Vol. III (mathematical correspondence), Unione Matematica Italiana, Ed. Cremonese, 1954, Rome.

E. Bortolotti, *La pubblicazione delle opere matematiche di Paolo Ruffini.* In: Atti del Congresso Internazionale dei Matematici: Bologna, 3–10 Settembre 1928 (VI). Nicola Zanichelli, 1929–1932, Bologna, vol. 6, pp. 401–406.

U. Bottazzini, *Mathematics in Unified Italy.* In: *Social History of nineteenth Century Mathematics,* H. Mehrtens, H. Bos, I. Schneider (eds.), Birkhäuser, 1981, Basel, pp. 165–178.

U. Bottazzini, *Review of "Il Circolo Matematico di Palermo" by A. Brigaglia and G. Masotto,* Mathematical reviews, MR658129 (84d:01077), 1984.

© Springer International Publishing AG, part of Springer Nature 2018
B. Bongiorno, G. P. Curbera, *Giovanni Battista Guccia,*
https://doi.org/10.1007/978-3-319-78667-4

U. Bottazzini, *Italy. Ettore Bortolotti.* In: J.W. Dauben, C.J. Scriba, (eds.), *Writing the History of Mathematics: Its Historical Development,* Birkhäuser, 2002, Basel, pp. 90–91.

U. Bottazzini, P. Nastasi, *La patria ci vuole eroi. Matematici e vita politica nell'Italia del Risorgimento,* Zanichelli, 2013, Bologna.

U. Bottazzini, L. Rossi, *Cremona, Luigi,* Dizionario Bibliografico degli Italiani, vol. 30, Treccani, 1984, Roma.

A. Brigaglia, *The Mathematical Circle of Palermo,* Sympos. Math., 27 (1986), 265–283.

A. Brigaglia, *Maurolico e le matematiche del secolo XVI,* In: C. Dollo (ed.), *Filosofia e scienze nella Sicilia dei secoli XVI e XVII,* Catania 1996, vol. I, pp. 15–27.

A. Brigaglia, *The Circolo Matematico di Palermo and its Rendiconti: the contribution of Italian mathematical community to the diffusion of international mathematical journals 1884–1914.* In: E. Ausejo, M. Hormigón (eds.), *Messengers of mathematics: European mathematical journals (1800–1946),* Siglo XXI España Ed., 1993, Madrid, pp. 71–93.

A. Brigaglia, *The first international mathematical community: the Circolo Matematico di Palermo.* In: K. Hunger Parshall, A.C. Rice, Adrian (eds.), *Mathematics unbound: the evolution of an International Mathematical Research Community, 1800–1945,* American Mathematical Society/London Mathematical Society, 2002, Providence, pp. 179–200.

A. Brigaglia, *Le scienze matematiche in Sicilia dal riformismo settecentesco all'unità nazionale.* In: L. Pepe (ed.), *Europa matematica e Risorgimento italiano,* Ed. CLUEB, 2012, Bologna, pp. 307–330.

A. Brigaglia, S. Di Sieno, *The Luigi Cremona Archive of the Mazzini Institute of Genoa,* Hist. Math., 38 (2011), 96–110.

A. Brigaglia, S. Di Sieno, *Luigi Cremona's Years in Bologna: From Research to Social Commitment.* In: S. Coen (ed.), *Mathematicians in Bologna 1861–1960,* Birkhäuser, 2012, Basel, pp. 73–104.

A. Brigaglia, G. Masotto, *Il Circolo Matematico di Palermo,* Edizioni Dedalo, 1982, Bari.

A. Brigaglia, P. Nastasi, *Due matematici siciliani della prima metà del '700: Girolamo Settimo e Niccolò Cento,* Società di Storia patria per la Sicilia Orientale, 1981, Catania.

W.H. Brock, R.M. Macleod, *The life and journals of Thomas Archer Hirst, FRS (1830–1892),* Hist. Math. 1 (1974), 181–183.

O. Cancila, *Palermo,* Editori Laterza, 1988, Palermo.

O. Cancila, *Storia dell'Università di Palermo dalle origini al 1860,* Editori Laterza, 2006, Roma-Bari.

O. Cancila, *I Florio. Storia di una dinastia imprenditoriale,* Bompiani, 2008, Milan.

G. Cacciatore, *Rapporti della Commissione italiana per l'osservazione dell'ecclisse del 22 dicembre 1870,* Osservatorio Astronomico di Palermo, 1872.

G. Capaccio, *Progetto di restauro del Palazzo Guccia a Palermo,* tesi di laurea, Faculty of Engeneering, 2010, University of Palermo.

L. Cardamone, *Il Circolo Matematico di Palermo,* L'Ora, October 4–5, 1962.

G. Castelnuovo (ed.), *Atti del IV Congresso internazionale dei matematici: Roma 6-11 Aprile 1908,* Reale Accademia dei Lincei, 1909, Rome.

G. Castelnuovo, *Luigi Cremona nel centenario della nascita,* Rendiconti della Reale Accademia dei Lincei, 12 (1930) pp. 613–618.

C. Cerroni, *Il carteggio Cremona-Guccia (1878–1900),* Mimesis, 2013, Milan.

S.D. Chatterji (ed.), *Proceedings of the International Congress of Mathematicians, August 3–11, 1994 Zürich,* Birkhäuser, 1995, Basel.

A. Chirco, *Memoria del 9 maggio 1943,* Fondazione Salvare Palermo, 2003, Palermo.

A. Chirco, M. Di Liberto, *Via Ruggiero Settimo ieri e oggi,* Dario Flaccovio Editore, 2002, Palermo.

I. Chinnici, *Gli instrumenti del Gattopardo,* Giornale di Astronomia, 23, (1997), 24–29.

I. Chinnici, *Giovan Battista Hodierna e l'astronomia,* Giornale di astronomia, 34 (2008), pp. 10–17.

I. Chinnici, D. Randazzo, *Old astronomical instruments on a movie set: the case of "The Leopard",* Bull. Scientific Instrument Soc., 109 (2011), 9–13.

Cicero, *Tusculan Disputations.*

C. Ciliberto, C. Pedrini, *L'Autonomia dell'Università da Cremona a Oggi,* Lettera Matematica Pristem, 61 (2007), pp. 30–37.

C. Ciliberto, E. Sallent Del Colombo, *Pasquale del Pezzo, Duke of Caianiello, Neapolitan mathematician,* Arch. Hist. Exact Sci., 67 (2003), pp. 171–214.

C. Ciliberto, E. Sallent Del Colombo, *Giovan Battista Guccia and the Circolo Matematico di Palermo,* Lettera Matematica, 3 (2015), pp. 177–188.

C. Ciliberto, E. Sernesi, *Some aspects of the scientific activity of Michele de Franchis,* In: *Opere di de Franchis,* Rend. Circ. Mat. Palermo (2) Suppl. No. 27, (1991), pp. 3–33.

Circolo Matematico di Palermo, *XXX anniversario della fondazione del Circolo Matematico di Palermo,* Adunanza solenne del 14 aprile 1914, Suppl. Rend. Circ. Mat. Palermo, 9 (1914). Also: Rend. Circ. Mat. Palermo (2) Suppl. No. 20, (1988), pp. 1–73.

S. Coen (ed.), *Mathematicians in Bologna 1861–1960,* Birkhäuser, 2012, Basel.

E.F. Collingwood, *A century of the London Mathematical Society,* J. London Math. Soc. 41 (1966), pp. 577–594.

Comptes-rendus de la 9ᵉ session Reims 1880, Association française pour l'avancement des sciences, Paris, 1881.

Comptes rendus du congrés international des mathématiciens, Oslo 1936, A.W. Brogers Boktrykkeri A/S, 1937, Oslo.

Congrès international de bibliographie des sciences mathématiques tenu à Paris du 16 au 19 juillet 1889. Procès-verbal sommaire, Exposition universelle, 1889, Paris.

L. Cremona, *Introduzione ad una teoria geometrica delle curve piane,* Memorie dell'Accademia delle Scienze di Bologna, Series I, 12 (1862) pp. 305–436.

L. Cremona, *Sulle trasformazioni geometriche delle figure piane,* Memorie dell'Accademia delle Scienze di Bologna, Series I, 2 (1863), pp. 621–630.

L. Cremona, *Sulle trasformazioni geometriche delle figure piane,* Memorie della Accademia delle Scienze di Bologna, Series II, 5 (1865), pp. 3–35.

C.J. Cunningham, *Piazzi, Giuseppe.* In: Th. Hockey, V. Trimble, Th.R. Williams, K. Bracher, R. Jarrell, J.D. Marché, F.J. Ragep (eds.), *Biographical Encyclopedia of Astronomers,* Springer, 2007, Berlin, pp. 902–903.

G.P. Curbera, *Mathematicians of the World: Unite! The international Congress of Mathematicians. A human endeavor,* A.K. Peters, 2009, Wellesley.

J.W. Dauben, *Georg Cantor: His mathematics and philosophy of the infinite,* Princeton Univ. Press, 1979, Princeton.

J.W. Dauben, C.J. Scriba (eds.), *Writing the History of Mathematics: Its Historical Development,* Birkhäuser, 2002, Basel.

H. Davenport, *Looking back,* J. London Math. Soc., 41 (1966), pp. 1–10.

M. De Franchis, *G.B. Guccia, Cenni biografici,* Rend. Circ. Mat. Palermo, 39 (1915), pp. i–x.

E. de Amicis, *Ricordi di un viaggio in Sicilia,* Giannotta Editore, 1908, Catania. Newly edited: Il Palindromo, 2014, Palermo.

M. De Franchis, *Il Circolo Matematico di Palermo dalla sua fondazione ad oggi.* In: *Atti della Società italiana per il progresso delle scienze, XVIII Riunione, 1929,* Rome, 1930, pp. 358–364.

A. De Morgan, *The London Mathematical Society. Speech of professor De Morgan at the first meeting of the society, January 16th, 1865,* Proc. London Math. Soc. 1 (1865), pp. 1–9.

E. D'Ovidio, *Luigi Cremona: cenno necrologico,* Atti della Reale Accademia delle Scienze di Torino, 38, 1903. Cited by: E. Bertini, *Della vita e delle opere di Luigi Cremona, Opere matematiche di Luigi Cremona,* Milano 1917, t. III, pp. V–XXII, p. XX. English translation: *Obituary Note: Life and Works of Luigi Cremona,* Proc. London Math. Soc. 1904, s. 2, vol. 1, pp. v–xviii, p. xvii.

F. San Martino De Spucches, *La storia dei feudi e dei titoli nobiliari di Sicilia dalla loro origine ai nostri giorni,* Boccone del povero, 1924, Palermo.

J. Deny, J.L. Lions, *Les espaces du type de Beppo Levi,* Ann. Inst. Fourier Grenoble, 5 (1953–54), pp. 305–370.

M. Di Carlo, M.A. Spadaro, *Commemorare a Palermo. Le Medaglie di Antonio Ugo*, Kalós Edizioni d'arte, 2014, Palermo.

J. Dieudonné, *History of functional analysis*, North-Holland, 1981, Amsterdam.

Y. Domar, *On the foundation of Acta Mathematica*, Acta Mathematica, 148 (1982), pp. 3–8.

C. Dollo (ed.), *Filosofia e scienze nella Sicilia dei secoli XVI e XVII*, Catania 1996, vol. I.

E. Duporcq (ed.), *Compte rendu du Deuxième Congrès Internationale des Mathématiciens*, Gauthier-Villar, 1902, Paris.

A. Durand, *Matematici parlamentari in Italia: uno sguardo alla politicizzazione di un'élite (1848–1915)*. In: L. Pepe (ed.), *Europa matematica e Risorgimento italiano*, Ed. CLUEB, 2012, Bologna, pp. 125–136.

A. Durand, *Matematici in politica. Nuovi scenari di ricerca?*, Lettera Matematica, 98 (2016), pp. 35–38.

M.R. Enea, *Francesco Gerbaldi e i matematici dell'Università di Palermo*, Pristem/Storia, 34–35 (2013), Milan.

G. Eneström, *Über die neuesten mathematisch-bibliographischen Unternehmungen. II. Die allgemeine mathematische Bibliographie des Herrn G. Valentin*. In: F. Rudio (ed.), *Verhandlungen des ersten Internationalen Mathematiker-Kongresses: in Zürich vom 9. bis 11. August 1897*, B.G. Teubner, 1898, Leipzig, pp. 285–286.

A. Lo Faso di Serradifalco, *Palermo 1713: numerationi delle anime di tutte le Parochie della Città di Palermo fatte nel mese di 9mbre 1713*, 3 vol., Ila-Palma, 2004, Palermo.

G. Foderà Serio, A. Manara, P. Siclo, *Giuseppe Piazzi and the discovery of Ceres*. In: W.F. Bottke, A. Cellino, P. Paolicchi, R.P. Binzel (eds.), *Asteroids III*, The University of Arizona Press, 2002, Tucson, pp. 17–24.

A. Gallo, *Elogio storico di Pietro Novelli da Monreale in Sicilia, pittore, architetto ed incisore*, Stamperia reale, 1830, Palermo.

L. Gariboldi, *Tacchini, Pietro*. In: Th. Hockey, V. Trimble, Th.R. Williams, K. Bracher, R. Jarrell, J.D. Marché, F.J. Ragep (eds.), *Biographical Encyclopedia of Astronomers*, Springer, 2007, Berlin, pp. 1119–1121.

P. Gario, *Resolution of Singularities of Surfaces by P. del Pezzo. A Mathematical Controversy with C. Segre*, Arch. Hist. Exact Sci., 40 (1989), pp. 247–274.

C. Gerini, N. Verdier (eds.), *L'émergence de la presse mathématique en Europe au 19ème siècle. Formes éditoriales et études de cas (France, Espagne, Italie et Portugal)*, College Publications, Collection «Cahiers de logique et d'épistémologie», Vol. 19, Oxford, 2014.

L. Giacardi (ed.), *Da Casati a Gentile. Momenti di storia dell'insegnamento secondario della matematica in Italia*, Agora Publishing, 2006, Lugano.

H. Gispert, *La France mathématique. La Société mathématique de France (1882–1914)*. Cahiers d'histoire & de philosophie des sciences, nouvelle série n. 34. Société francaise d'histoire des sciences et des techniques & Société mathématique de France, 1991, Paris.

H. Gispert, *Une comparaison des journaux français et italiens dans les années 1860–1875*. In: C. Goldstein, J. Gray, J. Ritter (eds.), *L'Europe mathématique: Histoires, Mythes, Identités*, Maison des sciences de l'homme, 1996, Paris, pp. 391–406.

H. Gispert, R. Tobies, *A comparative study of the French and German Mathematical Societies*. In: C. Goldstein, J. Gray, J. Ritter (eds.), *L'Europe mathématique: Histoires, Mythes, Identités*, Maison des sciences de l'homme, 1996, Paris, pp. 409–430.

L. Godeaux, *Les transformations birationelles du plan*, Mémorial des sciences mathématiques, vol. 122, Gautier-Villars, 1953, Paris.

C. Goldstein, J. Gray, J. Ritter (eds.), *L'Europe mathématique: Histoires, Mythes, Identités*, Maison des sciences de l'homme, 1996, Paris.

J. Goodstein, *The Volterra Chronicles: the life and times of an extraordinary mathematician 1860–1940*, American Mathematical Society/London Mathematical Society, 2007, Providence.

J.J. Gray, *Algebraic Geometry in the late Nineteenth Century*. In: D.E. Rowe, J. McCleary, *The History of Modern Mathematics, Vol. 1: Ideas and Their Reception*, Academic Press, 1989, Boston, pp. 376–379.

J. Gray, *Nineteenth-century mathematical Europe(s)*. In: C. Goldstein, J. Gray, J. Ritter (eds.), *L'Europe mathématique: Histories, Mythes, Identités,* Maison des sciences de l'homme, 1996, Paris, pp. 347–360.

J. Gray, *The Hilbert challenge: A perspective on twentieth century mathematics,* Oxford University Press, 2000, Oxford.

J. Gray, *A history of prizes in mathematics*. In: J. Carlson, A. Jaffe, A. Wiles (eds.), *The Millennium Prize Problems,* American Mathematical Society, 2006, Providence, pp. 3–27.

J. Gray, *Worlds Out of Nothing. A course in the History of Geometry in the 19th Century,* Springer, 2007, London.

G. Guccia, *Programma del saggio filosofico che darà il chierico Giuseppe Guccia novizio della Congregazione dell'Oratorio di S. Filippo Neri, nel Maggio 1837,* Roberti, 1837, Palermo. Biblioteca Centrale della Regione Siciliana, Palermo, MISC. A 523.

G. Guccia, *Osservazioni sulla vertenza Guccia e Demanio,* Tipografia Amenta, 1888, Palermo. Biblioteca Centrale della Regione Siciliana, Palermo, MISC. C 15.6, 15.7.

G. Guccia, *La quistione religiosa in Italia risoluta a Campo dei Fiori o il vero cristianesimo,* Tipografia Diretta da Sani Andó, 1891, Palermo. Biblioteca Centrale della Regione Siciliana, Palermo, 5.8.A.137.

G. Guccia, *Ancora una parola sulla convenienza e possibilità di una Riforma Religiosa,* Pensiero Italiano, 16 (1894), Biblioteca Centrale della Regione Siciliana, Palermo, Fondo Tumminelli, IX.E.51.5.

G.B. Guccia, *Sulla conduttura delle acque potabili: lettera alla giunta speciale di sanità,* Ed. Di Cristina, 1886, Palermo. Biblioteca Scienze Giuridiche, coll. BL M** 16408 U, University of Palermo.

G.B. Guccia, *Sulla conduttura delle acque potabili: lettera II alla giunta speciale di sanità,* Ed. Di Cristina, 1886, Palermo. Biblioteca Scienze Giuridiche, coll. BL M** 16408 T, University of Palermo.

A. Guerraggio, P. Nastasi, *Italian Mathematics Between the Two World Wars,* Springer, 2006, Berlin.

A. Guerraggio, P. Nastasi, *Roma 1908: il Congresso internazionale dei matematici,* Bollati Boringhieri, 2008, Torino.

A. Guerraggio, P. Nastasi, *L'Italia degli scienziati. 150 anni di storia nazionale,* Mondadori, 2010, Milano, pp. 54–69.

A. Guerraggio, G. Paoloni, *Vito Volterra,* Franco Muzzio Editore, 2008, Padua (in Italian). In English, published by Springer, 2012, Heidelberg.

G.H. Hardy, *Obituary Notice: Gösta Mittag-Leffler, 1846–1927,* Proc. Royal Society of London A, 119 (1928), pp. v–viii.

T.L. Heath, *The method of Archimedes. A supplement to the Works of Archimedes,* Cambridge University Press, 1897, Cambridge.

D.R. Heath-Brown, *Edmund Landau: Collected Works,* Bull. London Math. Soc. 21 (1989), 342–350, pp. 347–348.

F. Henryot (ed.), *L'historien face au manuscrit: Du parchemin à la bibliothéque numérique,* Presse universitaire de Louvain, 2012.

D. Hilbert, *Sur les problèmes futurs des mathématiques*. In: E. Duporcq (ed.), *Compte rendu du Deuxième Congrès Internationale des Mathématiciens,* Gauthier-Villar, 1902, Paris, pp. 58–114.

E.W. Hobson, E.A.H. Love (eds.), *Proceedings of the Fifth International Congress of Mathematicians,* Cambridge University Press, 1913, Cambridge.

Th. Hockey, V. Trimble, Th.R. Williams, K. Bracher, R. Jarrell, J.D. Marché, F.J. Ragep (eds.), *Biographical Encyclopedia of Astronomers,* Springer, 2007, Berlin.

O.J.R. Howarth, *The British Association for the Advancement of Science: a retrospect 1831–1921,* 1922, London.

K. Hunger Parshall, *Mathematics in National Contexts (1875–1900): An International Overview*. In: S.D. Chatterji (ed.), *Proceedings of the International Congress of Mathematicians, August 3–11, 1994, Zürich,* Birkhäuser, 1995, Basel, pp. 1581–1591.

K. Hunger Parshall, A.C. Rice, Adrian (eds.), *Mathematics unbound: the evolution of an International Mathematical Research Community, 1800–1945*, American Mathematical Society/London Mathematical Society, 2002, Providence.

A. Jackson, *Chinese acrobatics, an old-time brewery, and the "Much needed gap": The life of Mathematical Reviews*, Notices of the American Mathematical Society, 44 (1997), pp. 330–337.

F. Klein, *The present state of mathematics*. In: E.H. Moore, O. Bolza, H. Maschke (eds.), *Mathematical Papers Read at the International Mathematical Congress held in Connection with the World's Columbian Exposition Chicago 1893*, American Mathematical Society, 1896, New York, pp. 133–135.

H.G. Koenigsberger, *Review of "A History of Sicily: Medieval Sicily 800–1713; Modern Sicily after 1713" by Denis Mack Smith*. The English Historical Review, 85, no. 336 (1970), pp. 560–562.

A. Krazer (ed.), *Verhandlungen des dritten Internationalen Mathematiker-Kongresses in Heidelberg*, B.G. Teubner, 1898, Leipzig.

R. La Duca, E. Perricone, *Saluti da Palermo 1890–1940, Cinquant'anni di vita della città attraverso la cartolina illustrata*, Dario Flaccovio Editore, 2007, Palermo.

G. Lanza Tomasi, *Giuseppe Tomasi di Lampedusa. Una biografia per immagini*, Sellerio, 1998, Palermo.

O. Lehto, *Mathematics without borders: A History of the International Mathematical Union*, Springer, 1998, Berlin.

B. Levi, *Luigi Cremona*, In: Commemorazione tenuta nell'Istituto Matematico "S. Pincherle" di Bologna nel I centenario della nascita, 1930. See: http://www.luigi-cremona.it/download/Scritti_biografici/Comm_Levi.pdf.

D. Mack Smith, *Medieval Sicily: 800–1713*, Chatto & Windus, 1968, London.

D. Mack Smith, *Modern Sicily after 1713*, Chatto & Windus, 1968, London.

L.M. Majorca Mortillaro, *Il palazzo Francavilla in Palermo*, Alberto Reber, 1905, Palermo.

A. Mango di Casalgerardo, *Il nobiliario di Sicilia*, A. Reber Editore, 1912, Palermo.

D. Maraini, *La lunga vita di Marianna Ucría*, Rizzoli, 1990, Milan. In English: *The Silent Duchess*, The Feminist Press at CUYN, 2000, New York.

L. Martini, *Algebraic research schools in Italy at the turn of the twentieth century: the cases of Rome, Palermo, and Pisa*, Hist. Math., 31 (2004), pp. 296–309.

Memoir of the life and writings of M. Piazzi, Director General of the Observatories of Naples and Palermo, Edinburgh J. Science, 6 (1827), pp. 193–199.

H. Mehrtens, H. Bos, I. Schneider (eds.), *Social History of nineteenth Century Mathematics*, Birkhäuser, 1981, Basel.

A. Millán Gasca, *Mathematicians and the Nation in the Second Half of the Nineteenth Century as Reflected in the Luigi Cremona Correspondence*, Science in Context 24 (2011), pp. 43–72.

C.L.E. Moore, *The Fourth International Congress of Mathematicians*, Bull. Amer. Math. Soc., 14 (1908), pp. 481–498.

P. Nabonnand, L. Rollet, *Éditer la correspondance d'Henri Poincaré*. In: F. Henryot (ed.), *L'historien face au manuscrit: Du parchemin à la bibliothéque numérique*, Presse universitaire de Louvain, 2012, 285–304.

P. Nastasi, R. Tazzioli, *I matematici italiani e l'internazionalismo scientifico (1914–1924)*, La Matematica nella Società e nella Cultura, Rivista dell'Unione Matematica Italiana, 6 (2013), pp. 355–405.

S. Oakes, A. Pears, A. Rice, *The Book of Presidents 1865–1965*, London Mathematical Society, 2005, London.

G.E. Ortolani, *Biografia degli uomini illustri della Sicilia*, 4 vol., Arnaldo Forni Editore, Napoli, 1817.

L. Pepe (ed.), *Europa matematica e Risorgimento italiano*, Ed. CLUEB, 2012, Bologna.

B. Pettineo, *The Circolo Matematico di Palermo and the Palermo Mathematical School in the last century*, Boll. Un. Mat. Ital. A (5) 18 (1981), pp. 357–368.

F. Renda, *Storia della Sicilia dalle origini ai giorni nostri*, 3 vol., Sellerio, 2003, Palermo.

A.C. Rice, *London Mathematical Society Historical Overview.* In: S. Oakes, A. Pears, A. Rice, *The Book of Presidents 1865–1965*, London Mathematical Society, 2005, London, pp. 1–16.

A.C. Rice, *The library of the London Mathematical Society*, Brit. Soc. Hist. Math. 9, Newsletter 27 (1994), pp. 37–39.

A. Rizzo, *Palermo, città antica e sottosuolo antropizzato: ipotesi progettuale nell'area del Bastione del Papireto*, tesi di laurea, Faculty of Architecture, 1998, University of Palermo.

L. Rollet, P. Nabonnand, *An Answer to the Growth of Mathematical Knowledge? The répertoire bibliographique des Sciences Mathématiques*, European Math. Soc. Newsletter 47 (2003), pp. 9–14.

D.E. Rowe, J. McCleary, *The History of Modern Mathematics, Vol. 1: Ideas and Their Reception*, Academic Press, 1989, Boston.

D.E. Rowe, *Review of "Mathematics unbound: The evolution of an international mathematical research community, 1800–1945" by K. Hunger Parshall, A.C. Rice (eds.).* Bull. Amer. Math. Soc. (N.S.), 40 (2003), pp. 535–542.

F. Rudio (ed.), *Verhandlungen des ersten Internationalen Mathematiker-Kongresses: in Zürich vom 9. bis 11. August 1897*, B.G. Teubner, 1898, Leipzig.

F. Rudio, *Über die Aufgaben und die Organisation internationaler mathematischer Kongresse.* In: F. Rudio *op. cit.*, pp. 31–37.

A. Scirocco, *Garibaldi: battaglie, amori, ideali di un cittadino del mondo*, Laterza, 2001, Bari.

C.A. Scott, *Compte Rendu du Deuxieme Congrès International des mathematiciens tenu a Paris by E. Duporcq*, Bull. Amer. Math. Soc., 9 (1903), pp. 214–215.

N. Schappacher, R. Schoof, *Beppo Levi and the Arithmetic of Elliptic Curves*, Math. Intelligencer 18 (1996), pp. 57–69.

C. Segre, *La Geometria d' oggidí e i suoi legami coll' Analisi.* In: A. Krazer (ed.), *Verhandlungen des dritten Internationalen Mathematiker-Kongresses in Heidelberg*, B.G. Teubner, 1898, Leipzig, p. 111.

A. Stubhaug, *Gösta Mittag-Leffler: A man of conviction*, Springer, 2010, Berlin.

P. Tacchini, *Eclissi totali di sole*, 1888, Rome.

E.M. Tomarchio, *Analisi Storico-Costruttiva del Palazzo Guccia a Palermo*, tesi di laurea, Faculty of Engeneering, 2010, University of Palermo.

G. Tomasi di Lampedusa, *Il Gattopardo*, Feltrinelli, 1957, Milan. In English: *The Leopard*, Pantheon Books, 1960, New York.

L.E. Turner, *Cultivating Mathematics in an International Space: Roles of Gösta Mittag-Leffler in the Development and Internationalization of Mathematics in Sweden and Beyond, 1880–1920*, Ph.D. thesis, 2011, Aarhus Universitet.

Un dîner mathématique, L'Enseignement Mathématique, 9 (1907), pp. 491–492.

G. Veronese, *Commemorazione del socio Luigi Cremona*, Rendiconti della Reale Accademia dei Lincei, 12 (1903), pp. 664–678.

A. Vitello, *Giuseppe Tomasi di Lampedusa*, Sellerio, 2008, Palermo.

V. Volterra, *Betti, Brioschi, Casorati, trois analystes italiens et trois manières d'envisager les questions d'analyse.* In: E. Duporcq (ed.), *Compte rendu du Deuxième Congrès Internationale des Mathématiciens*, Gauthier-Villar, 1902, Paris, pp. 43–57.

V. Volterra, *Le matematiche in Italia nella seconda metà del secolo XIX.* In: G. Castelnuovo (ed.), *Atti del IV Congresso internazionale dei matematici: Roma 6–11 Aprile 1908*, Reale Accademia dei Lincei, 1909, Rome, vol. I, pp. 55–65.

S.A. Walter et al. (eds.), *Henri Poincaré Papers*, http://henripoincarepapers.univ-lorraine.fr/chp.

B.H. Yandell, *The Honors Class*, A.K. Peters, 2002, Natick.

Index

© Springer International Publishing AG, part of Springer Nature 2018
B. Bongiorno, G. P. Curbera, *Giovanni Battista Guccia*,
https://doi.org/10.1007/978-3-319-78667-4